Critical Social Challenges in the Tourism Industry:

Labor, Commodification, and Drugs

Barış Çıvak
Anadolu University, Turkey

Engin Bayraktoğlu
Anadolu University, Turkey

A volume in the Advances in Hospitality, Tourism, and the Services Industry (AHTSI) Book Series

IGI Global
PUBLISHER of TIMELY KNOWLEDGE

Published in the United States of America by
IGI Global
Business Science Reference (an imprint of IGI Global)
701 E. Chocolate Avenue
Hershey PA, USA 17033
Tel: 717-533-8845
Fax: 717-533-8661
E-mail: cust@igi-global.com
Web site: http://www.igi-global.com

Library of Congress Cataloging-in-Publication Data

Names: Çıvak, Barış, 1990- editor. | Bayraktaroğlu, Engin, 1986- editor.

Title: Critical social challenges in the tourism industry : labor, commodification, and drugs / edited by Barış Çıvak and Engin Bayraktaroğlu.
Description: Hershey, PA : Business Science Reference, 2023. | Includes bibliographical references and index. | Summary: "This book focuses on the controversial social and cultural aspects of tourism in the context of critical paradigm and contributes to the extant literature by demonstrating various aspects of the industry"-- Provided by publisher.
Identifiers: LCCN 2023040624 (print) | LCCN 2023040625 (ebook) | ISBN 9781668492550 (hardcover) | ISBN 9781668492567 (paperback) | ISBN 9781668492574 (ebook)
Subjects: LCSH: Tourism--Social aspects. | Tourism--Employees. | Commodification. | Tourists--Drug use.
Classification: LCC G156.5.S63 C75 2023 (print) | LCC G156.5.S63 (ebook) | DDC 306.4/819--dc23/eng/20230918
LC record available at https://lccn.loc.gov/2023040624
LC ebook record available at https://lccn.loc.gov/2023040625

This book is published in the IGI Global book series Advances in Hospitality, Tourism, and the Services Industry (AHTSI) (ISSN: 2475-6547; eISSN: 2475-6555)

British Cataloguing in Publication Data
A Cataloguing in Publication record for this book is available from the British Library.

All work contributed to this book is new, previously-unpublished material.
The views expressed in this book are those of the authors, but not necessarily of the publisher.

For electronic access to this publication, please contact: eresources@igi-global.com.

Advances in Hospitality, Tourism, and the Services Industry (AHTSI) Book Series

ISSN:2475-6547
EISSN:2475-6555

Editor-in-Chief: Maximiliano Korstanje, University of Palermo, Argentina

MISSION

Globally, the hospitality, travel, tourism, and services industries generate a significant percentage of revenue and represent a large portion of the business world. Even in tough economic times, these industries thrive as individuals continue to spend on leisure and recreation activities as well as services.

The Advances in Hospitality, Tourism, and the Services Industry (AHTSI) book series offers diverse publications relating to the management, promotion, and profitability of the leisure, recreation, and services industries. Highlighting current research pertaining to various topics within the realm of hospitality, travel, tourism, and services management, the titles found within the AHTSI book series are pertinent to the research and professional needs of managers, business practitioners, researchers, and upper-level students studying in the field.

COVERAGE

- Service Design
- Service Training
- Cruise Marketing and Sales
- Service Management
- Casino Management
- Customer Service Issues
- Travel Agency Management
- Sustainable Tourism
- Tourism and the Environment
- International Tourism

IGI Global is currently accepting manuscripts for publication within this series. To submit a proposal for a volume in this series, please contact our Acquisition Editors at Acquisitions@igi-global.com or visit: http://www.igi-global.com/publish/.

The Advances in Hospitality, Tourism, and the Services Industry (AHTSI) Book Series (ISSN 2475-6547) is published by IGI Global, 701 E. Chocolate Avenue, Hershey, PA 17033-1240, USA, www.igi-global.com. This series is composed of titles available for purchase individually; each title is edited to be contextually exclusive from any other title within the series. For pricing and ordering information please visit http://www.igi-global.com/book-series/advances-hospitality-tourism-services-industry/121014. Postmaster: Send all address changes to above address.

Titles in this Series

For a list of additional titles in this series, please visit:
http://www.igi-global.com/book-series/advances-hospitality-tourism-services-industry/121014

Cases on Traveler Preferences, Attitudes, and Behaviors Impact in the Hospitality Industry
Giuseppe Catenazzo (AUS American Institute of Applied Sciences, Switzerland & ICN Business School, France)
Business Science Reference • © 2023 • 348pp • H/C (ISBN: 9781668469194) • US $215.00

Women's Empowerment Within the Tourism Industry
Gül Erkol Bayram (Sinop University, Turkey) Syed Haider Ali Shah (Bahria University, Pakistan) and Muhammad Nawaz Tunio (The University of Sufism and Modern Sciences, Bhitshah, Pakistan)
Business Science Reference • © 2023 • 602pp • H/C (ISBN: 9781668484173) • US $235.00

Prospects and Challenges of Global Pilgrimage Tourism and Hospitality
S.K. Gupta (Centre for Mountain Tourism & Hospitality Management, HNB Garhwal University, India) Lilibeth C. Aragon (College of International Tourism and Hospitality Management, Lyceum of the Philippines University, Philippines) Pankaj Kumar (Department of Tourism & Hospitality Management, Mizoram University, India) Madhurima (Department of Environmental Science, Mizoram University, India) and Rajesh Ramasamy (Department of Tourism & Hospitality Management, Mizoram University, India)
Business Science Reference • © 2023 • 310pp • H/C (ISBN: 9781668448175) • US $250.00

Handbook of Research on Innovation, Differentiation, and New Technologies in Tourism, Hotels, and Food Service
Gonçalo Poeta Fernandes (CITUR, Polytechnic Institute of Guarda, Portugal & CICS, Universidade NOVA de Lisboa, Portugal) and António Silva Melo (CiTUR, Polytechnic Institute of Porto, Portugal & CIDTFF, University of Aveiro, Portugal)
Business Science Reference • © 2023 • 400pp • H/C (ISBN: 9781668469859) • US $295.00

Exploring Niche Tourism Business Models, Marketing, and Consumer Experience
Maria Antónia Rodrigues (CEOS.PP, ISCAP, Polytechnic of Porto, Portugal) and Maria Amélia Carvalho (ISCAP, Polytechnic of Porto, Portugal)
Business Science Reference • © 2023 • 300pp • H/C (ISBN: 9781668472422) • US $250.00

For an entire list of titles in this series, please visit:
http://www.igi-global.com/book-series/advances-hospitality-tourism-services-industry/121014

IGI Global
PUBLISHER of TIMELY KNOWLEDGE

701 East Chocolate Avenue, Hershey, PA 17033, USA
Tel: 717-533-8845 x100 • Fax: 717-533-8661
E-Mail: cust@igi-global.com • www.igi-global.com

Table of Contents

Detailed Table of Contents

Critical theory is a comprehensive, interdisciplinary framework for analyzing social, cultural, and political phenomena. It originated in the Frankfurt School in the 1930s by Max Horkheimer, and its leading proponents include Herbert Marcuse, Theodor Adorno, and Jürgen Habermas. Measuring the reflection of critical theory on tourism can be a complex task due to the interdisciplinary and multifaceted nature of critical theory and the tourism industry. This chapter will cover three critical topics. Firstly, the author will explore bibliometrics as a research tool for measuring scientific productivity in tourism journals through publications and citations. Secondly, he will delve into network analysis, which provides visualization of the relationships between nodes and components that shape the social structure of the tourism industry, including power dynamics. Lastly, the author will discuss using scientific methods to gain insight into the critical paradigm as a framework for understanding the tourism perspective. In the conclusion section, they will summarize and expand upon these findings.

Most villages of Coastal Ghana have lost their traditional livelihoods and their labor never entered global supply chains, i.e., a local working class never formed. There is a consensus in tourism literature, pointing towards tourism as a vehicle for development. However, tourism requires transformation of labor which cannot be seen in isolation from existing social and economic structures. In this chapter, the authors take a point of departure in data collected at a tourist resort in Ghana to understand why tourism has not created prosperity for the local community. Since young employees interviewed, as well as owners of the resort, expressed tension between the culture of their workplace and the culture of their communities, this case suggests that not merely transformation of labor, but transformation of communities is called for if tourism is to support livelihoods in rural Ghana.

The purpose of this chapter is twofold: first, to examine and explain the existing theory and practice of workplace integration in the organization, and second, to propose alternative mechanisms that could contribute to positive social changes in hospitality industry organizations. Critical research methodology is used to increase explanation in workplace integration research. Employers' and employees' perceptions and attitudes toward workplace integration are explored through interviews and focus group discussions in four hotels. Fourteen thematic categories (or demi-regularities) are identified as being able to affect the process of workplace integration and generate social impact in the hotel industry. The demi-regularities are explained through abductive reasoning and as a result, eleven workplace integration practices are formulated. Finally, ten practices, or 'mechanisms,' to encourage workplace integration in hotel companies are approved through retroduction.

Tourism has a strategic position in terms of production and employment, and stands out as a labor-intensive and feminized sector when examined in terms of employment. Asia-Pacific and Caribbean, Europe, and America come to the fore as geographies where female employment is very intense. This chapter strives to examine women's employment patterns in the tourism sector in the world, both in terms of regions and in the period before and after the pandemic. It is observed with the study that horizontal and vertical discrimination in the sector is quite evident. While women concentrate mostly in labor-intensive, low-paid, and informal jobs, their presence in jobs requiring high qualifications is extremely low. On the other hand, precarious employment of women as unpaid family workers is quite common. The Covid 19 pandemic has also adversely affected the female workforce. As the pandemic has increased inequalities in the sector, the hard-earned gains of women have also been eroded.

This chapter draws a conceptual framework for the exploitation and domination practices in tour guiding. Despite the emphasized significance of their role in the tourism industry, tour guides have been reported to be the victims of unfair practices from the stakeholders. Therefore, an extensive literature review was carried out, and the author offered their insider perspective based on their nine-year tour guiding experience. The findings in the body of tour guiding literature were scrutinized, and conceptual discussions were offered to provide an insight into the 'inside story' of tour guides. Eventually, some practical implications were noted, and the need to remediate the conditions of tour guides was confirmed due to their crucial role in the industry.

Tourism activities contribute to the economic development of countries through the transformation of values such as nature, culture, and history into attractive elements for tourists. However, the continuous growth of the tourism industry brings along numerous environmental and social issues, alongside the economic benefits it generates. To the extent that the tourism industry can perceive a country's culture, natural resources, historical sites, people, and even animals as commodities that can be used for capital gain. In this context, the study primarily evaluates the commodification of human and non-human labor in this industry. Subsequently, all tourism activities that exploit migrants, women, children, and animals, subjecting them to poor working conditions or sexual objectification, are examined through a cause-and-effect relationship. Regarding the concept of labor, a critical assessment is conducted based on the social reality where women, children, and animals are potentially seen as commodities in the tourism sector, and the topic is extensively addressed with examples.

The chapter begins by defining the concept of authenticity and discusses different types of authenticity. The concept of commoditization is then introduced, followed by a review of the literature on commoditization in terms of its positive and negative aspects. As a result, it seems obvious that a broad consensus on topics such as definition, classification, effects, precautions, or management decisions to be made on the idea of authenticity in tourism and the commodification of culture or cultural elements is not yet achievable. While it continues to be discussed in the relevant literature whether tourists today seek authenticity, it can be noted that certain cultural elements have become commodities and started moving away from authenticity in tourism destinations. Some cultural elements that provide a competitive advantage have become ordinary; and this situation is mostly accepted by local people who acquire an income from tourism. Moreover, when the relevant literature is reviewed, there are also studies claiming that commodification positively affects cultural authenticity.

wisdom, and criticism with critical pedagogy. These are: (1) increasing critical tourism studies in terms of quality and quantity, (2) increasing critical courses in the education curriculum and enriching their content, (3) creating freedom and developing dialogue, and (4) combining different teaching methods.

Preface

The ever-changing conditions of social interactions are pushing academics to find more creative ways to evaluate them. A reflection of these efforts shows itself in putting critical thinking, which also constitutes the nature of science, at the center of tourism research. In tourism research, critical thinking is gaining more and more attention. In an era where studies within the critical paradigm are scarce and newly developed, this project aims to fill a significant gap and serve as a crucial reference for scholars and practitioners seeking a deeper understanding of the social, economic, and cultural impacts of tourism.

The book adopts a critical perspective, shedding light on power relations, domination, and oppression within the tourism industry. By exposing these dynamics, the critical paradigm seeks to liberate both tourist destinations and employees from exploitative conditions. From examining the social and spatial effects of tourism to addressing critical topics such as consumerism, commodification, authenticity and critical pedagogy. This comprehensive study offers a fresh and thought-provoking analysis of the field.

With a focus on labor transformation, the formation of the working class, and the employment of disadvantageous groups, the book uncovers the intricate labor processes and interactions within the tourism industry. Furthermore, authors explore important aspects such as tourist-employee interactions, and the use of drugs and psychedelics in the context of tourist mobility. This book is ideal for business executives and managers, governmental and non-governmental organization leaders, students and teaching fellows of higher education, librarians, researchers, and academicians, offering a comprehensive examination of critical issues within the industry.

Chapter 1 aims to investigate critical research in tourism by using bibliometric techniques in three steps. First, the authors explore bibliometrics as a data pool to measure scientific productivity in tourism journals through publications and citations. Second, they mapped network analysis, which enables the visualization of the relationships between nodes and components that shape the social structure of the tourism industry, including power dynamics. Finally, a discussion was made to discuss using scientific methods to gain insight into the critical paradigm as a framework for understanding the tourism perspective.

Chapter 2 focuses on the labor transformation in tourism via a case study conducted in Ghana. As a result of the study, locals point out a tension between the culture of their workplace and the culture of their communities. The problem of developing new livelihoods for villages that lost their traditional way of life must be addressed as a community problem and not as a transformation of a group of individuals. This aspect is pivotal for the successful implementation of community participation tourism.

Chapter 3 investigates the workplace integration of employees in the hospitality industry. At first, the author examines and explains the existing theory and practice of workplace integration in the organization, and second, proposes alternative mechanisms that could contribute to positive social changes in hospitality industry organizations.

Chapter 4 examines women's employment in the tourism sector by regional analysis. The authors observed that horizontal and vertical discrimination in the sector is quite evident. While women are mainly employed in labor-intensive, low-paid and informal jobs, their rates of employment in high-skilled jobs are extremely low.

Chapter 5 aims to determine the practices of exploitation and domination over professional tour guides. Therefore, tour guides are mainly accepted as the victims of unfair practices of the stakeholders such as tour operators or travel agencies. The author points out some practical implications about the need for remediating the conditions of tour guides.

Chapter 6 discusses the commodification of labor through humans and non-human animals employed in the tourism industry. The authors gave examples of tourism activities which exploit immigrants, women, children and animals and expose them to bad working conditions or sexual objectification with a cause-effect relationship. As a result, authors noted based on the social reality in which women, children and animals are seen as potential commodities in the tourism sector especially in particular tourism types.

Chapter 7 aims to revisit the concepts of authenticity and commodification via critical thinking. As a result, the authors mentioned that a broad consensus has not been reached yet on the definition, classification, effects, precautions or management decisions on the idea of authenticity and the commodification of culture or cultural elements in tourism.

Chapter 8 discusses commodification in gastronomic elements by focusing on the concept of food porn. The author first discusses socio-cultural and economic reasons that cause the commodification of food phenomenon and explains the concept of food porn with its sub-dimensions.

Chapter 9 mainly focuses on the complex and controversial world of drugs and psychedelics in tourism by examining the resurgence of interest in psychedelic substances, their implications for tourist destinations, the therapeutic potential of psychedelics, and the associated health risks and ethical considerations.

Chapter 10 explains the roots of critical pedagogy and the ways and benefits of critical tourism pedagogy. By criticizing a tourism education designed for the sector, it tries to open a space for students' enlightenment and liberation through critical pedagogy. The boundaries of critical pedagogy, which is a radical and willing pedagogy for transformation, are drawn to develop critical consciousness that will lead to the conversion of the individual, the learning environment, and society in general.

This book, entitled *Critical Social Challenges in the Tourism Industry: Labor, Commodification, and Drugs,* focuses on the controversial social and cultural aspects of tourism in the context of critical paradigm and contributes to the extant literature by demonstrating various aspects of the industry. Chapters reveal the importance of critical thinking that scholars and practitioners should adopt to open new perspectives and respond to ever-changing conditions, especially considering the hospitality and tourism industry. We would like to thank the researchers that contribute to the aims of the book, by filling the prominent gaps in the literature, with their unique experience and knowledge.

Barış Çıvak
Anadolu University, Turkey

Engin Bayraktaroğlu
Anadolu University, Turkey

Chapter 1
Reflection of Critical Theory Within the Contemporary Tourism Industry

Hamid Derviş
https://orcid.org/0000-0002-9069-571X
Kastamonu University, Turkey

ABSTRACT

Critical theory is a comprehensive, interdisciplinary framework for analyzing social, cultural, and political phenomena. It originated in the Frankfurt School in the 1930s by Max Horkheimer, and its leading proponents include Herbert Marcuse, Theodor Adorno, and Jürgen Habermas. Measuring the reflection of critical theory on tourism can be a complex task due to the interdisciplinary and multifaceted nature of critical theory and the tourism industry. This chapter will cover three critical topics. Firstly, the author will explore bibliometrics as a research tool for measuring scientific productivity in tourism journals through publications and citations. Secondly, he will delve into network analysis, which provides visualization of the relationships between nodes and components that shape the social structure of the tourism industry, including power dynamics. Lastly, the author will discuss using scientific methods to gain insight into the critical paradigm as a framework for understanding the tourism perspective. In the conclusion section, they will summarize and expand upon these findings.

DOI: 10.4018/978-1-6684-9255-0.ch001

INTRODUCTION

Evaluating critical theory's impact, influence, and effectiveness can be challenging due to its intangible and subjective nature. However, several methods and criteria are available to assess its application and effect on the tourism industry. The critical paradigm has been a fundamental concept in social sciences since its inception, encompassing numerous fields such as sociology, psychology, communication, arts, and linguistics. Nevertheless, its implementation has yet to be investigated in modern tourism as a scientific domain. Critical theory is a comprehensive, interdisciplinary framework for analyzing social, cultural, and political phenomena. It originated in the Frankfurt School in the 1930s by Max Horkheimer, and its leading proponents include Herbert Marcuse, Theodor Adorno, and Jürgen Habermas. Measuring the reflection of critical theory on tourism can be a complex task due to the interdisciplinary and multifaceted nature of critical theory and the tourism industry. However, some techniques and indicators can be used to assess the extent to which critical theory is applied and its impact on tourism.

According to literature, constructivism suggests that individuals gain knowledge through experiences, observations, and interactions. Scientists actively collect data, create theories, and interpret results to understand the world. Post-constructivism, on the other hand, takes a critical and reflective approach to knowledge construction. It emphasizes the importance of power, discourse, and social context and encourages researchers to challenge dominant narratives, question assumptions, and consider knowledge production's social and political implications. This approach aims to facilitate social change and transform power relations in society. Two methods described in the above context, critical theory and its importance and effect on tourism governance, will be discussed.

Although critical theory manifests in many aspects of tourism, ranging from environmental impact, labor, tourist-employee interaction, sex and tourism, drugs, psychedelic tourism, and social effects of tourism, to name a few, each of these approaches highlights the importance of understanding power relations, social structures, and dominant ideologies in shaping social phenomena. We strive to approach the topic holistically, from a critical paradigm point of view, quantitatively and qualitatively. Critical theory is a paradigm that is usually intangible and subjective; direct measurement can be a challenging task. One can assess a paradigm's impact, influence, or effectiveness using various approaches. Some techniques and indicators can be used to determine the extent to which critical theory is applied and its effect on tourism. According to research, educational systems are divided into two parts: one focuses on science-related subjects like Engineering and Management studies, while the other focuses on social studies. The science-oriented part encourages using data and methods in the scientific process, promoting constructivism. In contrast, the latter can produce a highly critical perspective that challenges dominant narratives and assumptions about scientific outcomes. This article will discuss two important concepts: critical theory and how it affects tourism governance.

This chapter will cover three critical topics. Firstly, we will explore bibliometrics as a research tool for measuring scientific productivity in tourism journals through publications and citations. Secondly, we will delve into network analysis, which provides visualization of the relationships between nodes and components that shape the social structure of the tourism industry, including power dynamics. Lastly, we will discuss using scientific methods to gain insight into the critical paradigm as a framework for understanding the tourism perspective. In the conclusion section, we will summarize and expand upon our findings.

Bibliometric Analysis of Literature on Tourism

Bibliometric analysis provides a quantitative approach to studying literature, enabling researchers to assess research productivity, impact, collaboration networks, and emerging trends. It is commonly used in scientometrics, research evaluation, and knowledge mapping to support evidence-based decision-making and strategic planning. Authors have investigated tourism industries from different perspectives, for example, publication output, Authorship Patterns, Citation Analysis, Research Themes, and Trends. Still, there needs to be a study investigating the implication of critical theory in contemporary tourism. Scholars studied a systematic review of online reviews research in hospitality and essential elements and their relationships, conceptual evolution and thematic structure, and possible future trends (Zheng et al., 2023). Zheng analyzed 7293 Weibo (microblog) posts from 1977 to 2022 using Python, MYSQL, AI sentiment, and Tableau. Furthermore, their studies revealed that the first topic on safety published on WoS appeared in 1977 and increased incrementally.

Researchers utilized a bibliometric approach to examine the ocean ecosystem's significance in promoting tourism growth, preserving jobs, and ensuring economic sustainability (Liang et al., 2022). In a separate study, Lercari (2023) discovered that ten specific journals encompass more than 33% of the literature on sandy beaches that attract tourists. Based on the keyword co-occurrence analysis, it can be confidently stated that the primary research areas on sandy beaches include the ecology of benthic fauna, the impact of human activities such as pollution and urbanization, and various topics related to dunes.

Singh et al. (2023) conducted a retrospective analysis of the Journal of Convention and Event Tourism, exploring its bibliometric attributes. Similarly, in 2022, Zamorano and Lizárraga used bibliometric and visualization methods to examine the impact of the COVID-19 pandemic on sustainable tourism. Meanwhile, John Tribe (2015) emphasized the significance of scrutinizing critical theory within the tourism industry. Tribe conducted a comprehensive content analysis of scientific articles on critical theory approaches to tourism using the CAB International abstracts database.

Critical theory can be evaluated as a research paradigm using various methodologies. One such method is bibliometric analysis, introduced by Pritchard in 1969. This quantitative approach measures and analyzes scholarly publications and research, enabling the evaluation of the impact of researchers, institutions, and specific fields of study. Bibliometric analysis can also identify trends, patterns, and gaps in research.

The following are some methods mainly used in the bibliometric analysis:

- *Citation analysis* provides insights into scientific works' influence, visibility, and impact and reveals knowledge dissemination and collaboration patterns.
- *Authorship analysis* examines authorship patterns and characteristics within scholarly publications. It analyzes the number of authors per paper, author affiliations, collaboration networks, and author productivity and helps identify prolific researchers, collaborations between institutions, and emerging trends in research collaboration (Osareh, 1996).
- *Mapping* research domains can be used by analyzing co-citation patterns and co-authorship networks; researchers can identify clusters of related research, influential publications, and collaborations between institutions or researchers. Maps help to understand the structure and dynamics of research fields and can aid in identifying emerging trends or interdisciplinary connections (Chen, 2004).

Eugene Garfield, an American scientist, and entrepreneur, played a significant role in developing information science. He established the Institute for Scientific Information (ISI) and gained recognition in the bibliometric and scientometric analysis research communities. Garfield is widely recognized for creating the Science Citation Index (SCI) and establishing ISI, now part of Clarivate Analytics, in 2016. The ISI database comprises the Science Citation Index, Social Sciences Citation Index, and Arts and Humanities Citation Index as its core components. Moreover, the database integrates other essential indexes for carrying out bibliometric analysis, monitoring citations, and assessing the influence of academic publications.

As of 2023, WoS reports that there are 118,323 publications in their core collection on tourism (see Fig. 1). However, it is essential to remember that bibliometrics has limitations. This includes potential biases, disciplinary differences in publishing practices, and a reliance on quantitative measures alone. To fully comprehend research impact and influence, it is recommended to supplement bibliometric analyses with other qualitative methods.

Figure 1. Year wise of publication of tourism-related articles in WoS

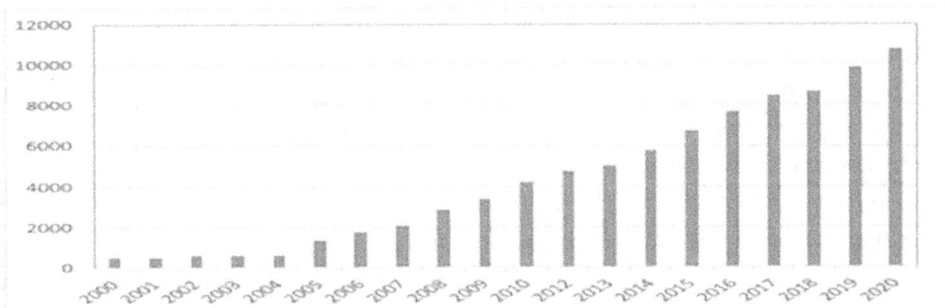

POWER DYNAMICS IN TOURISM

Power dynamics in tourism refer to the unequal distribution of power and influence among various actors involved in the tourism industry and the resulting consequences on local communities, cultures, and environments. Zhang and Xu (2023) studied the power relationship in the context of the ethnic tourism industry between host and guest and power in symbolic interactionism. These power dynamics shape tourism practices, decision-making processes, resource allocation, and benefits distribution. Examining power dynamics in tourism reveals how power is exercised, who holds it, and how it impacts different stakeholders. Here are some critical aspects of power dynamics in tourism:

Corporations and Tour Operators

Large multinational corporations and tour operators have much power in the tourism industry. They control everything from the supply chain to marketing channels and even destination development. This power allows them to influence tourism policies, determine who visits, and set prices. Unfortunately, this often harms local communities and smaller tourism businesses. Corporations having too much power can lead to exploitation, marginalization, and a lack of local control over tourism development. A study by Sigala (2008) examined the role of supply chain management in promoting socio-economic, cultural, and environmentally sustainable policies within tour operators, including TUI. The study found that better integration and implementation of these policies could have positive impacts.

Governments and Institutions

The role of government bodies and institutions in shaping tourism policies, regulations, and infrastructure development must be considered. They can allocate resources, grant licenses, and negotiate contracts with tourism stakeholders. However, it is crucial to acknowledge that power imbalances may exist within governments. Sometimes, decisions may be biased towards influential groups or prioritize economic interests at the expense of social and environmental considerations.

Governments and institutions significantly impact community involvement and decision-making processes, which can result in exclusion and inequality. This is evidenced by Kapera's (2018) survey on the impact of local government on tourism sustainability in southern Poland, as well as Gezici's (2006) research on the role of the government in coastal and inland tourism development in Türkiye. It is crucial to address this issue and ensure that communities have a fair and equal say in the decision-making processes that affect their lives.

Local Communities

Tourism destinations must recognize the negative consequences of ignoring local communities, such as losing control over resources and disrupting traditional ways of life. Unfairness in society and the economy cannot be tolerated. In a study conducted by Simmons (1994) in Huron County, Ontario, Canada, empowering residents by involving them in decision-making gave them a voice. Similarly, in the Dinsho area of the Bale Mountains National Park, Ethiopia, Wondirad and Ewnetu (2019) found that locals recognized the potential for extra income from tourism but struggled with negotiating fair contracts, accessing benefits, and preserving cultural heritage. Destinations heavily reliant on tourism are particularly susceptible to power imbalances resulting from economic dependence.

Tourists and Consumer Behavior

Tourists possess a significant amount of power as consumers in the tourism industry. Combined, they can significantly influence destination choices, shape the market, and promote sustainable practices. They must prioritize responsible tourism practices by supporting local businesses, respecting cultural norms, and prioritizing sustainability. However, when tourists prioritize low prices over social and environmental concerns, they can perpetuate negative impacts. Cohen et al. (2014) identified crucial factors that affect consumer behavior in tourism, including decision-making, values, motivations, self-concept, personality, expectations, attitudes, perceptions, satisfaction, trust, and loyalty. Therefore, tourists must recognize their power as consumers and use it to promote responsible and sustainable tourism.

Community Empowerment and Alternative Models

To tackle the power imbalances in tourism, it is imperative to establish initiatives that empower local communities and promote community-based tourism. Community cooperatives, social enterprises, and fair trade tourism are alternative models prioritizing community ownership, local control, and sustainable practices. Their ultimate objective is to minimize negative impacts and ensure that the benefits of tourism are distributed equitably, thus maximizing its advantages for host communities. Lee and Chien (2008) utilized an econometric approach to investigate the effect of tourism revenue on Taiwan's GDP and presented a sustainable tourism development model.

The above topics are crucial in implementing a sustainable contemporary tourism industry in a globalized world. Understanding and addressing power dynamics in tourism is crucial for promoting sustainable development, social justice, cultural preservation, and responsible tourism practices. It allows for a more equitable distribution of benefits, increased community empowerment, and the preservation of natural and cultural heritage. By recognizing power dynamics, stakeholders can work towards transforming the tourism industry into a more inclusive, responsible, and sustainable sector.

POWER DYNAMICS AND SOCIAL NETWORK ANALYSIS

There is a relationship between power dynamics and social network analysis (SNA). Social network analysis is a methodology used to study and analyze the patterns of relationships and interactions among individuals, groups, or organizations. It provides insights into how social connections and structures influence various social phenomena, including power dynamics. There are some ways in which power dynamics and social network analysis intersect:

Mapping Power Structures

Social network analysis can help map and visualize power structures within a social system. By examining the connections and relationships between actors, SNA can identify influential individuals, groups, or organizations within a network. Power dynamics can be understood by analyzing the centrality, connectivity, and influence of specific actors or groups, highlighting power disparities and dependencies.

Identifying Power Brokers

Social network analysis can identify individuals or organizations that act as power brokers or intermediaries within a network. These actors often hold strategic positions in the network, connecting different groups or bridging diverse social clusters. They can leverage their position to control information flow, resources, and decision-making processes, exerting influence and shaping power dynamics.

Analyzing Power Dependencies

Social network analysis can reveal patterns of power dependencies among actors within a network. It helps uncover who relies on whom for resources, information, or support. This analysis can unveil power imbalances and asymmetrical relationships, where some actors depend more on others, resulting in power differentials and potential exploitation.

Assessing Influence and Control

Social network analysis can assess actors' influence and control over others within a network. It examines factors such as centrality, degree of connectivity, and the ability to shape information flow or decision-making processes. This analysis can highlight power dynamics, showing which actors have the most influence and control over resources, opportunities, or outcomes.

Understanding Power Relations

Social network analysis provides insights into the nature of power relations within a network. It can reveal the presence of cohesive subgroups, cliques, or power clusters, where certain actors have more robust connections and mutual support. It can also uncover power asymmetries, such as power-law distributions, where a few actors have disproportionately more power and influence than others.

Researchers and practitioners can better understand power dynamics within social systems by employing social network analysis. This methodology helps uncover hidden power structures, dependencies, and relationships, shedding light on how power operates and influences social interactions. It provides a valuable tool for examining power dynamics in various contexts, including organizations, communities, and social movements. Social network analysis (SNA) can be applied to studying tourism to understand the social relationships, interactions, and information flow among various actors within the tourism system. For example, investigating the following networks can contribute to the understanding of social networks in the context of tourism:

Destination Networks

Social network analysis can be used to study the relationships and interactions between different actors in a tourism destination. This includes analyzing the connections between accommodation providers, tour operators, transportation services, local businesses, community organizations, and government agencies. By mapping the destination network, researchers can identify critical actors, information flows, collaboration patterns, and power dynamics within the destination.

Tourist Networks

Social network analysis can also focus on tourist connections and interactions. By studying social media data, online travel communities, or visitor surveys, researchers can identify patterns of social interactions, information sharing, and influence among tourists. This analysis can provide insights into how tourists exchange recommendations, form travel groups, share experiences, and influence each other's travel decisions.

Collaboration and Cooperation

Social network analysis can help identify opportunities for collaboration and cooperation among tourism stakeholders. By mapping the relationships between different actors, it becomes possible to identify potential partnerships, knowledge-sharing networks, and areas where cooperation can lead to mutual benefits. This analysis can inform destination management strategies and facilitate the development of sustainable and resilient tourism networks.

Innovation and Diffusion

Social network analysis can contribute to understanding the diffusion of innovations and new practices within the tourism industry. By examining the spread of information, ideas, and best practices through social networks, researchers can identify influential actors who drive innovation adoption and bottlenecks or barriers to diffusion. This analysis can support the implementation of innovative and sustainable tourism practices.

Power and Influence

Social network analysis can illuminate power dynamics and influence within tourism networks. By examining the connections and relationships between actors, researchers can identify influential actors who shape decision-making processes, control resources, or drive tourism development. This analysis helps understand the tourism system's power structures, dependencies, and potential inequalities.

Community Engagement and Empowerment

Social network analysis can inform community engagement strategies in tourism development. By mapping the relationships between local communities and tourism stakeholders, it becomes possible to identify the level of community involvement, social capital, and potential for community empowerment. This analysis can support participatory approaches, local ownership, and community-based tourism initiatives.

Social network analysis provides a valuable framework for understanding the tourism industry's complex social relationships and interactions. It offers insights into collaboration opportunities, innovation diffusion, power dynamics, and community engagement.

SOCIAL NETWORK ANALYSIS AS A RESEARCH TOOL

Although network analysis is an old theory rooted in mathematics and graph theory, it has emerged since the advent of information technology in the 2000s as a powerful tool for studying complex systems, such as social networks, transportation systems, and information networks. Social network analysis is employed in various disciplines, including sociology, anthropology, psychology, organizational studies, public health, marketing, and computer science. It offers valuable insights into social structures, social capital, collaboration, innovation, influence, and the dynamics of social systems. Analyzing social networks helps uncover hidden relationships, understand social phenomena, and inform decision-making in various domains. Two fundamental components of a network are nodes and relationships among them.

In social network analysis, actors or nodes represent individuals, organizations, or entities within a social system. These can be people, companies, groups, or any other unit of analysis. Relationships or edges represent the connections or interactions between nodes. These connections can be of various types, such as friendships, collaborations, communication, transactions, or any other form of social ties. Network data refers to information about nodes and their relationships. This data can be collected through databases, surveys, interviews, observations, existing records, or online sources. In this study, the data was collected from the Web of Science. Several measures are used to calculate and understand the characteristics of a network.

One of the phenomena of social network analysis (SNA) is the small world phenomenon. In the context of the small world concept, a network refers to a collection of nodes or individuals and the connections or links between them. These connections can represent various types of relationships, such as friendships, professional relationships, communication channels, or any other form of interaction between individuals. The concept became popularized through the "Milgram experiment" conducted by social psychologist Stanley Milgram in the 1960s. The experiment aimed to test the theory that two random individuals could be connected through a chain of acquaintances (Watts, 2003). Participants were given a package and instructed to pass it on to someone they knew personally who they thought would be closer to the target recipient. The packages eventually reached the target individual, and the average number of intermediaries required was around six. Overall, the small world concept underscores the notion that human social networks exhibit a surprising degree of interconnectedness, allowing for efficient communication, diffusion of information, and the potential for social influence to propagate rapidly.

Researchers can identify the key nodes or actors and the flow of information, resources, or other entities within them. In addition, network analysis can be used to identify factors such as patterns of influence, prestige, and centrality which are instrumental to the functioning of a system. Sociologists using Gestalt theory pioneered SNA methodology in the early 1920s. Jacob Moreno and Jennings (1938) were the first scientists to use SNA in the social sciences. Network centrality measures the importance or influence of a node or set of nodes by quantifying the relative importance of a node based on its position in the network. Commonly used measures include centrality (e.g., degree centrality, betweenness centrality), density, clustering coefficient, and network distance (Darvish & Tonta, 2016).

According to WoS (As of 2023), many social network studies within tourism and hospitality studies as a research field exist. Studies range from different perspectives on tourism; for example, analyzing the element of tourism: tourists, service-oriented organizations, tourism resources, destinations, and public institutions as nodes in a network (Casanueva et al., 2016), social network analysis of wildlife tourism micro entrepreneurship by interviewing 37 microentrepreneurs from North Carolina's Pamlico Sound Region (Morais et al., 2019), applying social network methodology to characterize on sport researchers phenomenon as six topological clusters: interaction networks, competition networks, intra-organizational networks, inter-organizational networks affiliation networks, and social environments (Wäsche et al., 2017).

One cannot understand the data trends, patterns, and relationships that may take time to be apparent from raw data. Data visualization techniques and bibliometric analysis complement each other to assess and analyze the information and data more actually and precisely. Network visualization software can be in a different format; command-based software, such as the Bibliometrix in R (Aria & Cuccurullo, 2017; Derviş, 2019), or interface-based such as Gephi (See Figure 2). In this section, we will explore the importance of data visualization and how it can enhance our understanding of data.

Co-word occurrences in network analysis provide a powerful tool for analyzing the intricate web of tourism-related concepts and their relationships. Researchers gain insights into emerging trends, tourist experiences, sustainable tourism practices, collaboration networks, and policy domains within the tourism industry by employing co-word analysis.

This knowledge contributes to informed decision-making, strategic planning, and the sustainable development of tourism destinations worldwide. Scholars utilized the co-word analysis to access the literature over time regarding the frequencies or co-occurrences of words in titles, abstracts, or, more generally, in text (Callon et al.,1983).

The network of co-word consists of 957 nodes and 2648 edges. For clarity in the graph, we filtered out nodes with less than 7 degrees. The degree of a node is calculated by its most occurrences with the adjacent node in the network, as shown in Figure 2. There are 7 clusters in the periphery and smaller clusters, even clusters of single nodes.

A network of co-words generated from journals within the WoS online database reveals that contemporary tourism is multifaceted attributes spanning from sexuality, fuzzy-set qualitative articles, consumer behavior, critical discourse, English as an international language, postmodernity, service experience, Harley Davidson, event marketing, Christmas, power, last but not least corporate social responsibility. Each topic mentioned above is a research field that a methodology can investigate quantitively, qualitatively, or both. Note that the collected data is in English documents.

In this endeavor, in the context of tourism, we will assess to what extent the intellectual structure of critical theory and tourism within scholarly communications sheds light on a critical paradigm as some phenomena by answering the following research questions:

(1) To what extent has the modern tourism industry engaged in critical theory?

Based on available inquired data from WoS, in contrast with other topics within tourism, there are few articles regarding critical theory as a research field.

(2) To what extent does the sociogram of the network formed from the Web of Science portate the "small world" phenomena?

The network of co-word analysis shown in Figure 2 is made of several components indicating a small-world phenomenon in the structure.

(3) To what extent does the empirical data reflected in the study shed light on leisure activity in the tourism industry?

Sub-components, including critical discourse, sexuality, and postmodernity, are mere examples of critical theory in tourism.

Figure 2. Example of network visualization of co-word (co-occurrence) of words appearing on critical theory within tourism as a research field using Gephi. Note: Each node represents a word, and each line depicts the relationship between each node.

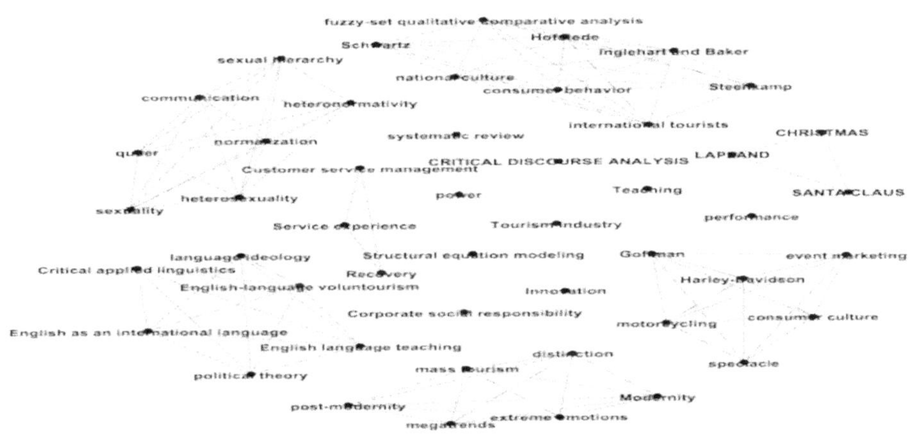

Researchers and practitioners can better understand power dynamics within social systems by employing social network analysis. This methodology helps uncover hidden power structures, dependencies, and relationships, shedding light on how power operates and influences social interactions. It provides a valuable tool for examining power dynamics in various contexts, including organizations, communities, and social movements.

Data Analysis and Results

The Web of Science Core Collection extracted seven hundred fifty-five complete records with cited references, including review articles followed by review articles, early access, proceeding papers, and editorial material. For the preliminary measurement of bibliometric analysis, Bibliometrix software was used. Bibliometrix is an open-source software that processes bibliographic data through a workflow: study design, data collection, data analysis, data visualization, and interpretation. Descriptive results of bibliographic data from 1991 to 2023 are shown in Table 1. The annual growth of publications is 3.93, a steady growth. The average number of citations per document is 23.76, and the average per year for each article is 3.39. Co-authored articles are 2.46.

Table 1. Main information about bibliographic data

Descriptive Data	
Author's Keywords	2944
Documents	755
Annual Growth Rate	% 3.93
International co-authorships	%28.87
Average citations per doc	23.76
Average citations per year per doc	3.395
Co-Authors per Doc	2.46

The number of publications is shown in Figure 3 which has increased steadily since 1991 until 2023. However, the number of citations was picked in 2003, and since then, it has kept on the same pace (See Figure 4).

Figure 3. Year-wise publication

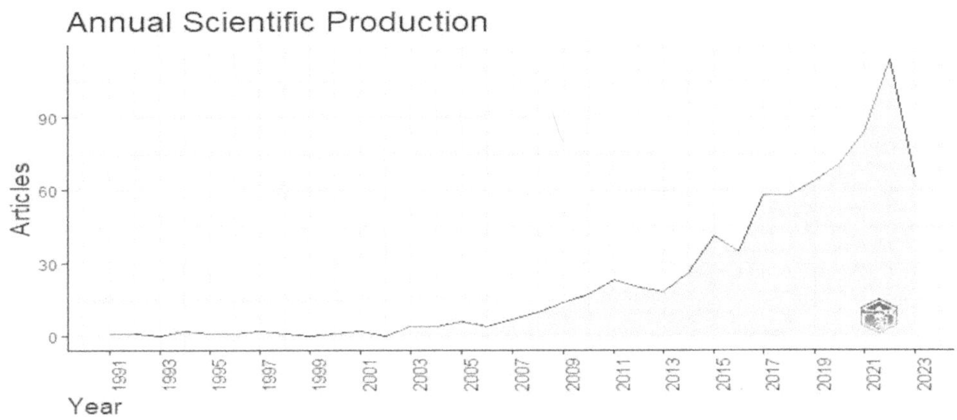

Figure 4. Year-wise citation

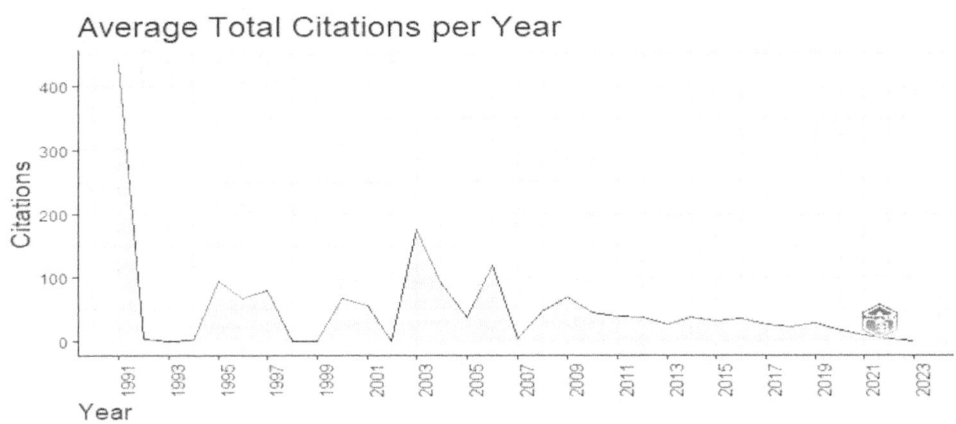

International co-authorships among scholars is almost %29 percent which determines the nature of tourism as a globalized business (see Figure 5). Almost all countries collaborate on scientific publications starting with China with the highest percentages, followed by the USA and the United Kingdom on the same level. Australia, Canada, Spain, New Zealand, Italy, Korea, and last but not least, India.

Figure 5. The most collaborative counties in tourism

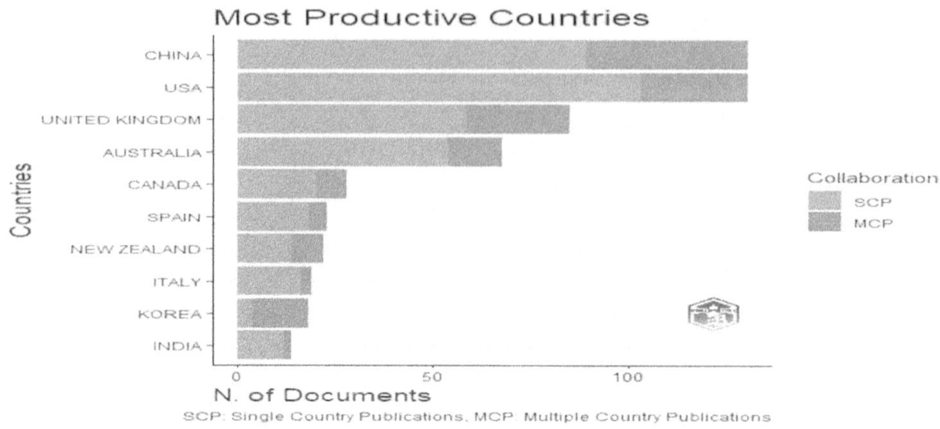

Since the study aims to access the reflection of critical tourism through scientific studies, investigating the discourse surrounding the literature sheds light on the research endeavor. A map of the keywords signifies the topics being investigated and further is a method used to examine and clarify the concepts and ideas underlying a particular subject or field of study (See Figure 6). Furthermore, it involves determining the scope and boundaries of the concept and understanding its relevance within the context of the research or discourse. (See Figure 7).

Figure 6. Two clusters of co-occurrences of the words in a bibliographic collection.

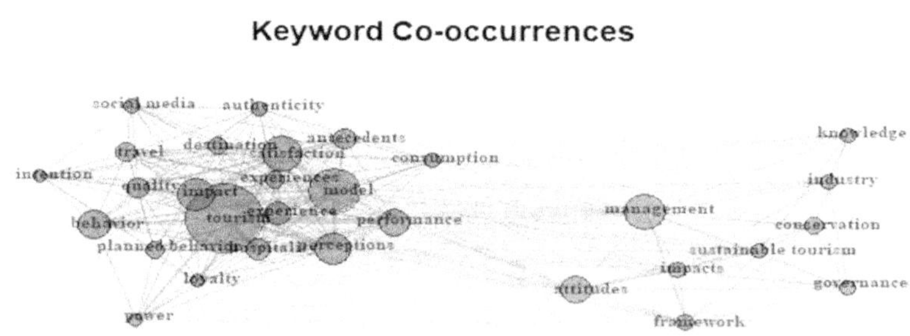

These general clusters' elements encompass a framework consisting of a bag of words, such as whiteness, critical race theory, on one side, leisure, spaces, and race, in the middle of the map, and trend on the left side of the map, finally last but not least, geographies, sex, death, reflexivity, sociology systematically examining and understanding core components and relationships of a concept.

Figure 7. A total of five clusters depicting the core component of the critical tourism

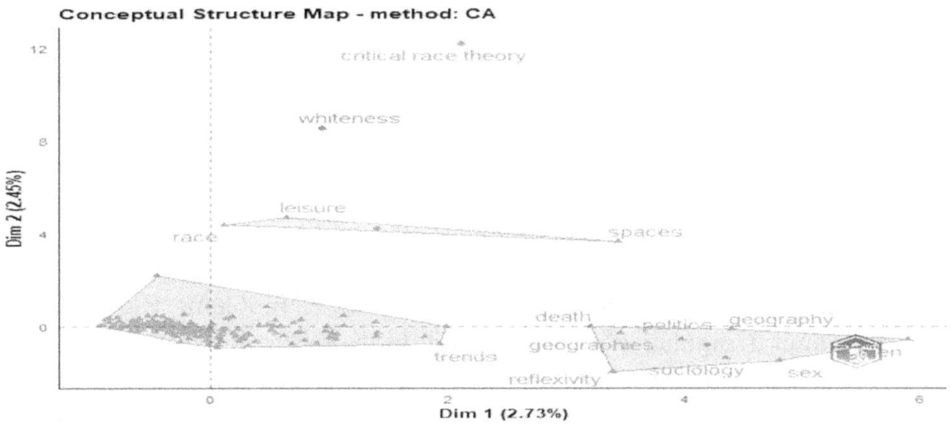

Analyzing the components and relationships of the concept is an essential part of conceptual analysis. It involves identifying the key elements or attributes that make up the concept and understanding how they relate. This step helps to establish a comprehensive and coherent understanding of the concept's structure.

In the next section, exploring the network of words, word or semantic networks are graphical representations that capture the relationships between words based on their co-occurrence patterns. In these networks, words are represented as nodes, and the connections or edges between nodes indicate the co-occurrence or proximity of words within a given text or corpus. Analyzing networks of words provides insights into semantic relationships, word associations, and the structure of language usage.

CRITICAL THEORY AND ITS IMPLICATION ON TOURISM

Critical tourism research adopts a rigorous approach to examining the tourism industry, encompassing its practices and effects. It entails questioning commonly held notions, power dynamics, and assumptions within the field. Critical tourism research aims to uncover and comprehend tourism's social, cultural, economic, and environmental drivers and outcomes.

The primary themes, concepts, and arguments present in the literature can be identified precisely through careful analysis of academic papers, research articles, and publications on critical theory and tourism. Results enable us to accurately gauge the extent to which critical theory perspectives are represented in discussions related to tourism.

Bibliometric analysis in tourism refers to applying bibliometric methods and techniques to study and analyze scholarly literature in tourism. It involves quantitative analysis of bibliographic data, such as publications, citations, and co-authorship, to gain insights into research trends, knowledge dissemination, and the impact of scholarly work within the tourism discipline. For example, 150 tourism-related articles have been published in the Web of Science (WoS) Core collection since 2012.

According to Kuhn, scientific communities create knowledge through a "paradigm." A paradigm is a globally recognized scientific achievement that provides a model for a community of researchers to solve problems for some time (Kuhn, 1970, p. 123). For the practical investigation of research problems, Kuhn suggests using paradigm triangulation as one method. In the tourism industry, the critical paradigm focuses on power dynamics, social inequalities, and environmental impacts associated with tourism practices. Critical tourism scholars argue that tourism is not unbiased or apolitical but is influenced by broader social, economic, and political contexts, affecting its development and impacts.

Collaboration across different fields is crucial in the critical paradigm as it emphasizes the importance of critical thinking about the social, labor, political, and economic inequalities in our world. Critical theory is a philosophical and social theory that aims to understand and critique societal structures, power dynamics, and ideologies. Critical theorists utilize various methods, including discourse and social network analysis, to examine and challenge dominant ideologies and power structures. Post-structuralism has significantly impacted various disciplines, including literary theory, cultural studies, sociology, and anthropology. It has influenced how scholars study power, language, subjectivity, and social construction. Post-structuralist ideas have also contributed to critical theories that question dominant ideologies, hierarchies, and oppressive structures. Critical theory encompasses different approaches and branches, with some relying more on structuralist ideas than others. For instance, the Frankfurt School, associated with critical theory, integrated structuralist concepts into their cultural, media, and ideology analysis. However, other branches of critical theory, such as post-structuralism and postmodernism, emerged as critiques and departures from structuralism, challenging its emphasis on stable structures and binary oppositions, such as guests versus hosts. The post-structuralist approach in tourism questions the dominant stories and power imbalances, aiming to uncover hidden meanings and areas of conflict. In contrast, structuralist perspectives rely on particular objects to analyze and represent the industry.

Applying the Foucauldian gaze (Cheong & Miller, 2000) to tourist attractions focusing on animals, the authors confidently advocate for a more ethical, sustainable, and compassionate approach to animal-related tourism. It is imperative for researchers and scholars to critically examine how animals are represented and treated in tourism settings. Responsible tourism practices prioritizing animal welfare, conservation, and respect for their rights and dignity should be confidently promoted. Conducting and sharing animal research with other qualitative and quantitative scientists is essential. Tomassini and Bertella (2023) suggest that more confident research should be conducted in this area.

In the study conducted by Ghaderi et al. (2023) on gender equality in tourism, 12 Iranian women were interviewed to gather insights on entrepreneurship for Asian women, with a specific emphasis on women in Middle Eastern countries. The study unequivocally identified eight crucial factors that contributed to their success: independence, gender roles and traditions, acceptance of the idea, family and social support, patriarchal society and government support, sponsorship and economic support, education, and entrepreneurs' networks.

Mura et al. (2022) applied a bibliometric analysis of critical studies published in tourism journals, books, and book chapters in four languages, for example, English, Spanish, Portuguese, and Italian, in the last 43 years from 1997 to 2020. Hoque et al. conducted a study in 2022 to investigate the effect of NGOs on poverty reduction in a rural Indigenous community in Bangladesh concerning tourism. The researchers employed qualitative methods based on critical theory, including case studies, semi-structured interviews, focus group discussions, and observations. It is worth noting that Thailand is a popular tourist destination for vacations. In 2023, Ocha explored the correlation between morality, religion, and sex tourism in Pattaya and Bangkok. Ocha (2023) investigated sex tourism in the correlation between morality and religion in Pattaya and Bangkok. The Philippines is home to many breathtaking tropical islands that have caught the interest of new immigrants. Ortega (2022) conducted a study on women, commonly referred to as Filipinas, who are in interracial relationships with foreign men. These women, who hail primarily from small villages in remote areas, establish businesses for themselves through relationships with immigrant men. Although this may seem empowering for Filipinas, it often leads to them being reduced to the label of "prostitutes." By examining these stories, we can see how gendered relationships and sexuality play a crucial role in urban transformation and transnational mobility, contributing to the diversity of the islands.

CONCLUSION

This chapter examines how critical theory is reflected in the present-day tourism industry. To achieve this, we utilized quantitative bibliometric methods and qualitative social network analysis techniques on scholarly literature related to tourism stored in the online database Clarivate Web of Science (WoS). The application of social network analysis can significantly benefit tourism researchers and practitioners. It has the potential to enhance destination management, promote sustainable tourism development, and encourage inclusive and participatory approaches within the industry. Social network analysis provides an interdisciplinary framework for studying social structures, relationships, and dynamics in tourism, and its practical applications and methodological advancements make it an increasingly important area of study.

Tourism industries are rapidly expanding, as evidenced by scientific articles in databases like WoS (although other databases also cover tourism-related topics). Through discourse analysis using technical and conceptual analysis methods, we can gain insight into the current cognitive development of modern tourism. Gephi and Bibliometrix software was utilized in this research. Although bibliometric studies started in information science, it is a popular tool in tourism. So far, we have utilized positivism theory to describe tourism's properties in the context of text data. However, critical theory proponents emphasize a discourse's hidden or latent variable in communication. It is crucial to conduct qualitative studies on various tourism-related topics, including gender, environmental impact, labor, tourist-employee interaction, sex and tourism, drugs, psychedelic tourism, the social impact of tourism, and ethics. That being said, it is essential to recognize that empirical findings from the literature should only be considered as one of many factors that influence research in the tourism field.

Utilizing a qualitative approach is imperative for conducting a thorough topic analysis. Critical theory encompasses a range of qualitative elements that shape its approach to social analysis and critique.

To conduct a thorough analysis of critical theory in tourism, researchers must boldly explore the following inquiries and methodologies:

- What is the proper attitude of local communities towards the ramifications of tourism development on their social and cultural identities?
- How do the power dynamics among stakeholders influence decision-making in tourism destinations?
- What discourses and depictions shape tourists' perception of authenticity in a destination?

Participant selection is an important part that can provide insights into the issues that aim to explore. This may include local community members, tourism workers, tourists, policymakers, industry professionals, and other relevant stakeholders and diverse perspectives that ensure the inclusion of marginalized or underrepresented voices. Data Collection Methods may take different shapes. Mainly, qualitative research methods are suited to capture the lived experiences, meanings, and social dynamics within the tourism context. Some commonly used methods are face-to-face interviews with participants to explore their perspectives, experiences, and interpretations of tourism phenomena or small groups of participants to engage in facilitated discussions about specific topics or issues related to critical theory in tourism. Another method would be participant observation, where one can immerse in the tourism setting as an observer to understand the interactions, power dynamics, and social practices in context. Also, content analysis of written or visual materials, such as policy documents, media representations, or tourist narratives, to gain insights into dominant discourses and power relations. Data analysis of the qualitative data using appropriate thematic, content, or discourse analysis methods is another qualitative method. Reflexivity and Ethical Considerations are critical qualitative research in tourism that requires researchers to critically reflect on their positions, biases, and potential impact on the research process. Consider how a researcher's perspectives and positionality may influence data collection, analysis, and interpretations. Researchers can gain deeper insights into the complexities of power, inequality, and social dynamics within the industry by studying the qualitative aspects of critical theory in contemporary tourism. By applying qualitative research methods, one can contribute to a sufficient understanding of critical issues, inform policy and practice, and empower marginalized communities within the tourism context. This chapter sheds light on critical theory, a philosophical and social theory that seeks to understand and critique societal structures and power dynamics. On the other hand, knowledge is not simply a neutral reflection of objective reality but is shaped by social, cultural, and historical processes.

As per Foucault (1977), knowledge is a potent form of power that finds expression through discourse. Discourse denotes a conversation amongst a group of people in society. The primary objective of critical tourism theory is to challenge and transform the existing power structures, social inequalities, and unsustainable practices prevalent in the tourism industry. It aims to attain a just, equitable, and sustainable future in tourism by critically analyzing and addressing these issues.

REFERENCES

Aria, M., & Cuccurullo, C. (2017). Bibliometrix: An R-tool for comprehensive science mapping analysis. *Journal of Informetrics*, *11*(4), 959–975. doi:10.1016/j.joi.2017.08.007

Bastian, M., Heymann, S., & Jacomy, M. (2009). Gephi: An open source software for exploring and manipulating networks. *International AAAI Conference on Weblogs and Social Media*. AAAI. 10.1609/icwsm.v3i1.13937

Callon, M., Courtial, J. P., Turner, W. A., & Bauin, S. (1983). From translations to problematic networks: An introduction to co-word analysis. *Social Sciences Information. Information Sur les Sciences Sociales*, *22*(2), 191–235. doi:10.1177/053901883022002003

Casanueva, C., Gallego, Á., & García-Sánchez, M. R. (2016). Social network analysis in tourism. *Current Issues in Tourism*, *19*(12), 1190–1209. doi:10.1080/13683500.2014.990422

Chen, C. (2004). Searching for intellectual turning points: Progressive knowledge domain visualization. *Proceedings of the National Academy of Sciences of the United States of America*, *101*(suppl_1), 5303–5310. doi:10.1073/pnas.0307513100 PMID:14724295

Cheong, S., & Miller, M. (2000). Power and Tourism. A Foucauldian Observation. *Annals of Tourism Research*, *27*(2), 371–390. doi:10.1016/S0160-7383(99)00065-1

Cohen, S. A., Prayag, G., & Moital, M. (2014). Consumer behavior in tourism: Concepts, influences and opportunities. *Current Issues in Tourism*, *17*(10), 872–909. doi:10.1080/13683500.2013.850064

Darvish, H., & Tonta, Y. (2016). Diffusion of nanotechnology knowledge in Turkey and its network structure. *Scientometrics*, *107*(2), 569–592. doi:10.100711192-016-1854-0

Derviş, H. (2019). Bibliometric Analysis using Bibliometrix an R Package. *Journal of Scientometric Res.*, *8*(3), 156–160. doi:10.5530/jscires.8.3.32

Foucault, M. (1977). *Discipline and punish*. Pantheon.

Gezici, F. (2006). Components of sustainability: Two Cases from Turkey. *Annals of Tourism Research*, *33*(2), 442–455. doi:10.1016/j.annals.2005.11.002

Ghaderi, Z., Tavakoli, R., Bagheri, F., & Pavee, S. (2023). The role of gender equality in Iranian female tourism entrepreneurs' success. *Tourism Recreation Research*, 1–15. doi:10.1080/02508281.2023.2168857

Hoque, B. L., Lovelock, B., & Carr, A. (2022). Alleviating Indigenous poverty through tourism: The role of NGOs. *Journal of Sustainable Tourism*, *30*(10), 2333–2351. doi:10.1080/09669582.2020.1860070

Kapera, I. (2018). Sustainable tourism development efforts by local governments in Poland. Sustainable Cities and Society, 40, 581–588. doi:10.1016/j.scs.2018.05.001

Kc, B., Morais, D. B., Peterson, M. N., Seekamp, E., & Smith, J. W.KC. (2019). Social network analysis of wildlife tourism microentrepreneurial network. *Tourism and Hospitality Research*, *19*(2), 158–169. doi:10.1177/1467358417715679

Kuhn, T. S. (1970). *The Structure of Scientific Revolutions*. University of Chicago Press.

Lee, C., & Chien, S. (2008). Structural breaks, tourism development, and economic growth: Evidence from Taiwan. *Mathematics and Computers in Simulation*, *77*(4), 358–368. doi:10.1016/j.matcom.2007.03.004 PMID:32288113

Lercari, D. (2023). Sandy beaches: Publication features, thematic areas and collaborative networks between 2009 and 2019. *Estuarine, Coastal and Shelf Science*, 281. doi:10.1016/j.ecss.2023.108211

Liang, J., Yin, Z., Yang, J., Li, Y., Xu, M., Li, J., Yang, M., & Niu, L. (2022). Bibliometrics and Visualization Analysis of Research in the Field of Sustainable Development of the Blue Economy (2006–2021). *Frontiers in Marine Science*, *9*, 936612. Https://Doi.Org/10.3389/Fmars.2022.936612. doi:10.3389/fmars.2022.936612

Milgram, S. (1967). The small world problem. *Psychology Today*, *1*(1), 61–67.

Moreno, J. L. (1934). *Who Shall Survive?* Beacon House.

Moreno, J. L., & Jennings, H. H. (1938). Statistics of social constellations. *Sociometry*, *1*(3), 342–374. doi:10.2307/2785588

Mura, P., & Wijesinghe, S. N. R. (2023). Critical theories in tourism – a systematic literature review. *Tourism Geographies*, *25*(2-3), 2–3, 487–507. doi:10.1080/14616688.2021.1925733

Ocha, W. (2023). (Diplomacy and International Studies, Rangsit University, Pathumthani, Thailand) Buddhism, gender, and sexualities: Queer spiritualties in Thailand. *Equality, Diversity and Inclusion*, *42*(5). doi:10.1108/EDI-01-2020-0025

Ortega, A. A. C. (2022). Bodies of transnational island urbanism: Spatial narratives of inclusion/exclusion of Filipinas in Philippine islands. *Urban Studies (Edinburgh, Scotland)*, *59*(16), 3365–3381. doi:10.1177/00420980211067941

Osareh, F. (1996). Bibliometrics, Citation Analysis and Co-Citation Analysis: A Review of Literature I. *Libri*, *46*(3), 149–158. doi:10.1515/libr.1996.46.3.149

Pritchard, A. (1969). Statistical bibliography or bibliometrics? *The Journal of Documentation*, *25*(4), 348–349.

Sigala, M. (2008). A supply chain management approach for investigating the role of tour operators on sustainable tourism: The case of TUI. *Journal of Cleaner Production, 16*(15), 1589–1599. doi:10.1016/j.jclepro.2008.04.021

Simmons, D. G. (1994). Community participation in tourism planning. *Tourism Management, 15*(2), 98–108. doi:10.1016/0261-5177(94)90003-5

Singh, A., Sibi, P. S., & Bashir, A. (2023). The Journal of Convention and Event Tourism: A retrospective analysis using bibliometrics. *Journal of Convention & Event Tourism, 24*(1), 87–108. doi:10.1080/15470148.2022.2150731

Tomassini, L., & Bertella, G. (2023). The human gaze at animals and the missing animal gaze in tourism studies. *Current Issues in Tourism*, 1–6. doi:10.1080/13683500.2023.2215920

Tribe, J. (2008). Tourism: A Critical Business. *Journal of Travel Research, 46*(3), 245–255. Advance online publication. doi:10.1177/0047287507304051

Wäsche, H., Dickson, G., Woll, A., & Brandes, U. (2017). Social network analysis in sport research: An emerging paradigm. *EJSS European Journal for Sport and Society, 14*(2), 138–165. doi:10.1080/16138171.2017.1318198

Watts, D. (2003). *Six Degrees: The Science of a Connected Age*. W. W. Norton & Company.

Wondirad, A., & Ewnetu, B. (2019). Community participation in tourism development as a tool to foster sustainable land and resource use practices in a national park milieu. *Land Use Policy, 88*, 104155. doi:10.1016/j.landusepol.2019.104155

Zamorano, M., & Lizárraga, C. (2022). The Impact of COVID-19 on the Sustainable Development Goals: Achievements and Expectations. *International Journal of Environmental Research and Public Health, 19*(23), 16266. doi:10.3390/ijerph192316266 PMID:36498340

Zhang, J., & Xu, H. (2023). The power dynamics in local–tourist interactions. *Annals of Tourism Research*. doi:10.1016/j.annals.2022.103509

Zheng, X., Huang, J., Wu, J., Sun, S., & Wang, S. (2023). Emerging trends in online reviews research in hospitality and tourism: A scientometric update (2000–2020). *Tourism Management Perspectives, 47*, 101105. doi:10.1016/j.tmp.2023.101105

KEY TERMS AND DEFINITIONS

Bibliometrics: It is a method of analyzing bibliographic data quantitatively. This data includes various forms of published literature, such as books, journal articles, conference papers, and scholarly documents.

Domination: It includes any violence towards tour guides and aims to sustain tour guides' labor for the benefit of others. Domination in tour guiding results from the unjust and mean practices of the employers, tour operators, and other stakeholders.

Exploitation: It includes the unfair practices that victimize the tour guides, such as low remuneration, the lack of job security and legal rights, hiring illegal labor, and forcing tour guides to maximize profits regardless of the probable dissatisfaction of tourists.

Gestalt Theory: Is a psychology theory that emphasizes comprehending how people perceive and experience the world as a unified entity rather than an assortment of individual components.

Illegal Guiding: This kind of labor appears as unauthorized individuals are employed by the tour operators to reduce costs. It is unlawful in destinations where the regulations for tour guiding services require licensed or certificated professionals to perform the job.

Power Dynamics: It refers to how power is spread, used, and discussed in social relationships, institutions, and systems. Power is the capacity to affect, manage, or mold the conduct, actions, and choices of others, and power dynamics dictate how power is exerted and how it affects people and groups in society.

Small World Phenomenon: The small world phenomenon refers to short chains of connected people in large networks, like social or human interaction networks acquaintances.

Social Network Analysis: Social Network Analysis (SNA) is a methodological approach that examines and analyzes the relationships between individuals, groups, or organizations within a social structure. It is a valuable tool for studying social network members' interactions, communication, and information flow patterns.

Stakeholders: These refer to external factors such as service providers and policymakers in the tourism industry.

Tour Guiding Service: It is performed by the tour guides and includes disseminating knowledge, narrating the values of the visited environment, and taking care of organizational/operational tasks in a destination.

Chapter 2
Labor Transformation Through the Prism of Communities:
The Case of a Tourist Resort in Rural Ghana

Charlotte Bruun
https://orcid.org/Orcid
University College of Northern Denmark, Denmark

Anna Hammershøy
https://orcid.org/0000-0001-9261-3071
University College of Northern Denmark, Denmark

Anne Lassen Zakaria
https://orcid.org/Orcid
University College of Northern Denmark, Denmark

ABSTRACT

Most villages of Coastal Ghana have lost their traditional livelihoods and their labor never entered global supply chains, i.e., a local working class never formed. There is a consensus in tourism literature, pointing towards tourism as a vehicle for development. However, tourism requires transformation of labor which cannot be seen in isolation from existing social and economic structures. In this chapter, the authors take a point of departure in data collected at a tourist resort in Ghana to understand why tourism has not created prosperity for the local community. Since young employees interviewed, as well as owners of the resort, expressed tension between the culture of their workplace and the culture of their communities, this case suggests that not merely transformation of labor, but transformation of communities is called for if tourism is to support livelihoods in rural Ghana.

DOI: 10.4018/978-1-6684-9255-0.ch002

INTRODUCTION

In 2018 we went to Ghana to study community participation tourism. We had spotted a small beach resort run by European owners. We expected to find a good example of how tourism can help develop rural villages in an area without many job opportunities. We did find that the resort owners financially supported local activities and seemed to prioritize what in their view was decency in operation. However, we noticed that their staff did not come from the local village, but from various other regions of Ghana. Something had gone wrong in their endeavour to develop their business in harmony with the local community. A community affected by poverty and lack of regular income and livelihood. In this chapter we search for an explanation to this paradox.

The original purpose of our study was to understand barriers to the otherwise obvious idea of development through community participation tourism. We therefore posed the question: how can local communities benefit from tourism? Our assumption was that at least the local community would benefit from the jobs available in the tourism sector, leading to tourism improving the pool of income available to the local community. In addition, we also hoped to search for secondary effects: Were there any signs that local produce or additional locally provided services were offered to tourists – things that might create a positive feedback loop between the local community and tourism resorts, with tourists getting experiences initiated and provided by locals and locals thus increasing their benefits from tourism beyond labor income. We also hoped to look for signs of additional "tertiary" effects with residents of the local community becoming more aware of the value of their community and surroundings – thus not only generating value for the tourists, but also for the community members themselves. However, we did not get around to studying the secondary effects, since we realized that not even the primary effect, i.e., the employment of local labor as unskilled workers in the tourism industry, had taken place.

Since, in our case, development through tourism had already failed in step one: hiring local labor, we turned our focus on transformation of labor for the tourism industry. Our research question became: *What hinders the transformation of community members from becoming paid employees in the tourism sector?* We started out investigating labor transformation *per se*, but quickly realized that communities in rural sub-Saharan Africa play a special role. Communities are not merely a collection of people living in the same area but are communities that are tightly connected through unwritten rules and structures. Communities in sub-Saharan Africa, and especially poor communities in rural areas, also hold values that are different from values held by typical communities in a Western context. This means that studies on labor, leadership, and job embeddedness done in a Western context are not immediately transferable to African contexts (in our case, Ghana). Therefore, beyond doing our own qualitative data collection, we needed to base our work on studies done in the sub-Saharan region. Since the Ghanaian government is purposing to create growth with (rural) jobs within tourism (Aryeetey et al., 2021), our research question is an important question to answer.

Community Participation Tourism as a Vehicle for Development

Promoting prosperity in the global south through tourism is not a novel idea. Growth in global tourism during the 1990s gave developing countries hope that tourism could be a way out of poverty for their marginalized populations (Ashley et al., 2000). However, the soundness of this tenet has been questioned (Adiyia, Vanneste & Van Rompaey, 2017; Sharpley 2020). There is evidence that economic dependency on tourism activities contributes to socio-economic inequalities, creates revenue leakages, encourages immigration of better qualified individuals instead of educating the local residents, and excludes the poor from the benefits which tourism may offer (Hummel & van der Duim, 2012; Adiyia et al., 2017; Medina-Muñoz et al., 2016). So far, economic growth in tourism destinations has not been inclusive, and the absolute number of poor has in fact increased (Medina-Muñoz et al., 2016; Bakker & Messerli, 2017).

Under the dominant sustainable tourism paradigm (UNWTO 2018a, 2018b), local communities are both a key stakeholder and the main beneficiary of tourism activities (Okello & Novelli, 2014; Spencely & Snyman, 2017; Jaafar et al., 2017; Burgoyne & Mearns, 2017). Communities are understood as human groups sharing a territory and involved in different but related aspects of livelihoods (Borrini, 1992 in Snyman, 2014).

There is evidence that community participation in tourism activities strengthens the community connections and creates a sense of belonging and trust among the members of the community. On the contrary, community apathy, disengagement or hostility can be a direct cause of failure of tourism initiatives (Twining-Ward et al., 2018). Therefore, it seems equitable that communities in tourist destinations should be given a prominent role in the decision-making related to tourism development. This approach is called community participation tourism (Okazaki, 2008).

However, there are multiple concerns about whether the principles of community participation in tourism activities can be implemented in developing countries, and particularly in poor communities. Community participation presupposes voluntary action from the citizens, and their active role as decision-makers. Citizen power can have a different degree in various communities and nations (Arnstein, 1971 in Tosun, 2000). Okazaki (2008) develops a typology of community participation in tourism, depending on the level of community participation and the level of power the community wields. Three scenarios can be identified: (1) citizen power, where communities exercise different degrees of power, (2) tokenism, where community consultation is formal and perfunctory, and finally, (3) non-participation of local communities in tourism development.

In the context of Sub-Saharan Africa, poverty is a significant challenge. The international baseline for extreme poverty is defined as an income of under US$1.90 a day. In Sub-Saharan Africa extreme poverty measures at 42% of the population (United Nations, 2023). In the case of Ghana, Eshun & Tichaawa (2020) conducted a study in two communities in Wechiau Hippo Sanctuary (WHS) and the Mole National Park (MNP). In Mole, the level of unemployment was 21%; and some 30% of the respondents earned no income. In Wechiau, 15% of the respondents earned less than 100 Cedis per month (Eshun & Tichaawa, 2020, p.319), which presently is about 8.5 US dollars.

The reality is that poor communities in rural areas have a limited capacity for empowerment and a lack of voice to handle issues which directly affect their dignity (Farrington et al., 1999; Adiyia et al., 2017). Farrington et al. (1999) in fact define poverty through absence of basic capabilities to achieve goals of participating in life of community and influencing decision-making. Poverty thus becomes a solid and so far, impenetrable barrier to community participation.

Apart from poverty, other barriers to community participation in tourism include low formal education, elite domination, lack of information and coordination between multiple stakeholders, and a traditional approach to livelihoods (Tosun, 2000; Okazaki, 2008; Okello & Novelli, 2014). Unfortunately, in developing countries, there is little evidence to date that community participation tourism development practices have gone beyond community consultation or manipulative strategies (Ashley et al, 2000; Mensah, 2017).

Ghana as a Tourist Destination

We chose to base our study in Ghana. The country boasts a relatively short but turbulent history according to Adu-Ampong (2019). In his study he presents a historical overview of Ghanaian tourism development policies and planning, from the country's independence in 1957, through its era of political instability and multiple coups-d'etat (1966–1981) to the current "democratic consolidation era" (p.129). During the past two decades Ghana has experienced "jobless growth" which has led government to focus on tourism as an "industry without smokestacks" to create jobs outside big cities (Aryeetey et al., 2021).

Yankholmes & McKercher (2015) suggest that the country lacks a clearly defined brand, due to the fact that Ghana Tourism Authority (GTA) brands the country as "a multi-purpose destination which offers a combination of nature based, heritage and cultural experiences along with traditional sea, sun and sand resorts" (p. 24). In geographical terms, tourism development has been asymmetrical, with most development concentrated in the Coastal and Central zones of Ghana (Adu-Ampong, 2019, p. 138). Akyeampong (2011) also points out that the expansion of tourism in Ghana has influenced the urban areas, whereas the impact on rural areas remains yet to be investigated.

Much of the heritage and cultural tourism in Ghana is inextricably linked to the country's colonial and slave trade past. There are several slave forts and castles along the coast. The Portuguese St. George castle in Elmina (also known as the Elmina castle) built in 1482, is one of the oldest European buildings in sub-Saharan Africa (Addo, 2011, p.408). In 1979 it was added to the World Heritage List and has since become the focal point for the slavery heritage tourism (Yankholmes & McKercher, 2015, p.23).

Another area of tourism development is nature-based tourism activities, including eco-tourism. Ghana ranks among the top 25% of African countries with the greatest wildlife diversity (Eshun & Tichaawa, 2019, p.1185). Eshun & Tichaawa (2019, 2020) explore the role of local communities in two ecotourism sites in Ghana, as well as the impact of the ecotourism activities on the wellbeing of the local residents. Their findings of the 2019 study showed that in terms of infrastructure, the researched community had not benefitted from ecotourism development in the area. Moreover, they demonstrated that most local residents had relatively little education and entrepreneurial training to be able to benefit from ecotourism business.

Adu-Ampong (2019, p. 127) also claims that currently, tourism is "one of the fastest growing sectors of the economy and is the fourth highest foreign exchange earner" in Ghana. In the period 2013-2017 annual growth rates within tourism is reported at 28% by Aryeetey et al. (2021). Interestingly, employment within the service sector has increased from 31% of the workforce in 2000 to 41% in 2021, whereas employment within agriculture has fallen from 53% to 39% (Data from World Bank). However, empirical studies suggest that locally available labor is not automatically transferred to the tourism industry (Fortainier & van Wijk, 2010).

The contribution of agriculture to the GDP only accounts for 20%, indicating that there is still much to be gained by transforming workers from agriculture to the service sector. Increasing productivity in agriculture and freeing labor for the tourism sector, could improve living conditions across Ghana (Data from World Bank).

It should be noted that a large fraction of the employed population (69% in 2019), represents "vulnerable employment" i.e., own-account workers with no formal work arrangements. Case studies reveal the continuous presence of smallholder farming and sharecropping in peasant communities in Ghana (Kidido & Baah, 2020). In rural Ghana there are few opportunities for formal jobs, and most of the rural population fall under the category of vulnerable employment, with small scale farming, fishing, and trading. Sharecropping with nonwage agricultural labor relationship where the landowner has obligations towards members of the kin. Sharecropping is beneficial for community members without land and no opportunities for formal (paid) work, and landowners are pressed into upholding their traditional obligations even though they may increase profits by hiring labor rather than sharecropping. But it also implies that productivity in agriculture remains low.

Case Organization and Methods Applied

Our study on community participation and labor transformation within tourism took place in Ghana. Our choice to focus on a tourist destination based in Ghana was motivated by the fact that tourism in the country had a relatively short but turbulent history. Furthermore, it was also due to our network, which was already strongly developed in the country.

To understand why it is so difficult for local communities in rural areas to benefit from tourism, we chose a small resort based in the Cape Coast area close to several tourist sites (the largest being Elmina and Kakun National Park). The village in our case held few opportunities for formal jobs, and most residents would fall under the category of vulnerable employment, with small scale farming, fishing, and trading.

Our case organization is based on a single pilot case-selection (Yin, 2009) of a small beach resort run by a European couple, placed in the outskirts of a small village in coastal Ghana. They have run the place for ten years and employed approximately 12 people at the time of collecting data in 2018. The resort is a small enclave where the tourists can eat their meals and have access to the beach. In the village there are no signs of tourism activity, but there are three other small resorts in the vicinity.

Overall, there seems to be limited qualitative research exploring tourism in Ghana (Adu-Ampong & Mensah, 2021). In Adu-Ampong and Mensah's comprehensive review of research studies on hospitality and tourism in Ghana, they demonstrate that quantitative methods seem to dominate: 53.78% of the articles in their study sample, compared to 34.03% qualitative methods (2021, p. 1102). Our study was explorative and qualitative in its design, as there were limited empirical studies taking on an anthropological qualitative approach in the study of community participation, when we started our data-collection in 2018 (Mensah, 2017). Furthermore, we applied a single case study approach due to the limited resources and options we had available during our data-collection period.

Using a single case-study might be critiqued for "putting all eggs in one basket" (Yin, 2009, p. 60). However, we were particular in adding voices and perspectives from the nearby hotels in addition to our chosen beach resort and therefore attempted an in-depth exploration of our selected case by investigating different perspectives of various stakeholders (Yin, 2009 p. 60). More specifically, and to address the potential weaknesses of the single explorative case-study approach, we applied a diverse range of methods such as semi-structured interviews, observations, informal talks within the community, as well as taking pictures, recording videos and keeping a logbook in which two researchers wrote daily observations, and compared notes (Yin, 2009). This allowed us to discuss the interviews and observations of the day, as well as our (different) assumptions[1].

We revisited our initial theoretical framework and methodology, which eventually led to the focus on labor transformation, in our exploration of reasons behind the lack of community participation in tourism activities. We are therefore informed by an abductive approach to the data, to let "the data speak", as well as to consider and revisit plausible explanations based on theories and other studies that could aid in understanding our case and the dilemmas we found and observed (Tavory &Timmermans, 2014). Data was collected over a period of one week. Overall, nine semi-structured interviews were carried out with staff members (who also lived in the local village), the assembly man, and the chief of the village, as well as the mentioned owners and manager at two neighboring hotels. Six of the interviews were recorded with permission from the interviewees, and subsequently transcribed within three weeks.

The interview guide included questions to the resort staff members about a perfect day at the beach for a tourist and for the village residents, in order to elucidate the level of expectation and whether these were shared by the local community and the staff. The respondents were chosen via a snowball technique, where the interviewees were picked via a third party, i.e., the beach resort manager and the assembly man in order to get access to the staff and community members. However, it also restricted us in choosing our own interviewees. For this reason, the community data were primarily based on observations and informal talks via our interpreter. We did, however, interview staff from the beach resort born or residing in the local community, a cleaning lady and the assembly man who did odd jobs at the resort, and thus managed to include more perspectives on the community.

The interviews mostly took place at the beach resort in the open restaurant area and lasted between 1 to 1.5 hours. The breeze and the noise from the ocean waves made it impossible for a third party to overhear the interview, although the interviewees could have been influenced by the fact that they were in a public place and at their workplace. The chief was interviewed at his palace in the presence of his stool members and via the assembly man. In terms of validity of the study, we do not claim the explorative single case-study to be unbiased, representative or generalizable (Flyvbjerg, 2006). However, by having a working awareness of bias, we embrace that no research starts with a "tabula rasa".

Initial Findings at the Case Organization

The owners of the resort bought the place with a dream of making a sustainable living from their business and supporting the local community by offering service jobs at their beach resort. From our research we found that the leadership style of the owners was very inclusive and aimed to promote a clan culture within the organization as opposed to more hierarchical leadership. This seemed to be the opposite approach of the next-door beach resort that catered to the high-end segment of Nigerian and British tourists, and had paid the local fishermen to move their boats, to make space for tourist speedboats which the owners wanted to invest in. The owners there had also built a wall around the hotel. At our case beach resort, there were no walls, and the beach area was public but maintained by the hotel. The owners seemed to guide their employees rather than commanding them. To give them a better idea of the tasks and behavior they preferred, they also took employees on a yearly trip to learn about service and tourism from the customer perspective. Their logic was: *How can you know how to provide service for tourists, if you never tried the role of guest yourself?* They seemed to have had success with this approach. Their employees appeared loyal and felt attached to the organization and the owners.

Despite their expressed intensions of building close relations to the village, we found that the owners, after struggling for a few years with hiring staff from the neighboring coastal village, finally abandoned the idea:

"Here everyone is a brother or a sister (not biologically). It's good to see the social connection they have (the community), but it is also very difficult for the people. There is a lot of pressure from the family. If you are working, you earn money, you are expected to help, and when you don't you are a bad person. Some time ago we decided that we did not want to have people from this village. The pressure on staff from the village, if you work (.....) is too much." (The owners of the resort)

If this is a typical situation in the tourism industry in rural Ghana, naturally it restrains the opportunities of not only labor transformation, but also community participation tourism. The typical employees at the resort were university students working towards getting the finances to finish their studies. However, none of the staff members studied hospitality or tourism. Interviews revealed that a factor putting stress on the staff members was family in need, demanding their savings for more immediate needs. Moving away from family for work appeared to reduce this strain. To better understand this factor, we need to understand the community, and some of the societal values and redistributive norms that characterize communities in Ghana as well as possible sources of conflict between work and community.

Although the employees at the resort had to rent accommodation in the village, we saw several signs that they did not feel welcome in the village. Neither did they express a desire to be included in the local community:

"There is jealousy from the locals towards you working at the hotel. They don't tell you. I stay for myself, not to get in trouble. I started hearing in town that I came to take another guy's position, as he was fired – and he was from the town." (Employee #1)

Another employee who has lived for 11 years in the village, and also moved his family from another region of Ghana to the area, adds:

"I am not one of them, so I am an outsider in the community. I am Twi – people here are Fante." (Employee #2)

Yet, a third employee reflects over his challenges of being an outsider to the village, and the cultural and educational differences he experiences in the community where he resides while working at the beach resort:

"When I first came it was difficult to make friends and understand the language. So, I made friends with other staff at first. So, I just went to work and back to the house in the village. There is jealousy from the locals towards you working at the hotel. […] They [the villagers] cannot wait to spend their money. What they get today they spend today. I think of tomorrow. They are used to living insecure lives. In my region, we learn from birth how to cope with life, cooking and planning your money. […] The chief here does not speak English, so he is not that educated. He does not put any impact. The chief in my area is an educationist, and head of the school too. […] Education is very important. […] So, when I see young ones not going to school, I get saddened." (Employee #3)

A fourth employee also touches on the challenges of hiring staff from the local community responding to the question: Would it be more difficult for the owners if the entire staff was from the local village?:

"They [employees] can bring personal family problems to work, and they will start fighting. […] When your family knows you are working, they start asking for help. If you don't help them, they say you are wicked – then you need to help them." (Employee #4)

To our surprise it turned out that only a few odd jobs at the resort were held by locals, and rather than including the village in the activities at the resort, there were several potential conflicts in their relation. The owners attempted to keep a good (working) relationship with the Chief by honoring the local traditions of regular visits and symbolic gifts to the chief, as well as attending funerals of any of his relatives. Moreover, they donated money to the local school, but did not feel accepted. Therefore, we decided to focus on interactions between tourism employment and the community. Focusing on the potential misfit between needs of the tourism industry and the competences of local labor would not explain our findings (Aryeetey et al., 2021; Amankwaa et al, 2022). The owners of the resort did not complain about lack of work skills of local employees. This was not a matter of the interactions between the individual employee and the employer, but of interaction between the employee, the community and the tourist resort. In order to seek further explanation to this complex issue, we turn our attention to understanding labor transformation: why had the owners of the beach resort employed migrant workers from other parts of Ghana and not from the local village? We must explore this to understand how community participation tourism can improve livelihoods in communities.

Labor Transformation and Work-Family Conflict

Tourism workers are widely portrayed as the critical dimension in the successful operation of tourism businesses. Moreover, shortage of skilled labor is considered one of the main explanations of the poor performance of the tourism industry as exemplified by studies in Sub-Saharan Africa (Ankomah, 1991; Aryeetey et al., 2021). According to Bakker (2019) insufficient human resource capacity is still a factor restricting tourism in Sub-Saharan Africa. It is quite apparent that the transformation of labor has not gone smoothly.

As a field between economics and sociology, labor transformation has its roots in the Fisher-Clark thesis stating that a rise in income causes a movement from agriculture to industry, and from industry to services (Singelmann, 1978). The thesis came with an assumption that "the penetration of capitalism would result in the development of a peasant proletariat dispossessed from their land and living of wage labor" (Blackwood, 1997). But the high number of vulnerable workers in Ghana, together with the fact that industry at its highest, which is today, only employ 19% of the workforce, tells us that most of the Ghanaian workforce never transformed to industry workers. This means that the labor to be transformed must come from people with no paid wage job experience and no worker mentality. This poses a challenge since the tourism industry is a rather demanding profession.

As argued by Aykaç (2007) labor transformation is also about transforming social and power dynamics. In her study she focuses on labor through the prism of the household, arguing that not only the employee is affected. Transforming small scale peasants into workers in the tourism industry is a complex process that affects the whole family. Aykaç (2007) describes three different transformation processes in Turkey: one geographical location where the workforce entering the tourism industry came from an identity as peasant household, one where they came as proletariat (paid workers) and a third where immigrant workers were hired. These three examples illustrate the complexity of the problem. Most of the households in the community of our study are subsistence-oriented small peasant households (Stauth, 1984). We did not meet any regular proletarian households, but rather what could be defined as semi-proletarian households (Aykaç, 2007). In semi-proletarian households wage income is generated primarily from migrant workers, or in the case of Ghana, remittances from family living abroad and is combined with subsistence production. In the semi-proletarian case, Aykaç describes the household as an income pooling unit for the reproduction of labor that can move to either of the poles (proletarian or subsistence-oriented households) depending on the current composition of the household. Because of the income pooling, it is not the individual member bringing home the wage that gets to decide how it is spent – that is for the household to decide. This may cause different types of conflict between family and work.

Aykaç's work illustrates that in a process of labor transformation one should not regard labor in isolation from social structures such as family but be aware of the work-family interface. Labor transformation may cause work-family conflict. Such conflicts are bi-directional (Mokomane, 2014). Family-to-work conflicts occur when experiences in the family or interpersonal conflicts within the family unit interfere with work life. Work-to-family conflict, on the other hand, occurs when experiences or demands at work interfere with family life. Work-family conflict is typically studied with a pre-understanding that work causes family conflict whereas family issues causing work conflict are less studied (Putnik et al., 2018; Annor & Burchell, 2017). However, in our study we found signs of both types of conflict.

Much research on the work-family interface focuses on individuals: Workers lacking resources to cope, and leaders that should consider the welfare of their employees. Conservation of resources (COR) theory is often applied in studying work-family stress. COR describes the motivation that drives humans to both maintain their current resources and to pursue new ones. Resources are defined as things that one values, specifically objects, states, and conditions. Studying the work-family conflict among hotel workers in Nigeria, Karatepe (2011) concludes that "managers should focus on hiring individuals high in job resourcefulness" referring to personal traits or resources as "emotional stability", "agreeableness" and "high-conscientiousness" to get a "pool of the right caliber employees" (p.15).

Nevertheless, personal resources cannot explain the pattern of local workers contra migrant workers, and in the view of Aykaç's income-pooling units, it need not be the individual worker that presents a challenge. As an example, stealing from one's employer need not be a sign of low moral standards of the individual employee – from the perspective of the kinship involved, it may be a sign of high moral standards. Even workers of "the right caliber", as Karatepe recommends hiring, may break down under structural pressures. Karatepe also recommends that management take measures to reduce work-family conflicts, e.g., family-friendly benefits. But this is exactly what the owners of our case resort did. They created a friendly atmosphere, they helped with school tuition for employees' children, and they made employees work shorter shifts as is common in Europe.

Despite observations of work-family conflicts, Aykaç's (Turkish) story has a happy ending. Households in rural areas are transformed and livelihoods are improved whilst remaining in and developing rural areas:

"In fact, although workers in the tourism industry become wage workers in a modern service sector, they also maintain their rural connections, whether by working in agriculture during the off-season, or by maintaining subsistence farming together with their tourism work. The analysis of households in this research explains the complexity of labor transformation and the coexistence of traditional forms of production and labor together with the market economy." (Aykaç, 2007)

It is interesting to transfer such a scenario to our study, and pose the question of whether the same development could have happened in the study area's coastal village, had the owners been more persistent in hiring and retaining local employees? When employees end up stealing from their employer despite their employer providing rather favorable work conditions, should they be given time to adapt to their new identity as paid workers, or might there be some underlying cultural and social structures that makes adaptation difficult?

Societal Values and Redistributive Norms

Our case suggests that transformation of labor is not merely a question of transforming competences, but also a question of transforming values. Values are not transformed in isolation but also concern organizational values and even societal values (Quinn & Rohrbaugh, 1983; Zeb et al., 2021). Forming a working class is not a question of making people work – it is about forming identities for the role of workers. In our case it is a question of the population of a village together forming an identity as employees (service providers) in the tourism industry.

The first question to be resolved is whether the family or household concept is directly transferable from a Western to a sub-Saharan context. Peters (2019) warns against using the modern Western concept of family to understand social structures in Africa. The Western nuclear family is very different from the sub-Saharan kinship, and it is not merely a question of the size of the family structure.

To understand kinship, one must take the cultural dimension into consideration. Whereas Western societies tend to be individualistic, sub-Saharan cultures tend to be collectivistic. This calls for a different work strategy in a Sub-Saharan context. According to Mokomane & Chilwane (2014), working in the West is about personal accomplishment and achievement, whereas in sub-Saharan societies, work is seen as a means of supporting the family. In a Western context people tend to view themselves as individuals, distinct from their family, which means that work-family conflict may arise because of conflict between the interest of "the self" and the needs of family. In contrast to this, according to Mokomane & Chilwane (2014), it implies that in a sub-Saharan context there is no conflict between "the self" and "the family" but only between "the work" and "the family".

Collectivism implies that African communities tend to be communal rather than individualistic. In its pure form, collectivism means holding the belief that an individual's benefit is at the cost of the remaining society. This also indicates that individuals take an internal focus always prioritizing the community over the external world. Governance in collectivist communities is hierarchical and dominated by sharing rules together with other rules. Redistributive norms work as an insurance in hunter-gatherer societies, and reminiscences of this logic are still strong in many African rural contexts – just as we saw in Aykaç's income-pooling households. The logic remains that if one individual gain economic success, it is at the cost of other members of society, and thus at the cost of collective prosperity (Platteau, 2014). This logic works well when a community's livelihood is traditional (smallholder farming) where such informal insurance networks allow the village to spread risks – and the individual to take chances (Fafchamps, 2004). But in other contexts, the logic may defer transition to other livelihoods.

With respect to work-family conflict, we may expect different types of conflicts in a collectivist culture. Just as one member's benefit is at the cost of the remaining community, one member's contribution will benefit the whole community. In an individualistic culture, working long hours may be perceived as egoistic by other family members since work is about self-realization. In a collectivist culture working long hours or migrating for work may be more acceptable to the family since it means providing for it. This also explains the different work ethics. In a Western context workers may prefer or be in a structure that allows them to work intense hours and then go home to the nuclear family. In some African contexts, workers tend to work longer, but less intense hours. In some cases, it might be acceptable to take a nap during work hours, although many African urban work settings influenced by Westernized work ethics would not allow this. In our case the owners implemented Western work ethics with shorter but more effective days. However, some of their employees expressed a desire for longer but less intense working hours.

Social structures in sub-Saharan Africa are often referred to as "extended family" – but this is an oversimplified image of a variety of different norms relating to different units, and more importantly in our context: norms that have changed with external events such as colonization, independence, changing political regimes and modern access to global media. Despite the dynamic nature of kinship relations, young people are still expected to take care of their elderly family members, and the less (economic) power the kinship holds, the stronger the "traditional way" becomes. In a proletarian household, income-pooling may be less important – but in the typical semi-proletarian household of rural Ghana, subsistence depends on the pooling. Thus, kinship allocates rights, responsibilities and obligations to people (Akanle & Olutayo, 2011 p.471) and is an important part of the governance structure in Sub-Saharan Africa. Today's remittances towards Sub-Saharan countries witness the importance of kinship obligations even today.

The different understanding of family, self and work made us conclude that reproducing the study of Aykaç (2007) is not relevant to our case. Instead, we chose to look at labor transformation through the prism of community, and search for potential conflict between work and community.

Transformational Leadership and Job Embeddedness

The labor transformation from the industrial or even agricultural to the service economy put strain on workers' habits, capacities and values (Chen et al., 2018; Booyens, 2022)'. Even more so in countries like India and countries in sub-Saharan Africa where transformation is taking place directly from the agrarian sector to the service sector. It is no longer enough to be obedient and perform the tasks demanded. In the service sector workers must perform tasks not specifically formulated by the management. They must foresee what their employer may want them to do in a given situation meeting the customer, i.e., be innovative.

Amankwaa et al. (2022) have studied how transformational leadership can encourage more innovative behavior among service workers. Transformational leadership is a leader-follower leadership style where leaders ensure that followers are aware of the vision of the organization and aspire employees to collaborate towards organizational interest. According to Amankwaa et al., "transformational leadership assumes that leaders affect followers by instilling in them a sense of trust so that followers appreciate and regard them positively" (p.512). When the owners take their employees on a yearly vacation trip to demonstrate what it is like to be a tourist, this can be seen as a typical example of transformational leadership – they do not tell the employees exactly what to do but teach them what service is. It also promotes innovative behavior among employees, which is important since employees must be able to understand and respond to the needs of tourists. The study of Amankwaa et al. (2022) is done in the hotel industry of Ghana, using the concept of "job embeddedness" as a mediator between leadership and innovative behavior.

Job embeddedness is a concept developed by Mitchell et al. (2001) to determine why employees leave or stay in their jobs. Job embeddedness is a broader concept than job satisfaction, and it was developed to help managers retain workers. What is important in our context is that the concept includes nonwork factors such as family attachment and work-family conflicts. Job embeddedness considers three factors; "(1) the extent to which people have links to other people or activities, (2) the extent to which their jobs and communities are similar to or fit with the other aspects in their life spaces, and (3) the ease with which links can be broken – what they would give up if they left [...]." (Amankwaa et al., 2022). These three factors are studied in two domains: the organization and the community (Table 1). In introducing the job embeddedness construct, Mitchel et al. (2001) suggest that it may be applied to other factors than job turnover, such as performance, absenteeism and general happiness. Later it has been adapted to study employees' innovation skills (Amankwaa et al., 2022).

From a perspective of labor transformation, the concept of job embeddedness is relevant because it may integrate the competences demanded of the individual worker in transforming from one set of work demands to another (e.g., from being a member of a subsistence household with small scale farming to becoming a service worker in the tourism industry), and the role expectations induced by social constructs.

From our perspective the concept of job embeddedness is particularly important because it divides job embeddedness into organizational embeddedness (whether one interacts with and shares values with co-workers) and community embeddedness (interaction and sharing values with other community members). Job embeddedness may be seen as a resource in the COR theory, as people will protect and develop their organizational and community embeddedness. Thus, the concept of job embeddedness may help us understand not only what it means for our case resort employees to be members of a kinship and its community, and what it means for them to be part of an organization supplying a good value fit and a network, but also how the two types of embeddedness may interact.

Studies of job embeddedness in a Western context based on Mitchell et al. (2001) have found that both organizational embeddedness and community embeddedness are necessary for people to be happy about their work and their life (Amankwaa et al., 2022). The two forms of embeddedness move in the same direction – i.e., there is no conflict between being embedded in a community and being embedded in a work organization, and both should be present.

Table 1. Based on Mitchell et al., 2001

Job embeddedness	FIT (values)	LINKS (network)	SACRIFICE (in leaving)
ORGANIZATION	I fit with the company's culture and needs	I interact with many co-workers regularly	People respect me at work
COMMUNITY	I love the place where I live	My family roots are in this community	People respect me in my community

Amankwaa et al. (2022) however do not find statistical significance for the community embeddedness in the data of their study, suggesting that communities play a different role in Ghana than in a Western context where transformational leadership and job embeddedness were first studied. From our perspective this is an interesting finding. The owners seem to have an implicit understanding of the theoretical concepts of transformational leadership and job embeddedness. But their employees feel no strong embeddedness in the community. They talk negatively about people living in the village and have no strong links with them. Rather than sacrificing anything by leaving the village where they currently work, to them it was a sacrifice to leave the village where they grew up, and to which they plan to go back.

One explanation why Amankwaa et al. (2022) do not find the importance of community embeddedness in the sub-Saharan Africa data may be that community plays a different role in this context. Grimm et al. (2013) point to the fact that "family and kinship ties" are different from "social networks", and "community" may cover both of these concepts. In a Western context community embeddedness is perceived as a set of "weak" links forming a social network – a network that we, as individuals, choose to connect to or not (thus the term weak link from Granovetter (1983). In some sub-Saharan African contexts, a community may primarily be viewed as a set of "strong" (blood) links given to the individual by birth and thus exogenous and only changeable at a high psychological, and possibly also economic cost (due to the risk sharing and insurance aspects of kinship ties).

Why is it difficult for employees of our case resort to connect to the community where they currently reside? In a Western context there may be a tendency to embed oneself in a new community, even if one still feels a bond with other communities. In an individual value context belonging to more communities is not problematic since relations do not go that deep and people typically have more connections. In a society with communal values, people prefer to have fewer but stronger relations, and conflicts between different communities would more easily arise (Ramesh & Gelfand, 2010). Belonging to both an organization where a clan culture is promoted and to a community with strong relations may create conflict – just as it may feel as betrayal to develop strong relations in a new community while still feeling a strong tie with the community of one's family. To investigate whether these theoretical ideas can explain why workers at our resort are not hired from the local village, we need to revisit our case.

Revisiting the Case: Transforming Communities?

One explanation why the best employees of the case resort come from other parts of the country may be that there is a potential conflict between organizational embeddedness and community embeddedness. If community values are not compatible with the organization, promoting labor transformation and innovativeness by community embeddedness is not an option. The community tends to be collectivist while the organization tends to be inspired by individualistic values. Furthermore, the hierarchical dimension differs. The European owners promoted and assumed equality as a desired objective, whereas the community may contrary have assumed hierarchical objectives, expecting submission to authority as a desired outcome (Fatehi et al., 2020). This was also confirmed in our interview with the local chief who strongly argued in favor of traditional teaching methods and curriculum at the local school, rather than promoting soft skills such as service mindedness and innovation[2].

Family and resembling family structures and community are important in a Ghanian context. One of the employees coming from a different part of Ghana stated that, "in Ghana it is important that you have a father of the family. Here our family is X & Y (the owners). In Ghana you need to know who is your father". Later he noted that "The staff and the owners, we all play together. Family like." But for an employer, taking the role of a father can be a double-edged sword. Later in the interview, in discussing his dreams for the future, the staff member noted that: "… nobody helped me yet". When asked: What do you need help for?, the interviewee responded:

Someone who would lead me, tell me what to do or help me to do it, because I don't know how it is. But I know my bosses have an NGO. If I get some money, if you can help me, then….

According to Beugré & Offodile (2001) the African cultural trait "respect for elders" is not a question of age, and not a one-sided respect. "As the younger respects the elder, the latter must, in return, take care of the former, provide him or her with advice and help him or her realize his or her full potential" (p.537). Perhaps this young person shows respect to the owners as "the elder" and might expect them to give advice and help him promote his career. On the contrary, the European owners are likely to regard their employees as independent decision-makers and see their own advice as not weighing more than the opinion of the young employee.

When it comes to the concept of job embeddedness, we see clear signs that employees feel well embedded in the organization but less embedded in the community, although different assumptions may affect the embeddedness:

Signs of organizational embeddedness:

Directors are friendly. We make jokes together. I feel like part of the community. I trust my boss, and my boss trusts me. Without trust you cannot do anything. (Employee #4)

The staff and the owners, we all play together – family like. When it is my day off, I still come here. (Employee #6)

Many staff members have been here many years: the supervisor has been here 10 years, others 3-4 years. […] The owners have their own way, so we take to what they want and move forward, but we also come with input. (Employee #8)

Signs of low community embeddedness:

Where do you see yourself in 5-10 years: Far away from here. I do not want family here. I want to go back where I came from. Sometimes people in the village think we earn a lot of money. (Employee #4)

I have a girlfriend. She does not live here. When my time is up, I'll go home. (Employee #6)

When Amankwaa et al. (2022) found that organizational embeddedness was a necessary mediator for innovative behavior, and in our case for transformation of labor to service jobs, our data shows similar tendencies to their findings. The owners had successfully embedded employees in the clan culture of their organization, but the employees were not well-embedded in the local community. On the contrary, our data suggested that it is easier for young people to embed themselves in the organization they work for, if they remove themselves from the community where they were brought up. Grimm et al. (2014) argue that

"kinship ties may become an important obstacle in the process of economic transition. Members of the kin system who achieve economic success in the modern sector may be confronted with sharing obligations by less successful fellows. […] The hypothesis is then that the need to meet such demands can adversely affect

the incentives of kin members to pursue and develop their economic activity in the modern sector." (p.128).

For young people who haven't yet started a family, leaving the community of one's kinship may be a way out of this dilemma, allowing them to make savings to pursue further education. Once they complete their education, our interviewees seemed ready to take on their kinship obligations, but maybe also ready to put less stress on the coming generations.

The fact that young workers move away from their kin and communities is a constraining dilemma for the idea of community participation tourism. To achieve the symbiosis between organizational and community embeddedness found in quantitative data from Western countries, our qualitative data suggest that organizations need to realize that they are dealing with communities and not just individual workers in isolation. But communities, consisting of a number of kinships, also need to change:

Kinship constructions are, however, not immutable. Unlike previous studies that showed the African kinship system as a given, imposed and constant, as a major way of life for Africans, kinship constructs are often deconstructable and reconstructable even in Africa [...]. What determines continuous kinship subscription is its "usefulness" to individuals and whether the kinship relations threaten the existence of the person and weakens or enhances their life chances. (Akanle & Olutayo, 2011)

The young migrant workers we met felt their life chances were weakened by their obligations towards their kin. They wanted education, but kinship obligations did not allow them to make savings. By relocating they deconstructed kinship. But that did not mean they gave up on their kin. They seemed to hope to reconstruct it later in life, when educated and ready to establish a family. Furthermore, they had to see their kinship structure as externally given; but as emphasized above and by Alber et al. (2010), kinship structures are not static but change over time - even in "rural areas in the hinterland of big towns" (Alber et al., 2010 p.45).

For community participation tourism to work in our context, one way could be to promote community-based leadership rather than circumventing completely governance structures in the local communities. When local youth migrate to work elsewhere, and local organizations prefer to hire migrant workers from other communities, local communities get to play no role at all. According to Haruna (2009), who is critical of the benefits of a "leader-follower" perspective on leadership, "community-based leadership has a better chance of succeeding because it taps into and builds on sub-Saharan Africa's culture of community, collaboration, and cooperation" (p.942). This, however, may imply a need to be more ambitious when it comes to tourism development in rural areas of sub-Saharan Africa. Furthermore, it also raises the question whether assuming equality as a preference in leadership in non-western contexts may collide with employees that expect and view hierarchical and paternalistic leadership styles as a positive and desired approach. Maybe community participation tourism is an oxymoron that should be replaced by a more ambitious concept of community tourism.

CONCLUSION AND FURTHER PERSPECTIVES

Ghana is currently struggling with a situation of jobless growth, where especially rural areas experience no growth while the capital Accra gets richer – but without providing more jobs for potential new inhabitants from rural areas. To overcome this situation "industries without smokestacks" are suggested as a way forward for rural areas (Aryeetey et al, 2021). The focus is on two areas of "smokestack free" industries: tourism and processing of agricultural produce.

The region of our case organization holds the potential to benefit from both these industries. The climate, beaches, wildlife and cultural heritage sites could form a solid base for tourism at a much larger scale than what is seen today. Fishing and agriculture in the area is still dominated by small-scale traditional methods, and this could give way to a second industry also benefiting from tourism: the supply of local produce and experiences related to agriculture and fishing. The two industries could enter a positive feedback loop where local communities would not merely benefit as low skilled labor in the tourism sector, but where community members could become entrepreneurs starting independent businesses thriving on the flow of tourists.

However, we witnessed no signs that our community was on the verge of entering such a positive feedback loop; neither did we observe them elsewhere in the region. During our stay we visited the Elmina castle – a UNESCO world heritage site with over 40.000 yearly visitors (Aryeetey et al, 2021). As international visitors we were not invited to visit the surrounding village. There were no cafés providing refreshments or vendors selling local produce targeted at tourists. Visitors arrived in buses, took the tour of the castle, and left.

Labor transformation originally dealt with the changes from primary to secondary and finally tertiary employment. Today we realize that labor transformation deals with the formation of identities. For the individual worker, moving between primary, secondary and tertiary sector may not seem like a big deal, and many people make several changes between the sectors during their work life. Individuals adapt to what is expected from them; obedience and hard work in the primary sector, skills and organizational understanding in the secondary sector and understanding customer needs and being innovative in the tertiary sector.

Communities, on the other hand, do not transform that easily. In communities it is not about the role you take on as a paid employee – it is about who you are as a person and about which role you take in the community. It is about the coevolution of individual and community values.

A community basing its subsistence on small-scale fishing and farming develops communal values since they have one fixed pool of resources to draw from. What one community member brings to the table negatively affects what another member may bring. Communal values ensure the survival of the community – not merely the survival of the individual.

A community basing its subsistence on industrial production, on the other hand, values the individual since individual entrepreneurship and ingenuity enlarges the pool of resources that the community can draw from. History still has to show the effect on values in communities basing their subsistence on services, but the concept of job embeddedness suggests that happiness and job satisfaction for the individual are important factors when your job is to provide service to others. Job embeddedness has two sides to it. To be happy and satisfied with your work, you must feel embedded in the organization, and you must feel embedded in the community where you live.

Our research question was: *What hinders the transformation of community members from becoming paid employees in the tourism sector*? In the literature on job embeddedness, we found a potential conflict that may prevent job transformation from traditional small-scale farming to the tourism industry. Quantitative studies in the tourism industry in Ghana suggested that community embeddedness played a different role in Ghana than the role identified in Western contexts. Whereas the quantitative study could not explain this abnormality, our qualitative study suggested an explanation. Young employees get caught between the values of their community and the values of their work organization. From their employers at the resort, employees were met with expectations of being loyal and satisfied with the good working conditions they had been given. From the community, employees were met with expectations of sharing this "luck" with the remaining community, even to the extent of being disloyal to the workplace.

This situation has turned into a deadlock with severe consequences for both the organization (the tourist resort) and for the community. First-order benefits for the community, i.e., the ability of some community members to secure a paid job at the resort, discontinued. Consequently, second-order benefits never came to be. The resort could not present local produce for their guests, and they could not refer their guests to a nice tourist friendly village to visit. Community participation tourism was not possible because the necessary labor transformation, when seen through a prism of communities, is not possible. Individuals can be transformed to service sector employees, but only when they leave their homes communities. Such individuals are not integrated in a new community but remain strangers in the village of the tourist resort. This is a major drawback for the aspect of community participation tourism in rural Ghana, and one for which there is no easy solution.

REFERENCES

Addo, E. (2011). European heritage and cultural diversity: The bricks and mortar of Ghana's tourism industry. *Journal of Contemporary African Studies*, *29*(4), 405–425. doi:10.1080/02589001.2011.607017

Adiyia, B., Vanneste, D., & Van Rompaey, A. (2017). The poverty alleviation potential of tourism employment as an off-farm activity on the local livelihoods surrounding Kibale National Park, western Uganda. *Tourism and Hospitality Research*, *17*(1), 34–51. doi:10.1177/1467358416634156

Adu-Ampong, E. A. (2019). Historical Trajectories of Tourism Development Policies and Planning in Ghana, 1957–2017. *Tourism Planning & Development*, *16*(2), 124–141. doi:10.1080/21568316.2018.1537002

Adu-Ampong, E. A., & Mensah, C. (2021). Tourism and hospitality research in the peripheries: Thematic focus and a research agenda from Ghana. *African Journal of Hospitality, Tourism and Leisure*, *10*(3), 1092–1109. doi:10.46222/ajhtl.19770720-151

Akanle, O., & Olutayo, O. (2011). Kinship construction variability among Nigerian international migrants: The context of contemporary Diaspora. *Human Affairs*, *21*(4), 470–480. doi:10.247813374-011-0044-2

Akyeampong, O. (2009). Tourism development in Ghana, 1957-2007. *Legon Journal of Sociology*, *3*(2), 1–23.

Akyeampong, O. A. (2011). Pro-poor tourism: Residents' expectations, experiences and perceptions in the Kakum National Park Area of Ghana. *Journal of Sustainable Tourism*, *19*(2), 197–213. doi:10.1080/09669582.2010.509508

Alber, E., Häberlein, T., & Martin, J. (2010). Changing Webs of Kinship: Spotlights on West Africa. *Africa Spectrum*, *45*(3), 43–67. doi:10.1177/000203971004500303

Amankwaa, A., Susomrith, P., & Seet, P.-S. (2022). Innovative behavior among service workers and the importance of leadership: Evidence form an emerging economy. *The Journal of Technology Transfer*, *47*(2), 506–530. doi:10.100710961-021-09853-6

Ankomah, P. K. (1991). Tourism skilled labor: The case of sub-Saharan Africa. *Annals of Tourism Research*, *18*(3), 433–442. doi:10.1016/0160-7383(91)90050-L

Annor, F., & Burchell, B. (2017). A cross-national comparative study of work demands/support, work-to-family conflict and job outcomes. *International Journal of Cross Cultural Management*, *18*(1), 53–72. doi:10.1177/1470595817746195

Aryeetey, E., Baffour, P. T., & Turkson, F. E. (2021). *Employment creation potential, labor skills requirements, and skill gaps for young people: Ghana case study*. African Growth Initiative at Brookings.

Ashley, C., Boyd, C., & Goodwin, H. (2000). Pro-poor tourism: Putting poverty at the heart of the tourism agenda. *Natural Resource Perspectives*, *51*. http://www.odi.org.uk/nrp/51.pdf

Aykaç, A. (2007). Labor transformation through the prism of the household: Evidence from the Turkish tourism industry. *New Perspectives on Turkey*, *37*, 87–113. doi:10.1017/S089663460000474X

Bakker, M. (2019). A Conceptual Framework for Identifying the Binding Constraints to Tourism-Driven Inclusive Growth. *Tourism Planning & Development*, *16*(5), 575–590. doi:10.1080/21568316.2018.1541817

Bakker, M., & Messerli, H. R. (2017). Inclusive growth versus pro-poor growth: Implications for tourism development. *Tourism and Hospitality Research*, *17*(4), 384–391. doi:10.1177/1467358416638919

Beugré, C. D., & Offodile, O. F. (2001). Managing for organizational effectiveness in sub-Saharan Africa: A culture-fit model. *International Journal of Human Resource Management*, *12*(4), 535–550. doi:10.1080/713769650

Blackwood, E. (1997). Women, Land, and Labor: Negotiating Clientage and Kinship in a Minangkabau Peasant Community. *Ethnology*, *36*(4), 277–293. doi:10.2307/3774038

Booyens, I. (2022). Social innovation for sustainable tourism development. In I. Booyens & P. Brouder (Eds.), *Handbook of innovation for sustainable tourism* (pp. 193–209). Edward Elgar. doi:10.4337/9781800372740.00020

Burgoyne, C., & Mearns, K. (2017). Managing stakeholder relations, natural resources and tourism: A case study from Ololosokwan, Tanzania. *Tourism and Hospitality Research*, *17*(1), 68–78. doi:10.1177/1467358416639088

Chen, Z, Huang, S, Liu, C, Min, M, & Zhou, L. (2018). Fit between Organizational Culture and Innovation Strategy: Implications for Innovation Performance. *Sustainability*, *10*(10), 3378. MDPI. doi:10.3390/su10103378

Cobbinah, P. B., & Darkwah, R. M. (2016). Reflections on tourism policies in Ghana. *International Journal of Tourism Sciences*, *16*(4), 170–190. doi:10.1080/15980634.2016.1212595

Eshun, G., & Tichaawa, T. M. (2019). Reconsidering participation for local community well-being in ecotourism in Ghana. *Geo Journal of Tourism and Geosites*, *27*(4), 1184–1200. doi:10.30892/gtg.27406-425

Eshun, G., & Tichaawa, T. M. (2020). Community participation, risk management and ecotourism sustainability issues in Ghana. *Geo Journal of Tourism and Geosites*, *28*(1), 313–331. doi:10.30892/gtg.28125-472

Fafchamps, M. (2004). *Market institutions in Sub-Saharan Africa: Theory and evidence*. MIT Press.

Farrington, J., Carney, D., Ashley, C., & Turton, C. (1999). Sustainable livelihoods in practice: Early applications of concepts in rural areas. *Natural Resource Perspectives*, *42*, 1–15.

Fatehi, K., Priestley, J. L., & Taasoobshirazi, G. (2020). The expanded view of individualism and collectivism: One, two, or four dimensions? *International Journal of Cross Cultural Management*, *20*(1), 7–24. doi:10.1177/1470595820913077

Flyvbjerg, B. (2006). Five Misunderstandings About Case-Study Research. *Qualitative Inquiry*, *12*(2), 219–245. doi:10.1177/1077800405284363

Fortainier, F., & van Wijk, J. (2010). Sustainable tourism industry development in sub-Saharan Africa: Consequences of foreign hotels for local employment. *International Business Review*, *19*(2), 191–205. doi:10.1016/j.ibusrev.2009.11.007

Granovetter, M. (1983). The Strength of Weak Ties: A Network Theory Revisited. *Sociological Theory*, *1*, 201–233. doi:10.2307/202051

Grimm, M., Gubert, F., Koriko, F. O., Lay, J., & Nordman, C. J. (2013). Kinship-ties and entrepreneurship in Western Africa. *Journal of Small Business and Entrepreneurship*, *26*(2), 125–150. Advance online publication. doi:10.1080/08276331.2013.771854

Haruna, P. F. (2009). Revising the Leadership Paradigm in Sub-Saharan Africa: A Study of Community-Based Leadership. *Public Administration Review*, *69*(5), 941–950. doi:10.1111/j.1540-6210.2009.02043.x

Hummel, J., & van der Duim, R. (2012). Tourism and development at work: 15 years of tourism and poverty reduction within the SNV Netherlands Development Organisation. *Journal of Sustainable Tourism*, *20*(3), 319–338. doi:10.1080/09669582.2012.663381

Jaafar, M., Rasoolimanesh, S. M., & Ismail, S. (2017). Perceived sociocultural impacts of tourism and community participation: A case study of Langkawi Island. *Tourism and Hospitality Research*, *17*(2), 123–134. doi:10.1177/1467358415610373

Karatepe, O. M. (2011). Job resourcefulness as a moderator of the work-family conflict – job satisfaction relationship: A study of hotel employees in Nigeria. *Journal of Hospitality and Tourism Management*, *18*(1), 10–17. doi:10.1375/jhtm.18.1.10

Kidido, J. K., & Baah, K. (2020). Sharecropping arrangement in the contemporary agricultural economy of Ghana: A study of Techiman North District and Sefwi Municipality, Ghana. *Journal of Planning and Land Management*, *1*(2), 50–62. doi:10.36005/jplm.v1i2.22

Medina-Muñoz, D. R., Medina-Muñoz, R. D., & Gutiérrez-Pérez, F. J. (2016). A Sustainable development approach to assessing the engagement of tourism enterprises in poverty alleviation. *Sustainable Development (Bradford)*, *24*(4), 220–236. doi:10.1002d.1624

Mensah, I. (2017). Benefits and challenges of community-based ecotourism in park-fringe communities: The case of *Mesomagor* of Kakum National Park, Ghana. *Tourism Review International*, *21*(1), 81–98. doi:10.3727/154427217X14866652018947

Mitchell, T. R., Holtom, B. C., Lee, T. W., Sablynski, C. J., & Erez, M. (2001). Why People Stay: Using Job Embeddedness to Predict Voluntary Turnover. *Academy of Management Journal*, *44*(6), 1102–1121. doi:10.2307/3069391

Mokomane, Z. (2014). Introduction. In Z. Mokomane (Ed.), *Work–Family Interface in Sub-Saharan Africa. International Perspectives on Social Policy, Administration, and Practice*. Springer. doi:10.1007/978-3-319-01237-7_1

Mokomane, Z., & Chilwane, D. (2014). A Review of Work–Family Research in Sub-Saharan Africa. In Z. Mokomane (Ed.), *Work–Family Interface in Sub-Saharan Africa. International Perspectives on Social Policy, Administration, and Practice*. Springer. doi:10.1007/978-3-319-01237-7_12

Okazaki, E. (2008). A Community-Based Tourism Model: Its Conception and Use. *Journal of Sustainable Tourism*, *16*(5), 511–529. doi:10.1080/09669580802159594

Okello, M. M., & Novelli, M. (2014). Tourism in the East African Community (EAC): Challenges, opportunities, and ways forward. *Tourism and Hospitality Research, 14*(1–2), 53–66. doi:10.1177/1467358414529580

Peters, P. E. (2019). Revisiting the Social Bedrock of Kinship and Descent in the Anthropology of Afrika. R.R. Grinker, R.R, Lubkemann, S.C., Steiner, C.B. & Gonçalves, E. (eds). A Companion to the Anthropology of Africa. John Wiley & Sons. doi:10.1002/9781119251521

Platteau, J.-P. (2014). Redistributive Pressures in Sub-Saharan Africa: Causes, Consequences, and Coping Strategies. Akyeampong, E., Bates, R.H., Nunn, N. & Robinson, J. (eds) Africa's Development in Historical Perspective, pp. 153 – 207. Cambridge University Press. doi:10.1017/CBO9781139644594.008

Putnik, K., Houkes, I., Jansen, N., Nijhuis, F., & Kant, I. (2018). Work-home interface in a cross-cultural context: A framework for future research and practice. *International Journal of Human Resource Management, 31*(13), 1–18. doi:10.1080/09585192.2017.1423096

Quinn, R., & Rohrbaugh, J. (1983). A Spatial Model of Effectiveness Criteria: Toward a Competing Values Approach to Organizational Analysis. *Management Science, 29*(3), 363–377. doi:10.1287/mnsc.29.3.363

Ramesh, A., & Gelfand, M. J. (2010). Will they stay or will they go? The role of job embeddedness in predicting turnover in individualistic and collectivistic cultures. *The Journal of Applied Psychology, 95*(5), 807–823. doi:10.1037/a0019464 PMID:20836585

Sharpley, R. (2020). Tourism, sustainable development and the theoretical divide: 20 years on. *Journal of Sustainable Tourism, 28*(11), 1932–1946. doi:10.1080/09669582.2020.1779732

Singelmann, J. (1978). The sectoral transformation of the labor force in seven industrialized countries, 1920-1970. *American Journal of Sociology, 83*(5), 1224–1234. doi:10.1086/226680

Snyman, S. (2014). The impact of ecotourism employment on rural household incomes and social welfare in six southern African countries. *Tourism and Hospitality Research, 14*(1–2), 37–52. doi:10.1177/1467358414529435

Spenceley, A., & Snyman, S. (2017). Can a wildlife tourism company influence conservation and the development of tourism in a specific destination? *Tourism and Hospitality Research, 17*(1), 52–67. doi:10.1177/1467358416634158

Stauth, G. (1984). Households, Modes of Living, and Production Systems. Smith, J., Wallerstein, I. & Evers, H.-D. (eds), Households and the World-Economy. Sage.

Tavory, I., & Timmermans, S. (2014). *Abductive Analysis: Theorizing Qualitative Research. Cicago*. University of Chicago Press. doi:10.7208/chicago/9780226180458.001.0001

Tosun, C. (2000). Limits to Community Participation in the Tourism Development Process in Developing Countries. *Tourism Management, 21*(6), 613–633. doi:10.1016/S0261-5177(00)00009-1

Twining-Ward, L. D., Li, W., Bhammar, H. M., & Wright, E. (2018). *Supporting Sustainable Livelihoods through Wildlife Tourism*. The World Bank Group. doi:10.1596/29417

United Nations. (2023). *Global Issues: Ending poverty*. UN. https://www.un.org/en/global-issues/ending-poverty, accessed on April 26, 2023.

UNWTO. (2018a). *Sustainable development*. UNWTO. http://sdt.unwto.org/content/about-us-5,

UNWTO. (2018b) *Sustainable Development of Tourism, One Planet – Sustainable Tourism Programme*. UNWTO. http://sdt.unwto.org/about-10yfp-stp,

Yankholmes, A., & McKercher, B. (2015). Understanding visitors to slavery heritage sites in Ghana. *Tourism Management, 51*, 22–32. doi:10.1016/j.tourman.2015.04.003

Yin, R. K. (2009). *Case study research: Design and methods* (4th ed.). SAGE.

Zeb, A., Hussain, K., Khan, F., & Safi, A. (2021). The competing value framework model of organizational culture, innovation and performance. *Business Process Management Journal, 27*(2), 658–683. doi:10.1108/BPMJ-11-2019-0464

ENDNOTES

1 The three authors have a background in tourism research, sociology and economics respectively, and thus did not have any unspoken preunderstanding. Thus, all aspects were open for discussion.

2 The Ghanian government appear to be aware of this potential conflict as they already in a educational reform in 1987 sought to "*de-emphasize the grammar-school type education*" (Aryeestey et al., 2021 p.57).

Chapter 3
Workplace Integration of Employees in the Hospitality Industry:
A Critical Perspective

Emil Nikolov Kotsev
https://orcid.org/0000-0001-6848-3051
University of Ruse, Bulgaria

ABSTRACT

The purpose of this chapter is twofold: first, to examine and explain the existing theory and practice of workplace integration in the organization, and second, to propose alternative mechanisms that could contribute to positive social changes in hospitality industry organizations. Critical research methodology is used to increase explanation in workplace integration research. Employers' and employees' perceptions and attitudes toward workplace integration are explored through interviews and focus group discussions in four hotels. Fourteen thematic categories (or demi-regularities) are identified as being able to affect the process of workplace integration and generate social impact in the hotel industry. The demi-regularities are explained through abductive reasoning and as a result, eleven workplace integration practices are formulated. Finally, ten practices, or 'mechanisms,' to encourage workplace integration in hotel companies are approved through retroduction.

DOI: 10.4018/978-1-6684-9255-0.ch003

INTRODUCTION

Human integration processes are very well studied at the national and international levels, but at the organizational level, they began to sharpen the interest of scholars only in the last two decades. However, it is debatable whether studies of workplace integration may have a place in this book. Normally, these studies are based on pragmatist philosophy and are carried out to increase individual and organizational efficiency. The process of integration is perceived as a positive phenomenon, desired by all parties concerned – both subjects (the influencing parties) and objects (the parties being influenced) of integration. Thus, some questions naturally arise. For example, is it possible for a study of workplace integration to adopt a critical stance? What could be wrong with workplace integration?

The answer to these questions requires specifying in advance what is meant by the term "workplace integration". In the scientific literature, there are different definitions of this concept, and it is often described as a *process of assimilation* (e.g. Florea, 2014; Berke-Berga, 2015). The addition of Roth et al. (2023) that it is a *two-way process of mutual assimilation* brings some clarity but does not fully reveal the essence of this process. That is why, for the needs of the present study, a relatively more thorough definition (Robeva-Stoyanova, 2023, p. 51) is used, according to which workplace integration is "a process of purposefully exerted managerial influences, in which individuals or groups are permanently committed to the goals of the organization and become a functionally indivisible part of it". More precisely, it concerns affective commitment, characterized by emotional engagement (Allen & Meyer, 1990).

To this definition, the existing organizational literature adds that people integration at work is a challenging process (Dglawi & Hassel, 2021) and deserves special consideration (Ploesser & Russell, 2009). Some experts are even more categorical, describing people integration as "the toughest challenge enterprises encounter today" (Tejaswi, 2010). With such definitive statements, it is normal to expect the research community to step into its role and offer businesses appropriate models and techniques for integrating all employees in the organization.

From this point of view, the tendency of scholars to focus on a limited number of social groups and individuals as the object of integration in the organization is at least puzzling. Most often, their interest is focused on refugees and migrants (see Hirst et al., 2021; Rajendran et al., 2017; Mahmud et al., 2014; Schmidt & Müller, 2021; McBrien et al., 2022; Ertorer et al., 2022), disabled (e.g., Ward & Baker, 2005; Freeman et al., 2005; Geisen & Harder, 2011; Geisen, 2014) and veterans (see Gonzalez & Simpson, 2021; Mael et al., 2022). Minorities based on gender and race complete this too short list of objects of integration that attract permanently the interest of the world research community (see Lee & Peccei, 2007; Stainback et al.,

2005). Thus, the majority of employees in organizations remain beyond the reach of integration attempts and outside the scope of scientific thought.

The present study identifies this weakness and "rings the bell" for attention to the real issues of business, in particular hospitality industry organizations. Since people are the most critical asset in service businesses (see Lasrado & Lasrado, 2018; Pham, 2020), and the hospitality industry is extremely labor-intensive (King, 2010; Elshaer & Marzouk, 2019) and dependent on garnering a fully engaged workforce (Thompson, 2016), the proposition that employee integration has proven especially critical in hotel companies (Emperger, 2004) can be accepted as logical. How does it correspond to the fact that the development of the theory of workplace integration of employees in hospitality industry organizations is still in its embryonic phase?

The lag of scientific knowledge in this area is more than obvious, but it is not the focus of the criticism expressed in this chapter. Previous research reports poor human resource practices (Kusluvan, 2003) throughout the hospitality industry, low organizational loyalty, insufficient employers' incentives to increase employees' sense of belonging (Mooney, 2014), and lack of identification with the employing organization (Lee-Ross, 2005; Lundberg et al., 2009). At the same time, it is clear that the industry needs dedicated employees, absorbed by their work (Karatepe & Olugbade, 2009). This makes the need for a more critical analysis of personnel practices in hotel companies, especially relevant. The objectivity of this analysis requires a good knowledge of the attitudes and behaviors of senior management regarding the integration of their employees.

It is clear that employees' perceptions of the work environment influence work commitment and engagement (Helms & Stern, 2001), but little is known about the specific characteristics of the work environment, influencing perceptions in the hotel industry. At the present moment, the extant hospitality literature does not provide a sufficient explanation for why many hotel companies fail to successfully integrate their employees. The research described in this chapter can be viewed as an attempt to illuminate some of the problems of workplace integration in the hospitality industry and to propose alternatives for their solution. The expectations are that these alternatives have the potential to provoke better relationships and communication between all interested parties – employers and employees, superiors and inferiors, white collars, and blue collars.

METHOD

Aim, Objectives, and Research Questions

The purpose of this chapter is twofold: first, to examine and explain the existing theory and practice of workplace integration in the organization, and second, to propose alternative mechanisms that could contribute to positive social changes in hospitality industry organizations.

The main objectives are as follows:

O1. To explore employers' and employees' perceptions and attitudes regarding workplace integration in four hotel companies.
O2. To explain O1 observations through abductive reasoning (theoretical redescription).
O3. To identify alternative mechanisms that could encourage employee integration and contribute to positive social changes in hospitality industry organizations.

Taking into account the existing conceptual framework and the objectives of the study, the following research questions are addressed:

Q1. What are the predominant perceptions and attitudes towards workplace integration of employees and employers in the four hotels studied?
Q2. How can existing concepts be redescribed to offer a better explanation of O1 observations?
Q3. What specific management practices could encourage employee integration in the four hotels studied?

Research Design and Data Collection

The research philosophy of this study is critical realism, since it recognizes that there is an objective reality, the perception of which is affected by social factors (O'Mahoney, 2016). Ontologically, critical realism believes that reality exists independently of our knowledge and identifies three domains of reality: the empirical, the actual, and the real (Bhaskar, 2013). Critical realist research aims at explaining structures and mechanisms rather than events (Robson, 2002) and assumes that there may be more than one feasible option to manage a situation (Price, 2016). Accordingly, the study employs a three-step critical research methodology – identification of demi-regularities, abduction, and retroduction (Fletcher, 2017). The research perspective follows an abductive and retroductive approach that combines elements from inductive and deductive research (Downward & Mearman, 2007).

The research design of the study is qualitative. Phenomenological and grounded theory approaches were used to describe and explain lived experiences and perceptions of participants. Empirical data was obtained through a combination of two research strategies. Since the study aims to examine and explain the existing theory and practice of workplace integration, descriptive research was involved. It also aims to propose alternative mechanisms, which implies an exploratory research strategy.

Semi-structured interviews and focus groups were used for primary data collection (Saunders et al., 2019). These were conducted with management teams of four wellness hotels (two Bulgarian and two Greek) as a part of a two-day in-company training course on communication skills and teamwork. The participants (N=37) formed four groups, each having 8-10 members. Most of the participants were female (83.8%), aged 30-49 (75.7%), and had three to five years of work experience as managers (54%).

Secondary data collection involved retrieving information on workplace integration from non-participant observations of organizational structures, scientific journals, reports, and books.

RESULTS

The Empirical Domain

As expected, the similarity of activities in the studied objects is reflected in their almost identical organizational structures. They can be described as 'flat' structures: directly subordinate to the general manager (in one of the hotels this position is called Managing Director) are several positions, including Hotel Manager, Food & Beverage Manager, Head Chef, Human Resource Manager, Spa Manager, Maintenance Supervisor, and Chief Accountant (called Financial Director in one of the hotels). Reception / front office managers and head housekeepers work within the span of control of the hotel managers.

Except for the head chefs, all other employees in managerial positions joined the training and participated in the research study. It seems that the content of the training had a positive effect on the willingness of the participants in the focus groups to share their opinions openly. Most of them expressed satisfaction with the conducted training and noted its importance for increasing their individual engagement and commitment, as well as for team cohesion in general. Here are just a few of their comments:

Head housekeeper (39):

We should do such training more often because they bring the team together... I don't remember the last time they asked for my opinion on one issue or another. Besides, I don't push myself to share so I don't suffer afterward (Laughs). I am flattered by the thought that I'm a good professional and sometimes I disagree with my boss, but guess what – I just keep quiet and perform. I don't want to get in trouble. The boss is the boss. He may know things that I'm not allowed to know.

Front office manager (35):

Of all that this training has given us, I find the opportunity for open communication between us to be the most beneficial. Working with bosses is cool. They asked me about my personal needs, can you imagine?... It was the first time we discussed our common goals as a team. We know the individual ones. We found many common values that make it worth the effort. Of course, I'll feel better knowing that my work is valued by the company (Starts to sing the ABBA song "Money, money, money").

Chief accountant (55):

Until now, I had not thought about how useful such meetings can be. We got closer to discussing and solving issues together. We probably would have become even closer if we knew our professions better... I understand the nature of the hotel business, and I think I'm doing my work well. Perhaps I could perform even better if I knew in detail the work of colleagues from other departments. And most likely they too would have achieved better results if they knew mine. For example, the Restaurant Manager would have finally provided adequate control procedures, and the Spa Manager might have been more careful about what consumables she ordered (Smiles).

The hotel manager (47):

Integration is a good thing, but in our business, there is rarely time to think about it. Customer satisfaction is the most important thing for us. My daily work routine is overwhelmed with constant firefighting... My long experience has taught me that if you want a job done well, you have to do two things. First, explain to them in detail what you expect, and second, constantly keep your eyes open. If they notice you are not there, the work stops. I know this will make integration difficult, but you have to constantly monitor and check them. Someone had said that verification is the highest form of trust. And he was right.

If this feedback can be somewhat expected, it is a different matter from the opinion expressed by the four general managers. During the individual interviews, they expressed identical attitudes, well summarized in the following comment by one of them:

Instead of making special efforts to integrate people, it is much better to think in advance, even during the selection of employees. You cannot integrate a person who does not want to integrate. That is why such a person should not be hired at all.

The results of the conducted interviews and focus group discussions reveal differences in employers' and employees' perceptions and attitudes regarding workplace integration. It seems that inferiors react positively and manifest readiness to the possibility of their further integration into the organization, while superiors (the general managers) do not see a particular need to make efforts for employee integration. However, observed attitudes and behaviors cannot fully reveal the causal mechanisms of social interactions. Going beyond the empirical domain of reality implies the identification of some common patterns (also called demi-regularities). They can provide research focus and "contribute to the generation of hypotheses about causal mechanisms" (Lawson, 1997, p. 14).

Identification of Demi-Regularities

The process of data analysis in critical realism begins with the identification of demi-regularities (Fletcher, 2017; Peter & Park, 2018), which is analogous to thematic coding (Kirca-Demirbaga, 2021). Several thematic categories (or demi-regularities) were identified in this study as being able to affect the process of workplace integration and generate social impact in the hotel companies:

1. Team building and sensitivity training ("*We should do such training more often because they bring the team together...*")
2. Open communication ("*Of all that this training has given us, I find the opportunity for open communication between us to be the most beneficial...*")
3. Appreciation of subordinates' ideas ("*I don't remember the last time they asked for my opinion on one issue or another... Sometimes I disagree with my boss, but guess what – I just keep quiet and perform. I don't want to get in trouble...*")
4. Interaction with the executer ("*Working with bosses is cool...*")
5. Attention to subordinates' personal needs ("*They asked me about my personal needs, can you imagine?*")
6. Identification of common goals ("*It was the first time we discussed our common goals as a team...*")

7. Emphasis on shared values (“*We found many common values that make it worth the effort...*”)
8. Innovative reward and recognition practices (“*I’ll feel better knowing that my work is valued by the company...*”)
9. Joint problem solving (“*We got closer discussing and solving issues together...*”)
10. Mutual learning within the group (“*We probably would have become even closer if we knew our professions better...*”)
11. Detailed specification of tasks (“*Explain to them in detail what you expect...*”)
12. Strict management control (“*I know this will make integration difficult, but you have to constantly monitor and check them*”)
13. Withholding information/knowledge hiding (“*The boss is the boss. He may know things that I’m not allowed to know*”)
14. Disrespecting the principle of delegated responsibility (“*Constantly keep your eyes open. If they notice you are not there, the work stops*”).

It is hypothesized that most of the demi-regularities listed above will have a positive impact on workplace integration, while the last four of them will affect adversely the degree of employee integration. Theoretical explanation of these thematic categories is done through abductive reasoning (Haig, 2005), following the principle that in critical realism the focus is on studying demi-regularities through abduction (Collins, 2019).

Abductive Reasoning

Abduction involves reasoning that goes beyond observation and experience to their possible explanations (Mäkinen, 2018) and can be defined as the process of forming explanatory hypotheses (Frankfurt, 1958). Based on the identified demi-regularities, in this study, fourteen hypotheses are formulated. Their brief theoretical rationale is presented below.

Hypothesis 1: Team building and sensitivity training encourage employee commitment and workplace integration in hotel companies.

Paradoxically, sensitivity training is not among the most used training methods in the hotel industry, although it is among the methods perceived as most effective (Furunes, 2005; Luo et al., 2019). Nevertheless, several recent studies report a positive impact of sensitivity/group training and team building on employee commitment in the hospitality industry (e.g. Coston & Salazar, 2011; Bies, 2019; Achieng, 2021; Sandjong, 2021) Therefore, this hypothesis is theoretically sound.

Hypothesis 2: Open communication encourages employee commitment and workplace integration in hotel companies.

Open communication is when people can freely express their ideas and thoughts to one another. This approach has positive effects on many aspects of organizational performance, but for the purpose of this study, it is important to note the unanimity with which scholars support the idea that fostering a sense of open communication stimulates employee commitment in the hospitality industry (see Dawson & Abbott, 2011; Rieu & Kamara, 2016; Taylor, 2019; Sandjong, 2021). Therefore, hypothesis 2 is theoretically sound.

Hypothesis 3: Appreciation of subordinates' ideas encourages employee commitment and workplace integration in hotel companies.

Paying attention to subordinates' ideas and their positive effect on workplace integration is not new to scholars (see Kotsev, 2008; Sarmento et al., 2016; Qian et al., 2020). This approach is not among the most widespread in the hotel sector, but here, too, research shows a clear and significant positive relationship between appreciation of subordinates' ideas and employee commitment (Marshall et al., 2015; Mansour et al., 2021; Tariq et al., 2023). Therefore, hypothesis 3 is theoretically sound.

Hypothesis 4: Interaction with the executer encourages employee commitment and workplace integration in hotel companies.

There are probably subordinates who prefer to work together with their superiors (as was the case with two of the participants in the focus groups), but there are at least as many people who prefer to work independently (Pagés, 2010; Simpson et al., 2022). More importantly in this case, the study of the available literature did not lead to information that confirms the existence of a relationship between 'interaction with the executer' and 'workplace integration'. Therefore, hypothesis 4 is not theoretically sound.

Hypothesis 5: Attention to subordinates' personal needs encourages employee commitment and workplace integration in hotel companies.

Paying attention to the individual needs of subordinates is among the key characteristics of transformational leaders. Its positive influence on employee engagement and commitment has been studied and proven by numerous scholars (e.g. Park, 2019; Mi et al., 2019; Sondaitė & Keidonaitė, 2020). It turns out that for successful workplace integration in the hospitality industry, satisfying the need for belonging is particularly important (Srivastava & Singh, 2020; Dar & Rahman, 2022). Therefore, hypothesis 5 is theoretically sound.

Hypothesis 6: Identification of common goals encourages employee commitment and workplace integration in hotel companies.

It is no surprise to anyone anymore that common goal identification and attainment is among the factors affecting employee commitment in the workplace (Estigoy et al., 2020). A closer look even shows that the relationship is bidirectional – workplace integration, in turn, "keeps the organization focused on common goals" (Fink, 1992, p. 9). There is some evidence that in hotel organizations the adoption of common goals can foster employee integration, as well (e.g. Pourbarkhordari et al., 2016; Folakemi et al., 2017; Pekerşen & Tugay, 2020; Saltik, 2023). Therefore, hypothesis 6 is theoretically sound.

Hypothesis 7: Placing emphasis on shared values encourages employee commitment and workplace integration in hotel companies.

The importance of creating a vision and mission based on shared values is pointed out in most leadership models (Davids, 2018). It is true that sometimes the strong emphasis on shared values in an organization may be a double-edged sword (Cha & Edmondson, 2006), but the positive effect it has on employee commitment is not doubted by scholars (e.g. Caldwell et al., 1990; Byrtek, 2000; Dolatabadi & Safa, 2010). In this regard, hotel organizations are no exception (e.g. Peyton, 2023; Deale & Lee, 2023). Therefore, hypothesis 7 is theoretically sound.

Hypothesis 8: The provision of flexible and innovative reward and recognition practices encourages employee commitment and workplace integration in hotel companies.

The implementation of more flexible and innovative reward and recognition practices affects the degree of employee integration (see Wright, 2004; Aladwan et al., 2015; Agustina et al., 2022), which Hendersen (1979) defines as the extent to which employees identify with the company and its goals. Numerous studies in the hospitality industry endorse these findings by arguing that employee recognition fosters employee engagement (see Ncube & Jerie, 2012; Chathoth et al., 2020; Chiwawa & Wissink, 2021) and commitment (see Ibrahim, 2020; Bae, 2021; Uzochukwu et al., 2023). Therefore, hypothesis 8 is theoretically sound.

Hypothesis 9: Joint problem-solving encourages employee commitment and workplace integration in hotel companies.

Joint problem-solving and its influence on employee commitment (see Walton, 1985; Caldwell, 1996; Kotsev, 2022) and workplace integration (e.g. Hyman & Mason, 1995; Rozuel & McGhee, 2012) has long attracted the interest of scholars of different generations. Furthermore, several studies in hospitality and tourism have reported a positive impact of employee involvement in decision-making and problem-solving on workplace integration (Lashley, 1996; Lashley, 1999; Hatjidis & Parker, 2018; Ugwu et al., 2022; Chathoth et al., 2022). Therefore, hypothesis 9 is theoretically sound.

Hypothesis 10: Mutual learning within the group encourages employee commitment and workplace integration in hotel companies.

An experiment conducted at the Colorado School of Mines with the participation of twenty-three students from various majors showed that mutual learning within a group broadens the horizons of its members and strengthens both their willingness and their ability to cooperate and interact (Sutton & Thompson, 1998). According to Bugdol (2015), mutual learning (also called cross-learning) and playing are useful employee integration methods. Although there is no evidence that similar research has been done in hospitality and tourism, it can be assumed that hypothesis 10 is theoretically sound.

Hypothesis 11: Too detailed specification of tasks discourages employee commitment and workplace integration in hotel companies.

After a thorough study of the available literature, no information was found that confirms or rejects the hypothesis that too detailed instructions and task specifications discourage employee commitment and workplace integration. Therefore, the initial formulation of hypothesis 11 is not theoretically sound.

Hypothesis 12: Strict management control discourages employee commitment and workplace integration in hotel companies.

Tight control is not suitable in a dynamic environment like the one in the hospitality industry (Khoa & Hoa, 2021) and has a negative impact on employee commitment (Walton, 1985). Regarding hospitality organizations, Ersoy and Ehtiyar (2021) recommend minimization or even withdrawal of strict control behaviors. Even before them, a similar opinion has been expressed by Naqvi (2017), who, however, assumes that the cultural factor might play a role. Nevertheless, hypothesis 12 could be viewed as theoretically sound.

Hypothesis 13: Withholding information discourages employee commitment and workplace integration in hotel companies.

According to Brooks (2012), intentional withholding of information has a negative impact on employee commitment. Bratley & Aloysius (2019) confirm his findings, reporting a significant relationship between employees' perceptions of transparency and affective commitment. Kremers (2018) expresses a similar opinion, stating that knowledge hiding in the organization leads to a decrease in affective commitment. All these publications recommend just the opposite – open communication, as a means of increasing employee commitment. However, this is exactly what hypothesis 2 states. Therefore, hypothesis 13 duplicates hypothesis 2 and does not need independent proof.

Hypothesis 14: Disrespecting the principle of delegated responsibility discourages employee commitment and workplace integration in hotel companies.

Numerous studies (see Kazlauskaite et al., 2006; Castro et al., 2008; Amehule, 2022) report a significant positive relationship between delegation of responsibility and employee commitment. Some hospitality industry researchers (e.g. Ahmed & Khalid, 2019) come to a similar conclusion. Therefore, hypothesis 14 is theoretically sound.

As a result of the abductive reasoning and contrary to the opinion of the four general managers, according to whom "you cannot integrate a person who does not want to integrate", eleven workplace integration practices were formulated and put through a process of retroduction.

Retroductive Reasoning

Retroductive reasoning derives from a broader understanding of conditions where data have been collected (Conlon et al., 2020) and involves testing the proposed mechanisms (Stutchbury, 2022) for employee integration. In this case, the testing implies presenting the proposed integration practices to the general managers and seeking feedback from them. Ten out of eleven proposed practices have received the general managers' approval. Their common recommendation was to combine two of the proposed practices ('less management control' and 'respecting the principle of delegated responsibility') into a more general one ('practicing employee empowerment'). What is more, after becoming familiar with the practices proposed, three of the general managers announced a positive shift in their opinion regarding the capability of influencing the degree of integration of an employee – they already assumed it was possible.

Finally, ten mechanisms to initiate or improve workplace integration were approved by the general managers, and, more importantly, by the three-step critical research methodology used in this study (identification of demi-regularities, abduction, and retroduction). These are as follows:

- Employee involvement in team building and sensitivity training;
- Open communication between management and employees;
- Paying attention to subordinates' ideas;
- Paying attention to subordinates' personal needs;
- Identification and attainment of common goals;
- Placing emphasis on shared values;
- Implementation of flexible and innovative reward and recognition practices;
- Employee involvement in decision-making and problem-solving;
- Team members' involvement in mutual learning within the group;
- Practicing employee empowerment;

CONCLUSION

In this chapter for the first time, critical research methodology was used to increase explanation in workplace integration research. Its first objective (O1) was achieved by exploring employers' and employees' perceptions and attitudes regarding workplace integration through interviews and focus group discussions. It became clear that employees perceive workplace integration as possible and necessary, while employers are more skeptical and, although they consider it needed, do not believe that this process can be influenced (answer of Q1). Contrary to their opinion, fourteen thematic categories (or demi-regularities) were identified as being able to affect the process of workplace integration and generate social impact in hotel companies. The demi-regularities were explained through abductive reasoning and as a result, eleven workplace integration practices were formulated (answer of Q2). Thus the second objective of the study (O2) was accomplished. Finally, ten practices, or 'mechanisms', to encourage workplace integration in hotel companies were approved through retroduction (O3 and Q3).

However, although Bohman's (2005) main criteria for an adequate critical study (to be explanatory about what is wrong with current reality and to identify actions to change it) are met here, the study is not without limitations. First, it covers a small number of hotel companies. It is unclear whether the survey would yield similar results in other organizations in the hospitality industry. Second, it disregards the impact of cultural diversity on individual perceptions and attitudes toward workplace integration. Third, the proposed mechanisms were approved by the employers, but have not been tested in practice and their reliability has not been properly proven yet.

Finally, the study enters too timidly into the three domains of reality (empirical, actual, and real) and leaves many questions unanswered. For example, what particular changes in organizational structure could encourage employee commitment and workplace integration in the hospitality industry? Or what other 'causal mechanisms' exist in the third, 'real' layer of reality? Is the employer's attitude towards workplace integration not one of the main 'causal mechanisms' for the current state? These and many other complex questions wait to be answered.

REFERENCES

Achieng, O. L. (2021). *Effect of training on employee commitment at Prideinn Azure Hotel*. School of Hospitality and Tourism Management.

Agustina, R., Yusuf, M., Sutiyan, O. S. J., Ardianto, R., & Norvadewi, N. (2022). Employee performance mediated quality of work-life relationship satisfaction on the job and organizational commitment. *Jurnal Darma Agung*, *30*(2), 589–605.

Ahmed, M. A., & Khalid, S. (2019). Empowering Leadership and Proactive Behavior: Mediating Role of Psychological Empowerment and Moderating Role of Leader-Follower Distance. *Abasyn University Journal of Social Sciences, 12*(1).

Aladwan, K., Bhanugopan, R., & D'Netto, B. (2015). The effects of human resource management practices on employees' organizational commitment. *The International Journal of Organizational Analysis*, *23*(3), 472–492. doi:10.1108/IJOA-11-2014-0822

Allen, N. J., & Meyer, J. P. (1990). The measurement and antecedents of affective, continuance, and normative commitment to the organization. *Journal of Occupational Psychology, 63*(1), 1–18. doi:10.1111/j.2044-8325.1990.tb00506.x

Amehule, S. (2022). Delegation of duty and employee commitment in hotels in Port Harcourt. *BW Academic Journal, 1*(1), 10.

Bae, S. (2021). Evaluating hospitality employees' various relationships and the effects on organizational commitment. *International Hospitality Review*. doi:10.1108/IHR-01-2021-0005

Berke-Berga, A., Dovladbekova, I., Sumilo, E., & Baumane-Vitolina, I. (2015). Development of employee integration: Share ownership perspective. *Proceedings of Economic Science for Rural Development, 39*, 79–88.

Bhaskar, R. (2013). *A realist theory of science*. Routledge. doi:10.4324/9780203090732

Bies, A. R. (2019). *Organizational Socialization Tactics and Learning in the United States Hotel Industry* [Doctoral dissertation, Purdue University Graduate School].

Bohman, J. (2005). Critical theory. Edward N. Zalta (ed.). Stanford Encyclopedia of Philosophy. Stanford Press.

Bratley, K. J., & Aloysius, S. M. (2019). Transparency in managerial practices and affective commitment. *The Journal of Business, 6*(2), 2019–2061.

Brooks, G. (2012). Misbehavior, its dimensions, and relationship to commitment in organizations. *Rethinking misbehavior and resistance in organizations, 19*, 237-257.

Bugdol, M., Jedynak, P., Bugdol, M., & Jedynak, P. (2015). Integration in Different Organizational Situations. *Integrated Management Systems*, 161-192.

Byrtek, G. J. (2000). Creating shared values: An exploratory examination of methodology. [Doctoral dissertation, Walden University].

Caldwell, D. F., Chatman, J. A., & O'Reilly, C. A. (1990). Building organizational commitment: A multi-firm study. *Journal of Occupational Psychology, 63*(3), 245–261. doi:10.1111/j.2044-8325.1990.tb00525.x

Caldwell, R. (1996). Creating commitment to change: From a closed to open communication style. In D. Hussey (Ed.), *The Implementation Challenge* (pp. 229–239).

Castro, C. B., Periñan, M. M. V., & Bueno, J. C. (2008). Transformational leadership and followers' attitudes: The mediating role of psychological empowerment. *International Journal of Human Resource Management, 19*(10), 1842–1863. doi:10.1080/09585190802324601

Cha, S. E., & Edmondson, A. C. (2006). When values backfire: Leadership, attribution, and disenchantment in a values-driven organization. *The Leadership Quarterly, 17*(1), 57–78. doi:10.1016/j.leaqua.2005.10.006

Chathoth, P. K., Chan, E. S., Harrington, R. J., Okumus, F., & Song, Z. (2022). The Role of Organizational Factors on Employee Engagement and Hospitality Service Co-creation. *Journal of China Tourism Research, 18*(4), 756–777. doi:10.1080/19388160.2021.1972068

Chathoth, P. K., Harrington, R. J., Chan, E. S., Okumus, F., & Song, Z. (2020). Situational and personal factors influencing hospitality employee engagement in value co-creation. *International Journal of Hospitality Management, 91*, 102687. doi:10.1016/j.ijhm.2020.102687

Chiwawa, N., & Wissink, H. (2021). Determinants of employee engagement in the South African hospitality industry during COVID-19 lockdown epoch: Employee perception. *African Journal of Hospitality, Tourism and Leisure, 10*(2), 487–499. doi:10.46222/ajhtl.19770720.113

Collins, R. (2019). *Developing a lean internationalisation strategy through open innovation, entrepreneurship and cross-cultural intelligence* [Doctoral dissertation, University of Newcastle].

Conlon, C., Timonen, V., Elliott-O'Dare, C., O'Keeffe, S., & Foley, G. (2020). Confused about theoretical sampling? Engaging theoretical sampling in diverse grounded theory studies. *Qualitative Health Research, 30*(6), 947–959. doi:10.1177/1049732319899139 PMID:31959073

Costen, W. M., & Salazar, J. (2011). The impact of training and development on employee job satisfaction, loyalty, and intent to stay in the lodging industry. *Journal of Human Resources in Hospitality & Tourism, 10*(3), 273–284. doi:10.1080/15332845.2011.555734

Dar, N., & Rahman, W. (2022). Interactional Justice and Deviant Workplace Behavior: The Mediating Role of Perceived Supervisor Support in Hotel Industry of KP-Pakistan. *Journal of Entrepreneurship, Management, and Innovation, 4*(1), 149–166. doi:10.52633/jemi.v4i1.121

Davids, J. (2012). Have a Heart: Challenges for Lead Vocational Teachers in the Changing VET Landscape. National Centre for Vocational Education Research Ltd.

Dawson, M., & Abbott, J. (2011). Hospitality culture and climate: A proposed model for retaining employees and creating competitive advantage. *International Journal of Hospitality & Tourism Administration*, *12*(4), 289–304. doi:10.1080/15256480.2011.614533

Deale, C. S., & Lee, S. H. (2023). Employee benefits in the hospitality industry: What do employees want and need? *Journal of Human Resources in Hospitality & Tourism*, 1–27. doi:10.1080/15332845.2023.2207998

Dglawi, M., & Hassel, M. (2021). *Integration of new employees: A study of integration challenges for international companies with homework environment*.

Dolatabadi, H. R., & Safa, M. (2010). The effect of directive and participative leadership style on employees' commitment to service quality. *International Bulletin of Business Administration*, *9*(1), 31–42.

Downward, P., & Mearman, A. (2007). Retroduction as mixed-methods triangulation in economic research: Reorienting economics into social science. *Cambridge Journal of Economics*, *31*(1), 77–99. doi:10.1093/cje/bel009

Elshaer, A. M., & Marzouk, A. M. (2019). *Labor in the tourism and hospitality industry: Skills, ethics, issues, and rights*. Apple Academic Press. doi:10.1201/9780429465093

Emperger, W. (2004). Customer Relationship Management in the Hospitality Industry. *Erişim Tarihi, 10*. https://www.hsyndicate.org/file/152001440.pdf

Ersoy, A. & Ehtiyar, R. (2021). Determination of managerial resourcefulness in the Turkish hospitality industry: a grounded theory study. *Journal of Hospitality and Tourism Technology*. doi:10.1108/JHTT-03-2021-0083

Ertorer, S. E., Long, J., Fellin, M., & Esses, V. M. (2022). Immigrant perceptions of integration in the Canadian workplace. *Equality, Diversity and Inclusion*, *41*(7), 1091–1111. doi:10.1108/EDI-02-2019-0086

Estigoy, E., Sulasula, J., & Guodu, X. A. (2020). Factors affecting employee commitment in the workplace: An analysis. *Journal of Education and Practice*, *11*(27), 160–171.

Fink, S. L. (1992). High Commitment Workplaces. Greenwood publishing group, NY.

Fletcher, A. J. (2017). Applying critical realism in qualitative research: Methodology meets method. *International Journal of Social Research Methodology, 20*(2), 181–194. doi:10.1080/13645579.2016.1144401

Florea, N. V. (2014). Good practices of integration for new employees in roman organizations. *Buletin Stiintific, 37*(1), 13–23.

Folakemi, O., Anthonia, A., & Dayo, A. I. (2017). Human resource management practices and organisational commitment: Employees' perspective in the Nigerian hospitality sector. *New Trends and Issues Proceedings on Humanities and Social Sciences, 4*(10), 246–255. doi:10.18844/prosoc.v4i10.3090

Frankfurt, H. G. (1958). Peirce's notion of abduction. *The Journal of Philosophy, 55*(14), 593–597. doi:10.2307/2021966

Freeman, D., Cromwell, C., Aarenau, D., Hazelton, M., & Lapointe, M. (2005). Factors leading to successful workplace integration of employees who have experienced mental illness. *Employee Assistance Quarterly, 19*(4), 51–58. doi:10.1300/J022v19n04_04

Furunes, T. (2005). Training paradox in the hotel industry. *Scandinavian Journal of Hospitality and Tourism, 5*(3), 231–248. doi:10.1080/15022250510014372

Geisen, T. (2014). Workplace integration through disability management. In *Handbook of Vocational Rehabilitation and Disability Evaluation: Application and Implementation of the ICF* (pp. 55–71). Springer International Publishing.

Geisen, T., & Harder, H. G. (Eds.). (2011). *Disability management and workplace integration: International research findings*. Gower Publishing, Ltd.

Gonzalez, J. A., & Simpson, J. (2021). The workplace integration of veterans: Applying diversity and fit perspectives. *Human Resource Management Review, 31*(2), 100775. doi:10.1016/j.hrmr.2020.100775

Haig, B. D. (2005). An Abductive Theory of Scientific Method. *Psychological Methods, 10*(4), 371–388. doi:10.1037/1082-989X.10.4.371 PMID:16392993

Hatjidis, D., & Parker, A. (2018). The role of relationship quality in raising organizational change acceptance: The case of the Greek hotel industry. *Journal of Human Resources in Hospitality & Tourism, 17*(3), 340–359. doi:10.1080/15332845.2017.1406284

Helms, M. M., & Stern, R. (2001). Exploring the factors that influence employees' perceptions of their organization's culture. *Journal of Management in Medicine, 15*(6), 415–429. doi:10.1108/EUM0000000006502 PMID:11811793

Henderson, R. M. (1979). *Employee integration and automation: a study at Caterpillar Tractor Company, Peoria, Illinois* [Doctoral dissertation, Massachusetts Institute of Technology].

Hirst, G., Curtis, S., Nielsen, I., Smyth, R., Newman, A., & Xiao, N. (2021). Refugee recruitment and workplace integration: An opportunity for human resource management scholarship and impact. *Human Resource Management Journal*, 1748-8583.12349. doi:10.1111/1748-8583.12349

Hyman, J., & Mason, B. (1995). Managing employee involvement and participation. *Sage (Atlanta, Ga.).*

Ibrahim, A. M. (2020). Psychological empowerment and organizational commitment among employees in the lodging industry. *Journal of Human Resources in Hospitality & Tourism*, *19*(3), 277–295. doi:10.1080/15332845.2020.1737766

Karatepe, O. M., & Olugbade, O. A. (2009). The effects of job and personal resources on hotel employees' work engagement. *International Journal of Hospitality Management*, *28*(4), 504–512. doi:10.1016/j.ijhm.2009.02.003

Kazlauskaite, R., Buciuniene, I., & Turauskas, L. (2006). Building employee commitment in the hospitality industry. *Baltic Journal of Management*, *1*(3), 300–314. doi:10.1108/17465260610690944

Khoa, B., & Hoa, L. (2021). The effect of knowledge management process on the employee commitment: Evidence from digital marketing industry. *Management Science Letters*, *11*(5), 1557–1564. doi:10.5267/j.msl.2020.12.017

King, C. (2010). "One size doesn't fit all": Tourism and hospitality employees' response to internal brand management. *International Journal of Contemporary Hospitality Management*, *22*(4), 517–534. doi:10.1108/09596111011042721

Kirca-Demirbaga, K. (2021). *A study on the character development and education of gifted pupils using a critical realist approach* [Doctoral dissertation, Durham University]. http://etheses.dur.ac.uk/14348/

Kotsev, E. (2008). Integration of the Individual and the Group in the Organization. A Grup, Bulgaria.

Kotsev, E. (2022). Knowing When and How to Trust Superiors' Decisions: Toward a Conceptual Model of Subordinate Managers' Behavior. *Organizacija*, *55*(1), 50–63. doi:10.2478/orga-2022-0004

Kremers, A. (2018). The relationship between knowledge hiding and turnover intentions: The mediating role of affective commitment and the moderating effect of commitment HR attributions. Human Resource Studies, 1-51. Tilburg University.

Kusluvan, S. (2003). *Managing employee attitudes and behaviors in the tourism and tourism industry*. Nova Science Publishers.

Lashley, C. (1996). Research issues for employee empowerment in hospitality organisations. *International Journal of Hospitality Management*, *15*(4), 333–346. doi:10.1016/S0278-4319(96)00025-4

Lashley, C. (1999). Employee empowerment in services: A framework for analysis. *Personnel Review*, *28*(3), 169–191. doi:10.1108/00483489910264570

Lasrado, F., & Lasrado, F. (2018). Awards and Assessments. *Achieving Organizational Excellence: A Quality Management Program for Culturally Diverse Organizations*, 121-141. Springer.

Lawson, T. (1997). *Economics and reality*. Routledge. doi:10.4324/9780203195390

Lee, H. J., & Peccei, R. (2007). Organizational-Level Gender Dissimilarity and Employee Commitment. *British Journal of Industrial Relations*, *45*(4), 687–712. doi:10.1111/j.1467-8543.2007.00644.x

Lee-Ross, D. (2005). Perceived job characteristics and internal work motivation: An exploratory cross-cultural analysis of the motivational antecedents of hotel workers in Mauritius and Australia. *Journal of Management Development*, *24*(3), 253–266. doi:10.1108/02621710510584062

Lundberg, C., Gudmundson, A., & Andersson, T. D. (2009). Herzberg's two-factor theory of work motivation tested empirically on seasonal workers in hospitality and tourism. *Tourism Management*, *30*(6), 890–899. doi:10.1016/j.tourman.2008.12.003

Luo, C. C., Wang, Y. C., & Tai, Y. F. (2019). Effective training methods for fostering exceptional service employees. *Journal of Hospitality and Tourism Insights*, *2*(4), 469–488. doi:10.1108/JHTI-01-2019-0013

Mael, F., Wyatt, W., & Iyer, U. J. (2022). Veterans to workplace: Keys to successful transition. *Military Psychology*, *34*(5), 516–529. doi:10.1080/08995605.2021.2016307

Mahmud, S., Alam, Q., & Härtel, C. (2014). Mismatches in skills and attributes of immigrants and problems with workplace integration: A study of IT and engineering professionals in Australia. *Human Resource Management Journal*, *24*(3), 339–354. doi:10.1111/1748-8583.12026

Mäkinen, A. (2018). Development of internal communication policy in growth enterprise. International Business, 1-75. Oulu Business School.

Mansour, A., Rowlands, H., Al-Gasawneh, J. A., Al-Qudah, S., Shrouf, H., & Al-smadi, L. A. (2021). The mediation effect of employees' affective organizational commitment: The impact of employees' perceptions of the social support for training programs on enhancing their organizational citizenship behavior in Jordanian hotels sector. *Academy of Strategic Management Journal*, *20*, 1–23.

Marshall, T., Mottier, E. M., & Lewis, R. A. (2015). Motivational factors and the hospitality industry: A case study examining the effects of changes in the working environment. [JBCS]. *Journal of Business Case Studies*, *11*(3), 123–132. doi:10.19030/jbcs.v11i3.9289

McBrien, B., O'Brien, F., Shetty, S. R., Eustace-Cook, J., & Byrne, G. (2022). Assessment of workplace integration of migrant nurses and midwives within international health care settings: A scoping review protocol. *JBI Evidence Synthesis*, *20*(7), 1799–1805. doi:10.11124/JBIES-21-00239 PMID:35131992

Meyer, S. B., & Lunnay, B. (2013). The application of abductive and retroductive inference for the design and analysis of theory-driven sociological research. *Sociological Research Online*, *18*(1), 86–96. doi:10.5153ro.2819

Mi, L., Gan, X., Xu, T., Long, R., Qiao, L., & Zhu, H. (2019). A new perspective to promote organizational citizenship behaviour for the environment: The role of transformational leadership. *Journal of Cleaner Production*, *239*, 118002. doi:10.1016/j.jclepro.2019.118002

Mooney, S. K. (2014). *How the intersections of age, gender, ethnicity and class influence the longevity of a hospitality career in New Zealand* [Doctoral dissertation, Auckland University of Technology].

Naqvi, F. (2017). The Relationship between Employee Empowerment and Organizational Commitment in the Hotel Industry and its Implications for the Management. *International Journal of Economic Research*, *14*(12), 17–35.

Ncube, F., & Jerie, S. (2012). Leveraging employee engagement for competitive advantage in the hospitality industry. A comparative study of hotels A and B in Zimbabwe. *Journal of Emerging Trends in Economics and Management Science*, *3*(4), 380–388.

O'Mahoney, J. (2016). Critical realism and qualitative research: An introductory overview. In C. Cassell, A. L. Grandy, & G. Cunliffe (Eds.), *Sage handbook of qualitative research methods*.

Pagés, C. (2010). The age of productivity. In *The age of productivity: Transforming economies from the bottom up*. Palgrave Macmillan US.

Park, C. K. (2019). Transformational leadership and employee engagement from the HRD perspective. *Korean Review of Corporation Management*, *10*(3), 19–36. doi:10.20434/KRICM.2019.10.10.3.19

Pekerşen, Y., & Tugay, O. (2020). Professional satisfaction as a key factor in employee retention: A case of the service sector. *Journal of Tourism and Services*, *11*(20), 1–27. doi:10.29036/jots.v11i20.123

Peter, S., & Park, L. S. C. (2018). Theoretical research: Changing research methodology: Two case studies of critical realism informing social work doctoral research. *Aotearoa New Zealand Social Work*, *30*(1), 65–70. doi:10.11157/anzswj-vol30iss1id426

Peyton, T., Gip, H., Pasamehmetoglu, A., & Guchait, P. (2023). How authentic leadership cultivates trust and desirable workplace behaviors in hotels: Commitment and leader-follower value congruence matters. *Journal of Human Resources in Hospitality & Tourism*, 1–28. doi:10.1080/15332845.2023.2207987

Pham, P. T. (2020). *The Impact of Servitisation on Businesses*. Metropolia University of Applied Sciences.

Ploesser, K., & Russell, N. (2009). The people integration challenge. In *Handbook of research on business process modeling* (pp. 274–298). IGI Global. doi:10.4018/978-1-60566-288-6.ch013

Pourbarkhordari, A., Zhou, E. H., & Pourkarimi, J. (2016). Role of transformational leadership in creating a healthy work environment in business setting. *European Journal of Business and Management*, *8*(3), 57–70.

Price, L. (2016). Using retroduction to address wicked problems. In *Crisis System* (pp. 109–129). Routledge. doi:10.4324/9781315563138-7

Qian, X., Zhang, M., & Jiang, Q. (2020). Leader humility, and subordinates' organizational citizenship behavior and withdrawal behavior: Exploring the mediating mechanisms of subordinates' psychological capital. *International Journal of Environmental Research and Public Health*, *17*(7), 2544. doi:10.3390/ijerph17072544 PMID:32276358

Rajendran, D., Farquharson, K., & Hewege, C. (2017). Workplace integration: The lived experiences of highly skilled migrants in Australia. *Equality, Diversity and Inclusion*, *36*(5), 437–456. doi:10.1108/EDI-11-2016-0094

Rieu, J. N., & Kamara, M. (2016). Determinants of Employee Commitment in Hotel Industry in Nairobi County: A Case of Rated Hotels in Nairobi Central Business District. *International Journal of Innovative Social Sciences & Humanities Research*, *4*(4), 1–21.

Robeva-Stoyanova, B. (2023). Integrating people in the organization: A neglected management function? *XIX. IBANESS Congress Series on Economics, Business and Management*, (pp. 50-54). Ibaness..

Roth, C., Breckner, A., Wensing, M., Mahler, C., Krug, K., & Berger, S. (2023). Barriers and enabling factors for workplace integration of internationally qualified nurses: A qualitative study of perceptions of German nurses. *Nursing Open*, *00*(8), 1–11. doi:10.1002/nop2.1760 PMID:37060232

Rozuel, C., & McGhee, P. (2012). Investigating the convergence of corporate social responsibility and spirituality at work. *Australian Journal of Professional and Applied Ethics*, *14*(1), 47–62.

Saltik, Z. (2023). Behavioral Approaches to Leadership. In *Leadership Approaches in Global Hospitality and Tourism* (pp. 40–60). IGI Global. doi:10.4018/978-1-6684-6713-8.ch003

Sandjong, F. L. D. (2021). *Training Strategies for Increasing Employee Job Satisfaction and Retention in the Hospitality Industry* [Doctoral dissertation, Walden University].

Sarmento, M., Pedro, M. I., & Pinto, E. (2016). Integration and socialization of new employees in the bank sector. *International Journal of Latest Trends in Finance & Economic Sciences*, *6*(2), 1098–1107.

Saunders, M., Lewis, P., & Thornhill, A. (2019). *Research methods for business students* (8th ed.). Pearson.

Schmidt, W., & Müller, A. (2021). Workplace universalism and the integration of migrant workers and refugees in Germany. *Industrial Relations Journal*, *52*(2), 145–160. doi:10.1111/irj.12320

Simpson, E., Conway, C., Erickson, E., Kramer, O., Oberbroeckling, M., O'Malley, C., & Ybarra, H. (2022). Examining the OT Student Perspective of Group Work in Graduate School. *The American Journal of Occupational Therapy*, *76*(Supplement_1), 76. doi:10.5014/ajot.2022.76S1-PO120

Sondaitė, J., & Keidonaitė, G. (2020). Experience of transformative leadership: Subordinate's perspective. *Business: Theory and Practice*, *21*(1), 373–378. doi:10.3846/btp.2020.11113

Srivastava, S., & Singh, S. (2020). Linking personal growth initiative and organizational identification to employee engagement: Testing the mediating-moderating effects in Indian hotel industry. *Journal of Hospitality and Tourism Management*, *45*, 79–89. doi:10.1016/j.jhtm.2020.07.005

Stainback, K., Robinson, C. L., & Tomaskovic-Devey, D. (2005). Race and workplace integration: A politically mediated process? *The American Behavioral Scientist*, *48*(9), 1200–1228. doi:10.1177/0002764205274816

Stutchbury, K. (2022). Critical realism: An explanatory framework for small-scale qualitative studies or an 'unhelpful edifice'? *International Journal of Research & Method in Education*, *45*(2), 113–128. doi:10.1080/1743727X.2021.1966623

Sutton, J., & Thompson, R. (1998). Multidisciplinary integration: A decision methodology and procedure for instruction. In *FIE'98. Proceedings of 28th Annual Frontiers in Education Conference,* (pp. 450-455). IEEE. 10.1109/FIE.1998.736891

Tariq, H., Abrar, M., & Ahmad, B. (2023). Humility breeds creativity: the moderated mediation model of leader humility and subordinates' creative service performance in hospitality. *International Journal of Contemporary Hospitality Management*. doi:10.1108/IJCHM-07-2022-0851

Taylor, C. Y. (2019). *Strategies to improve employee engagement in the hospitality industry* [Doctoral dissertation, Walden University].

Tejaswi, M. J. (2010). People Integration a Challenge: Experts. *The Times of India*.

Thompson, R. J. (2016). *An investigation of work environment characteristics and work engagement of limited-and full-service hotel property front-line employees*, [Doctoral dissertation, The University of Southern Mississippi].

Ugwu, K. E., Okoroji, L. I., & Chukwu, E. O. (2018). Participative decision making and employee performance in the hospitality industry: A study of selected hotels in Owerri Metropolis, Imo State. *Management Studies and Economic Systems*, *4*(1), 57–70. doi:10.12816/0053651

Uzochukwu, A. C., Nwankwo, E. C., & Okafor, S. O. (2023). Effect of reward system on employee's productivity in Nigeria hospitality industry. *World Journal of Advanced Research and Reviews*, *17*(01), 361–373. doi:10.30574/wjarr.2023.17.1.0035

Walton, R. E. (1985). From control to commitment in the workplace. *Harvard Business Review*, (March), 1–9.

Ward, A. C., & Baker, P. M. (2005). Disabilities and impairments: Strategies for workplace integration. *Behavioral Sciences & the Law*, *23*(1), 143–160. doi:10.1002/bsl.631 PMID:15706600

Wright, A. (2004). *Reward Management in Context.* CIPD Publishing.

KEY TERMS AND DEFINITIONS

Abductive Reasoning: reasoning that goes beyond observation and experience to their possible explanations.

Demi-Regularity: A semi-predictable pattern of behavior.

Employee Commitment: An emotional attachment to and involvement with an organization.

Employee Engagement: The degree to which an employee invests cognitive, emotional, and behavioral energies toward positive organizational outcomes.

Retroductive Reasoning: A broader understanding of the context and conditions where data have been collected, which involves testing of proposed mechanisms.

Workplace Integration: A process of purposefully exerted managerial influences, in which individuals or groups are permanently committed to the goals of the organization and become a functionally indivisible part of it.

Chapter 4
Female Employment Patterns in the Tourism Sector: A Global Review

Hatice Işıl Alkan
Faculty of Economics and Administrative Sciences, Ondokuz Mayıs University, Turkey

Bora Alkan
Ondokuz Mayıs University, Turkey

ABSTRACT

Tourism has a strategic position in terms of production and employment, and stands out as a labor-intensive and feminized sector when examined in terms of employment. Asia-Pacific and Caribbean, Europe, and America come to the fore as geographies where female employment is very intense. This chapter strives to examine women's employment patterns in the tourism sector in the world, both in terms of regions and in the period before and after the pandemic. It is observed with the study that horizontal and vertical discrimination in the sector is quite evident. While women concentrate mostly in labor-intensive, low-paid, and informal jobs, their presence in jobs requiring high qualifications is extremely low. On the other hand, precarious employment of women as unpaid family workers is quite common. The Covid 19 pandemic has also adversely affected the female workforce. As the pandemic has increased inequalities in the sector, the hard-earned gains of women have also been eroded.

DOI: 10.4018/978-1-6684-9255-0.ch004

INTRODUCTION

Tourism is a vital sector for several economies around the world. From a macroeconomic point of view, tourism is a substantial sector that increases the revenue of countries, develops employment potential with job opportunities, reduces unemployment, and closes the balance of payments deficits. Moreover, it stands out as a key sector that makes significant contributions to growth and development, especially for developing countries. In addition to these features, tourism is a sector that has back-and-forth connections with many sectors, that is, an added value created in the tourism sector also affects the economic size of different sectors in the country. For example, the food and beverage sector, which is a sub-sector of tourism, is in close relationship with the agriculture sector. The travel and accommodation sector is linked to both the agricultural and industrial sectors. In other words, the development of the tourism sector in a country paves the way for the development of other economic sectors. The tourism sector, which has significant contributions to global GDP and employment, is a key sector in certain regions of the world. Asia-Pacific and the Caribbean are regions where the total contribution of tourism to GDP and employment is considerably high (WTTC, 2022).

Women are disadvantageous compared to men all over the world in terms of participation in the labor market. When the basic indicators of the labor market are examined, it is observed that female employment is low and female unemployment is high compared to males all over the world (ILO, 2022a). However, the presence of women in the labor market is a very important factor that paves the way for women's economic empowerment, welfare, and future. Women's employment has a positive impact not only on the welfare and future of themselves but also on the future and welfare of their children and family members. At this point, the tourism sector comes to the fore as a very important sector because the sector is largely feminized. Moreover, international statistics (such as WTO&ILOSTAT) reveal that in many countries, the rate of female employment in the tourism sector is higher than the female employment rate in the general economy. Therefore, the tourism sector has a notable role in empowering women and ensuring gender equality. It should be underlined that tourism is central not only to achieving gender equality but also to many of the sustainable development goals targeted to be achieved by the end of 2030 globally.

The main aim of this study is to examine the employment patterns of women in the tourism sector globally in the pre-and post-Covid-19 period, based on comprehensive international statistics and literature. In the study, regional differences in employment in the tourism sector were also tried to be examined in detail. Since there is no study in the literature that examines women's employment in the tourism sector in such detail in the axis of current data, it is thought that this study will fill an important gap in the literature. In this framework, the first section of the chapter discusses the contribution of the tourism sector to the global economy and global employment. The second part examines patterns of female employment in the tourism sector. The third part investigates the impact of the Covid-19 pandemic on female employment in the tourism sector. The study ends with the conclusion section in which the findings are summarized.

Overview of the Contribution of the Tourism Sector to the Global Economy and Global Employment

The tourism sector is one of the significant sectors of the world economy with its weight in the global GDP and share in employment. In the pre-pandemic period, the contribution of the tourism and travel sector to the global GDP is around 10%, the related sector accounts for 1 in 4 of all new jobs created in the world, as well as 10.3% of all jobs. The sector creates substantial socio-economic impacts, especially for women, youth, and minorities (WTTC, 2022).

Table 1. Contribution of the travel&tourism sector to the global GDP

	2019	2020	2021
Travel&Tourism GDP (% of the global economy)	10.3	5.3	6.1
Amount	9.630 billion USD	4.775 billion USD	5.812 billion USD

Source: WTTC (2022)

The contribution of the travel and tourism sector to GDP and employment differs from region to region. The Caribbean, Asia Pacific, and Europe are prominent regions in this scope (Table 2). Tourism is an activity carried out in many different places nowadays. Besides urban, coastal-sea tourism, one of the activities that contributes significantly to the economic development of the regions is rural tourism. Rural tourism, being an important driving force in terms of rural development, is worthy notably for developing regions. As it is known, the population living in rural in developing countries is considerably higher than in developed countries and agriculture is almost a shelter sector for rural. Moreover, poverty is mostly a problem in rural (Casteneda et al., 2018). Hence, rural tourism stands out as a sector that contributes significantly to the economy of rural as support to agriculture for increasing production and employment capacity. Rural tourism is a significant tool in combating rural poverty, preventing child labor, hazardous and dangerous working conditions, and low pay, and ensuring gender equality in rural (ILO, 2011). The scope of rural tourism is quite broad, encompassing many activities such as ecotourism, hiking, climbing, riding, adventure, sport, and health tourism. This sector, which makes a significant contribution to family income in rural areas and is important for regional development, is developing gradually (Irshad, 2010, p. 2). Therefore, rural tourism is an essential activity to support women who can find limited employment opportunities other than agriculture in rural.

Table 2. Contribution of the travel and tourism sector to GDP and employment by regions, 2019

	Total contribution of travel and tourism to GDP	Total contribution of travel and tourism to employment
Europe	9.2%	9.9% (37.8 mn jobs)
North America	8.9%	11.0% (25.5 mn jobs)
Caribbean	13.9%	15.2% (2.8mn jobs)
Asia Pasific	9.8%	9.9% (184.7 mn jobs)
Middle East	8.4%	8.9% (6.9 mn jobs)
Africa	6.8%	5.7% (25.5 mn jobs)
Latin America	8.2%	8.0% (17.1 mn jobs)

Source: WTTC (2022)

The Covid 19 pandemic, which started in Wuhan, China in November 2019 and spread rapidly all over the world, has deeply affected the tourism industry. With the restrictions on mobility and the closing of the borders, the tourism economy has been hit at the core. The pandemic has greatly affected not only international tourism activities but also domestic tourism. Although tourism entered into a recovery process towards 2022 with the removal of travel restrictions, Russia's war against Ukraine increased the uncertainties and negatively affected the sector. In addition, tourism companies had to cope with rising energy, food& input prices, labor shortages, and skills gap after the pandemic. Therefore, the contribution of tourism to the global economy and employment has decreased due to the pandemic, the war, and several macroeconomic problems in recent years. Various OECD countries with a sizeable tourism sector in the pre-pandemic period (such as Island, Mexico, and Portugal; respectively 8.1%, 8%, and 8.1% of GDP) faced significant reductions in the sector's contribution to GDP. Besides this, regions, where tourism has a great contribution to the local economy, have also been strikingly affected by the pandemic. Corsica in

France, the Ionian Islands in Greece, and the Balearic and Canary Islands in Spain are prime examples (OECD, 2022, p. 18).

The effects of the pandemic and other mentioned developments on different regions are also divergent. As can be observed in Table 3, the contribution of tourism to GDP and employment decreased dramatically in all regions as of 2020 and entered a recovery period as of 2021. In the post-pandemic period, the contribution of the sector to GDP decreased the most in the Asia-Pacific region, while the contribution of the sector to employment decreased the most in North America. The region with the fastest recovery in contribution GDP and employment in 2021, is the Caribbean (Table 3).

Table 3. Contribution of the travel and tourism sector to GDP and employment in the post-pandemic period

	The total contribution of travel and tourism to GDP 2020	**The total contribution of travel and tourism to GDP 2021**	**The total contribution of travel and tourism to employment 2020**	**The total contribution of travel and tourism to employment 2021**
Europe	-47.1%	+28%	-12.5%	+4.7%
North America	-45.8%	+22.3%	-34.7%	+8.2%
Caribbean	-53.2%	+36.6%	-25.8%	+15.2%
Asia Pasific	-58.8%	+16%	-18.9%	+6.2%
Middle East	-49.8%	+15.9%	-24.7%	+7.6%
Africa	-47.1%	+23.5%	-22.9%	+8.2%
Latin America	-41.4%	+26.5%	-22.7%	+8.0%

Source: WTTC (2022)

The tourism sector, being a labor-intensive sector, is an outstanding sector for those who enter the labor market for the first time and for disadvantaged groups. In other words, the tourism sector is an important source of employment for low-skilled workers, minorities, immigrants, unemployed youth, and women. Tourism businesses are generally privately owned or family-run businesses. Therefore, small and medium-sized enterprises are quite common in the sector. Especially the accommodation sector is the sector in which part-time, seasonal, and youth employment are intense (WTO & ILO, 2014, p. 16).

Patterns of Female Employment in the Tourism Sector

The employment opportunities of women in the world are quite limited compared to men. According to ILO, while the employment rate of women (15+) in the world was 43.8% in 2021, the employment rate of men (15+) is 67.9%. Women's unemployment rates in the world have also been higher than men's since the past. While the female unemployment rate is 6.1% in 2022, the male unemployment rate is 5.7%. (ILO, 2022a). Although female employment rates in the world vary from region to region, women are less involved in the labor market and are more discriminated in all regions. Industrial production is a mode of production in which female employment is very limited. It is seen that women's labor is concentrated in industrial production, especially in regions where export-oriented industrialization is high. The main reasons why women's labor is preferred in industrial production are that women work cheaper than men, women are more effective in repetitive women, and women are more compliant and obedient in the workplace. With globalization, the increase in flexible production techniques has led to the spread of subcontracted forms of production and home-based production in women's employment (Dedeoğlu, 2004, pp. 96-98). With the spread of flexible working methods in the world, especially in the 1990s, the share of standard employment in total employment has decreased, and informal employment has become widespread. Since women are seen as more prone to this type of work, they have become primarily employed in newly created informal jobs. While the majority of women in informal employment are unpaid family workers, temporary workers, and self-employed, few are employers (Toksöz, 2011, pp. 154-155). From a sectoral perspective, women are mostly employed in the services sector all over the world, and the tourism sector is one of the prominent sectors that provide employment to women. In this context, the tourism sector can be considered as a sector that contributes to gender equality and the empowerment of women. Globally, more than half of the workforce in the tourism sector, 54%, is women as of 2018. In the sector, women are mostly employed in low-skilled jobs, and they are rarely represented at the professional level. This situation shows that although the tourism sector constitutes an important employment framework for women, the quality of women's participation is poor. In addition, women's earnings are lower than men's earnings in the sector (WTO, 2019, pp. 31-35). ILOSTAT and WTO data reveal that female employment in the tourism sector is considerably higher than female employment in the broader economy in certain countries (Table 4).

Table 4. Comparison of UNWTO data on employees in tourism by sex with ILOSTAT data on overall employees by sex, 2005–2016

	Women employees in tourism (%)	Women employees in the broader economy (%)
Czech Republic	55	46.1
Slovenia	54.2	48.3
Canada	51.4	49.4
Slovakia	51	47.2
Brazil	45.7	45.5

Source: UNWTO and ILOSTAT (2020)

Tourism is a relatively labor-intensive industry. The sector employs women and youth more intensively than other sectors. Globally, the vast majority of workers in the industry are under the age of 35 (half of whom are 25 and below). Migrant workers make up a significant portion of workers in the industry. Therefore, the tourism sector can be considered as a sector that creates jobs, especially for women and young individuals, mainly through small and medium-sized enterprises (ILO, 2017, pp. 10-12). The concentration of the tourism sector on female labor certainly contributes positively to the empowerment of women. Various studies in the literature confirm this assertion. A study which targets focuses on women's empowerment in varied Arab countries; Egypt, the United Arab Emirates, and Oman determines that female employment in tourism and women's entrepreneurship are outstanding predictors of women's empowerment in the sector. Moreover, the results of the study indicate that the effects are different in the mentioned countries. Among the three Arab countries included in the study, the United Arab Emirates is the primary regarding the impact of women's empowerment on tourism development, followed by Egypt and then Oman (Abou-Shouk et al., 2020). Another study focuses on women empowerment in a panel of 24 selected European countries, over a period of 1990–2015 reveals that tourism and financial factors promote gender equality in education and employment (Nassani et al., 2019). Research analyzing the influence of tourism on gender equality in 36 Asian countries over the period 2006–2018, verifies that the positive impact of tourism on gender equality is significant. While this effect is quite high in Eastern and Southeastern Asian countries, these countries are followed by Western and Central Asian countries and then South Asian countries (Zhang & Zhang, 2020). Another study conducted in a region known for its poverty in China (namely Haiyuan County, Ningxia Hui Autonomous Region, in northwest China) underlines that the tourism sector not only empowers women economically but also psychologically and politically. According to the study, working in tourism strengthened women socially and advanced them in terms of education (Su et al., 2020).

In the sector, women are mostly employed in low-qualified and low-paid jobs (mainly in housekeeping and customer contact areas). They are underrepresented in qualified areas such as managerial and technical positions. Hence, it should be underlined that tourism is a sector including horizontal and vertical segregation and gender inequalities (ILO, 2017, pp.10-12). Many studies in the literature underline the notable segregation in the tourism sector. Research on tourism industries in 21 APEC member economies reveals that there is an ongoing segregation in terms of women employed in the tourism sector. The study also emphasizes national cultural barriers and intersectionality impacting female employment and progression in the tourism sector (Hutchings et al., 2020) Another study investigating the glass ceiling in tourism management asserts that covert discrimination in tourism businesses is much more common than overt discrimination. The study includes senior managers in hotels and travel businesses in the tourism sector. According to the results the idea of women's great family orientation, the belief that men are more competent in the field of work, and exclusionary practices pave the way for this discrimination (Carvalho et al., 2019).

The jobs undertaken by women in the tourism sector are quite similar to the jobs they undertake in the household. For this reason, women are mostly employed as cleaners, housekeepers, receptionists, and travel agents. Related jobs are the expanded version of the care tasks women undertake in the household. On the other hand, some jobs are masculinized and women have little access to these occupations. Men are concentrated in jobs that require more information and technology use. For example, airplane pilots are mostly men (Figueroa-Domecq et al., 2020). Besides, the stewardesses are mostly women. As can be understood from this example, while women undertake expanded and differentiated versions of labor-intensive jobs such as cooking, cleaning, and care services in the sector, men concentrate on technological knowledge, machine use, and skill-based jobs. Therefore, the categorization of "women's work and men's work", which stands out in the horizontal segregation, is quite obvious in the tourism sector. Glass ceiling syndrome, which is examined within the scope of vertical segregation, is another problem as women are underrepresented in managerial and professional positions.

One of the prominent features of employment in the tourism sector is informal employment. The fact that tourism activities are seasonal and mainly carried out by small-sized enterprises intensifies informality regarding employment in the sector. The intense informal employment in the sector triggers social insecurity for those employed in the sector (Çelik Uğuz & Kaya, p. 524). According to the World Bank, women in tourism are more likely to undertake seasonal, part-time, and informal jobs. It is also common for women to be victims of sexual exploitation in the industry. The propertylessness of women and the limited access to financial services are among the factors that prevent women from entrepreneurship (WB-IFC, 2017, p. 12). Another distinguishable pattern of female employment in the tourism sector is part-time employment. Tourism is a sector where the part-time employment of women is intense compared to the general economy. The seasonal nature of the tourism sector increases the presence of part-time employment especially for women and the density of part-time employment creates significant wage differences (Calvet et al., 2021, pp. 10-11). In EU-28 countries, 1 out of every 5 women employed in the tourism sector is employed part-time. For men, the corresponding rate is about 6% (WTO, 2019, p. 107). It should also be noted that part-time work, which is an employment pattern in which women can associate their household responsibilities with paid work, is often a vulnerable employment pattern. Another outstanding feature of female labor in the tourism sector is unpaid labor. Especially in family businesses in the sector, women's labor is largely unpaid. This way of working also reflects the gender stereotypes within the family. In family businesses, women undertake jobs similar to the ones they undertake in the household (for example, care and cleaning) in an informal way of working. Therefore, women in family businesses are largely deprived of social security (EIGE, 2023).

In terms of earnings, it is observed that the gender wage gap is high in the tourism sector. The wage gap is particularly striking in the hotel sector. Various studies confirm this situation in the literature. Guimaraes and Silva examined the earnings of women and men in the tourism sector in Brazil and found that the hourly earnings of men were 35% more than women. They concluded that there is pay discrimination in the Brazilian tourism sector and that female labor is less valued than men despite the same job characteristics (Guimaraes & Silva, 2016, pp. 440-447). Another research determines a significant difference in the income of women and men working in the hospitality sector in the United States. The research shows that "being a woman was associated with a loss of income of $620, relative to a mean income of $11,271 per year across the entire hospitality industry in the U.S." (Fleming, 2015, pp. 180-189). In addition to the mentioned studies, many studies reveal that there is a gender wage gap in the tourism sector that cannot be explained by the difference in worker qualifications (Kortt et al., 2018; Muñoz-Bullón, 2009; Oliver & Sard, 2020). The pay gap and the discrimination in the sector is frequently emphasized outside the literature. For instance, it was even reported in the press that male housekeeping workers at Hotel Best Tenerife (in Spain) received 4.6 times more bonuses than females in 2017, despite receiving the same wages and being at the same professional level (PUBLICO, 2017).

Although the tourism sector is dominantly a feminized sector, the intensity of female employment in the sector differs from region to region. When the distribution of female employment in tourism in the world is examined, it is observed that the highest number of women are employed in the Asia Pacific region. When analyzed proportionally, it is determined that the rate of female employment (53%) in the Asia Pacific region is very close to the world average (54%) (Table 5).

Table 5. Number and percentage of persons employed by sex in 'accommodation and food services activities, 2009–2018

Region	Number of countries included in the sample	Number of persons employed (× 1,000)	Number of women employed (× 1,000)	Female workers (%)
Americas	33	28,040	16,034	57
Africa	37	6,602	4,881	69
Asia and the Pacific	29	58,669	31,047	53
Europe	45	16,264	8,580	53
Middle East	13	2,897	310	9
World	157	112,500	60,867	54

Source: WTO (2019); ILOSTAT (2018)

Socio-cultural norms and ideologies in the region directly affect women's employment in the tourism sector. Despite being heavily employed in numbers, women in the Asia Pacific region mostly work in part-time, temporary, and seasonal jobs. It is striking that women in the sector are underrepresented in managerial status. Migrant labor is intense in the region and it has been determined that migrant women are mostly employed on a part-time and contract basis. The priority expected from women in Asia and the Pacific region is to fulfill their family and care responsibilities, and the tourism sector in the region provides women with the opportunity to contribute to the family and earn income with flexible job opportunities (WTO, 2022, pp. 24-25).

Table 6. Share of female employees in tourism and in the broader economy in Asia and the Pacific region

Countries	Female share in tourism %	Female share in broader economy %
Australia	55	46.7
China	58.9	43.9
Indonesia	57.3	39.6
Japan	62.4	44.3
Korea, Democratic People's Republic o	64	47.3
Malaysia	49.8	38.4
Singapore	51.5	41.7
New Zealand	61.5	47.7
Total (ILOSTAT)	51.7	36.1

Source: WTO (2022, p.28)

Europe is one of the regions where the tourism industry is an important source of employment for women. On the continent, the tourism sector employed more women (58%) than men in 2020. It has been determined that women are mostly employed in tourism agencies (64%) and tour operators (60%) in the sector. On the other hand, about 3 out of 10 women working in the sector work part-time and this rate is 1 in 10 for men (EUROSTAT, 2023). Although tourism mainly employs women in the region, discrimination, and the wage gap are important problems in the labor market. It is determined that the hourly gross wage of women in accommodation and food services is lower than that of men in European Union countries. Moreover, women are largely employed in low-skilled jobs and have little chance of advancement in their careers (Obadic, 2016, p. 823)

The American continent is another continent where employment in tourism is predominantly female. The female share in travel&tourism employment in the USA in 2019 is approximately equal to the female employment rate in the general economy (46.4% and 46.9%) (WTTC, 2019). The female employment rate in the tourism sector in Canada is higher than the female employment rate in the general economy and women are (48.4%) more likely to have part-time jobs than men (36.9%) (STATCAN, 2019). In Brazil, one of the South American countries, tourism is an important sector that enables women to enter the labor market. However, it should be emphasized that informal employment of women is quite intense in this sector. Accommodation, food and beverage, and travel agencies are the prominent sub-sectors in which female employment is dominant. Although it tends to decrease, there are wage gaps between men and women (Costa et al, 2013, p. 56; IPEA 2008). As Vomick et al. (2023) emphasized, tourism significantly affects women's integration into the economy and employment in the services sector in South American and Caribbean countries.

The economic contribution of the tourism sector in Africa is quite limited when compared to other continents (Table 2). Although employment in the tourism sector on the continent is limited, women constitute a significant part of employment in the sector (Table 4). Young women constitute a significant part of those working in the accommodation and food and beverage sector in Africa (UNCTAD, 2023). Moreover, African women play a pivotal role in the African travel industry, for instance approximately one-third of employees in Ethiopian Airlines are women (Travelpulse, 2023). Tourism stands out as a very worthy aspect of empowering women in the region. Gender inequality is widespread in the tourism sector on the continent. While managerial and professional jobs are the responsibility of men, women work in less qualified jobs. For example, in Ghana and Nigeria, more than half of businesses are owned by men (Kimbu et al, 2019).

The representation of women in the tourism sector is marginal in the Middle East, one of the geographies where women's labor force participation is the lowest in general. Contrary to the general situation in the world, employment in tourism in the Middle East is predominantly male. Socio-cultural and gender norms, working conditions, poor image of the sector, low wages, and weak advancement opportunities cause marginal employment of women in the tourism sector in the Middle East. Female employment is concentrated in the hospitality sector in the region. Migrant workers are predominantly employed in low-skilled jobs, and women are not sufficiently represented in management levels in the region. The participation of women in decision-making processes in the sector is limited. In fact, tourism is a sector with significant economic returns for Middle Eastern countries. Egypt and Jordan are among the countries where the contribution of tourism to GDP is significant. However, even in countries where tourism is developed, the employment of women in the sector is quite low (WTO, 2020, pp. 9-26). As can be seen in Table 7, the employment of women in the tourism sector in the Middle East is generally less than employment in the general economy. It is noteworthy that the only exception is Qatar.

Table 7. Share of women employees in tourism and in the broader economy in the Middle East, 2018

Countries	Women employees in tourism %	Women employees in broader economy %
Egypt	4.8	20.5
Iraq	5.7	10.5
Jordan	2.9	16.2
Lebanon	11.2	23.4
Qatar	24.7	13.5
Saudia Arabia	8.1	19.8
United Arab Emirates	12.8	16.8
Total	8.3	16.4

Source: WTO (2020, p. 27)

A notable share of female participation in the labor market belongs to women's entrepreneurship. Entrepreneurship is also an area that includes gender inequalities. Gender-based barriers are the main barriers women face in their entrepreneurial activities. Because of gender inequalities, it is difficult for women to start and run a business compared to men. When a woman succeeds in establishing a business, it is observed that women's businesses differ from those of men. Women's businesses are generally smaller than men's and have a greater presence in the informal economy. When these difficulties are combined with the care responsibilities of women, the burden on women entrepreneurs increases (ILO, 2023). When tourism employment is examined in terms of female entrepreneurship and female employers, World Bank reveals that the travel and tourism sector contains twice as many female employers compared to other sectors. Latin America is the leader in this field, with more than half of the tourism businesses in the region being run by women. In Panama and Nicaragua, more than 70% of tourism businesses are run by women, compared to 20% for other sectors. On the Asian continent, the prominent countries are Indonesia, Malaysia, the Philippines, and Thailand, where more than half of the businesses are run by women (WB-IFC, 2017).

The Impact of the Covid 19 Pandemic on Female Employment in Tourism Sector

The Covid 19 pandemic, which is defined as the plague of the twenty-first century, emerged in Wuhan, China in 2019 and then spread all over the world in a short time, and has affected the tourism sector on a vast scale. In fact, it can be said that the tourism sector has been one of the sectors most heavily affected by the pandemic due to restrictions such as closing the borders and minimizing inter- and intra-country mobility. As can be seen in Table 1, the contribution of the travel and tourism sector, which contributed 10.3% to the global GDP in 2019, decreased to 5.3% in 2020. While 334 million were employed in the tourism sector in 2019, this figure decreased to 272 million in 2020. In other words, there was a decrease of 18.5%. Since many jobs in the tourism sector are seasonal, take place in the informal economy, and unpaid family work is very common in tourism, the actual losses are actually well above these figures. The global pandemic has put thousands of small and medium-sized businesses that employ women predominantly in the tourism sector at risk, causing significant employment losses. The epidemic affected the countries whose economy depends on tourism more negatively, hence, significant revenue and employment losses came to the fore in the relevant countries (UNWOMEN, 2022, p. 2).

The pandemic has not only dragged the health sector into crisis and brought millions of deaths, but also caused unemployment, the closure of millions of workplaces, loss of income, increased informality, and job insecurity. Women, youth, and immigrants have been disproportionately affected by the pandemic in the tourism sector. Travel restrictions decreased international tourism arrivals globally. In 2020, there was a 74% decrease in international tourism arrivals compared to the previous year. This effect is much more severe in the Asia Pacific, as the reduction rate in the relevant geography is 84% (Table 8).

Table 8. International tourist arrivals, 2019-2020

Region	2019	2020	Change
Americas	219 mn	69 mn	-69%
Europe	746 mn	221 mn	-70%
Africa	70 mn	18 mn	-75%
Middle-East	65 mn	16 mn	-75%
Asia-Pacific	360 mn	57 mn	-84%

Source: UNWTO (2020)

The Covid 19 pandemic has affected women in tourism much more negatively, as the tourism industry is largely a feminized industry. As emphasized in the second part, female employment in tourism is more intense than female employment in the overall economy in various countries such as the Czech Republic, Australia, China, Indonesia, and Japan. Therefore, the impact of Covid 19 on female employment in these countries has been much more severe. If we take the example of Austria; while women were employed by 55% of all occupations (both full-time and part-time) in the country before the pandemic, this rate dropped to 53% after the pandemic. Full-time employment of women in the tourism sector has also decreased significantly in the post-pandemic period. For instance, women's full-time employment decreased to 52% in Australia. Moreover, since female labor in the tourism sector is largely informal and invisible, female employees in the sector could not benefit from the relief packages announced by the governments (UNWOMEN, 2022, p. 4). For example, since the majority of women in India are working in the unorganized sector, these women could not benefit from the support packages announced by the government (THENEWSMINUTE, 2020). In Small Island Developing States (SDIS), whose economy is heavily dependent on tourism and whose tourism sector is heavily feminized, women have been the group most severely affected by the pandemic. In SDIS, such as the Dominican Republic, Kiribati, Tonga, and Mauritius, women have very high employment rates in low-skilled jobs, especially in the food and beverage sector. In contrast, women are scarce in high-skilled jobs such as travel agents and tour operators. These countries are among the countries most adversely affected not only by the pandemic, but also by climate change, and gender is the biggest factor creating inequality (UNWOMEN, 2022, p. 5).

The most adversely affected group of workers by Covid 19 has been those who are informally employed globally. This group was quickly pushed out of the labor market as they lacked social security. ILO data highlights that 5 billion workers in the world are adversely affected by Covid 19, and 1.6 billion of this number are those who are informally employed. Those who are employed informally are concentrated in sectors where women are predominant (ILO, 2020a, 2020b). Since the tourism sector is heavily feminized and informal employment intensive, the impact of Covid 19 on female workers in countries whose economy is dependent on tourism is quite negative. For example, Mexico is an example of a country whose economy is dependent on tourism and where informal employment is intense. In Mexico, Covid 19 has directly affected those working in the informal tourism sector, increasing the vulnerability of this group and causing significant social inequality (Monterrubio, 2021). In a study that dealt with the effects of the pandemic on the tourism sector in Spain from a gender perspective, it was revealed that the pandemic had negative permanent effects on employment and these effects were more severe for women. According to the findings of the study, women are more vulnerable to the shocks of Covid 19 (Claudio-Quiroga et al, 2022).

As emphasized before, tourism is an important source of employment, especially for migrant workers. When examined globally, millions of female migrants work in the tourism sector. Closing the borders during the pandemic has caused millions of people to be unemployed, and especially low-skilled workers employed in high-income countries have been adversely affected by this situation. For example, one-sixth of tourism workers in the European Union (EUROSTAT, 2018) and one-fifth of tourism workers in the United States are foreign-born (NAE, 2018). The number of workers with temporary work passports in the tourism and hospitality sector in Australia is strikingly high (Warren, 2014). Due to Covid 19, millions of workers have either been laid off, have been on leave, or their salaries have been reduced. Many migrant workers could not go to the host countries for work due to travel restrictions. Therefore, millions of job losses have occurred in the tourism sector, and the income levels of those who can continue to work in the sector have eroded significantly. This situation not only affected the migrant workers but also put their families under significant financial pressure. As is known, remittances from migrant workers are an important source of income, especially in developing countries. However, with the pandemic, this flow has also been significantly eroded and millions of families have become impoverished as they are deprived of this resource (Dempster&Zimmer, 2020). According to Asian Development Bank data, the Covid 19 pandemic has had a direct negative impact on the lives of 91 million migrant workers originating from the Asia-Pacific region. It should be emphasized that migrants originating from Asia-Pacific make up one-third of the world's total migrant workers. Remittances to Asia decreased by at least 31.4 billion in 2020. A stop in remittance flow has pushed millions of Asian citizens into poverty. Especially in Central and Western Asia and the Pacific region, the related impoverishment was felt more sharply (Takenaka et.al.,2020).

Thus, the Covid 19 pandemic has deepened pre-existing inequalities in the labor market in the tourism sector. The negative effects of the pandemic are more severe on women, youth, minorities, and immigrants, who are defined as vulnerable groups and are predominantly involved in the informal sector. Since the small and medium-sized enterprises, where women are heavily employed in the tourism sector, are more active in the informal economy, these enterprises could not benefit from the aid & support packages announced by the governments after the pandemic. In summary, Covid 19 has caused the loss of women's hard-won gains in terms of gender equality in the tourism sector and deepened inequality (ILO, 2022b).

CONCLUSION

Tourism is a sector that has significant contributions to global GDP and employment. It is also a key sector for many developing countries. When observed regionally, the contribution of tourism to GDP and employment is higher in Caribbean and Asia Pacific countries compared to other regions. On the other hand, tourism is a very important economic resource not only for developing countries but also for developed geographies. The contribution of tourism to income and employment is quite high in developed geographies such as Europe and North America. In other words, the tourism sector has noteworthy economic contributions for both developed and developing countries.

Being a labor-intensive sector, tourism is a substantial step toward entry into the labor market. It is one of the leading ways to be included in the labor market, especially for disadvantaged groups like women, youth, and immigrants. The tourism sector is largely a feminized sector as WTO (2019) data reveal that more than half of the workers in the sector are women. In addition to many developing countries, in countries such as the Czech Republic, Australia, and Japan, female employment in the sector is higher than female employment in the general economy. Considering that women are less involved in the labor market than men globally and unemployment is a bigger problem for women, the role of the tourism sector in achieving gender equality and empowering women globally comes to the fore. However, despite providing significant employment opportunities for women, women in the sector are mostly employed in low-paid jobs that require low qualifications. In other words, gender inequalities are evident in the labor market in sector; women are underpaid compared to men and are very limitedly represented in high-skilled jobs. Most of the enterprises in the tourism sector are small and medium-sized, and this structure significantly increases the informal employment of women. Especially in family businesses, women work mostly informally in jobs similar to their household responsibilities. This situation paves the way for the vulnerable employment of women. In addition to the deprivation of social security, unpaid employment of women is also very common in the sector, especially in small family businesses. Apart from the scale structure of the enterprises, another factor that triggers the informal employment of women is the seasonality of the jobs in the sector. World Bank data indicate that women are largely employed in the tourism sector on an informal, part-time, and seasonal basis. Part-time employment is a type of employment that allows women to combine their household responsibilities with paid work, however, it should be noted that part-time jobs are generally low-paid and largely devoid of social security, especially in developing countries. When analyzed in terms of earnings, it is determined that the gender wage gap in the sector is high. In many countries in the world, women's earnings in the tourism sector are lower than men's, and this cannot be explained only by the difference in human capital (Kortt et al., 2018; Muñoz-Bullón, 2009; Oliver & Sard, 2020). When female employment in tourism is analyzed on the axis of regions, the country with the highest female employment in absolute terms is the Asia Pacific. In many countries in Asia Pacific, the employment of women in tourism is proportionally higher than employment in the general economy. The continents of Europe and America are also continents where tourism is largely feminized. Although the contribution of tourism to GDP and employment in Africa is limited in comparison with most regions, women are heavily employed in the tourism sector in this continent. Especially in the accommodation and food and beverage sector, the share of female employees is quite high. The region with the lowest share of female employees in the tourism sector is the Middle East. This region has the lowest female employment rate generally. The prevailing socio-cultural structure and patriarchal norms in the region are important obstacles to women's inclusion in the labor market (WTO, 2020)(WTO, 2022). Tourism has a privileged position in terms of women's entrepreneurship, which paves the way for women's empowerment and employment. As known, women entrepreneurs in the world are quite limited compared to men. On the other hand, World Bank data reveals that women entrepreneurs in the travel and tourism sector are approximately twice as high as those in other sectors, and Latin American countries are premier. Women's entrepreneurship in tourism is also high in several Asian countries (WB-IFC, 2017). Briefly, tourism is also a prominent sector in terms of women's entrepreneurship.

Finally, tourism is one of the sectors most severely affected by the Covid 19 pandemic globally. Travel restrictions implemented with the closure of borders brought both domestic and international travel and tourism activities to a standstill. The activities of many enterprises in the sector have stopped and millions of jobs have been lost. As tourism is a highly feminized sector, female workers have been adversely affected by the pandemic. On the other hand, the intense informality and social insecurity of female employment left women vulnerable to the crisis. ILO data (2022b) reveals that many informal workers in the sector could not even benefit from state aid & support because their labor was invisible. In summary, Covid 19 has exacerbated existing inequalities in the sector, while significantly eroding the gains made so far.

REFERENCES

Abou-Shouk, M. A., Mannaa, M. T., & Elbaz, A. M. (2021). Women's empowerment and tourism development: A cross-country study. *Tourism Management Perspectives, 37*. https://doi.org/ doi:10.1016/j.tmp.2020.100782

Calvet, N.A., Conde, C.I., Ballart, A.L., & Almela, M.S. (2021). *Gender inequalities in tourism labor market*. Alba sud editorial.

Carvalho, I., Costa, C., Lykke, N., & Torres, A. (2019). Beyond the glass ceiling: Gendering tourism management. *Annals of Tourism Research, 75*, 79–91. doi:10.1016/j.annals.2018.12.022

Castañeda, A., Doan, D., Newhouse, D., Nguyen, M. C., Uematsu, H., & Azevedo, J. P. (2018). A new profile of the global poor. *World Development*, *101*, 250–267. doi:10.1016/j.worlddev.2017.08.002

Çelik Uğuz, S., & Kaya, İ. (2016). Informal employment in tourism. In U. Yedipınar & E. Yıldız (Eds.), *Global Issues and Trends in Tourism*. St. Kliment Ohridski University Press.

Claudio-Quiroga, G., Gil-Alana, L. A., Gil-López, A., & Babinger, F. (2022). A gender approach to the impact of COVID-19 on tourism employment. *Journal of Sustainable Tourism*. doi:10.1080/09669582.2022.2068021

Costa, H. A., Costa, C., Breda, Z., Çacador, S., Carvalho, I., & Vıeira, R. (2013). Women in Brazil with higher education in tourism: Preliminary results of the Gentour Project. *Revista Turismo & Desenvolvimento (Aveiro)*, *19*, 53–66.

Dedeoğlu, S. (2004). Dünya ekonomisi ve hayatta kalmanın feminizasyonu: Toplumsal cinsiyet, kalkınma ve küreselleşme. In S. Dedeoğlu & T. Subaşat (Eds.), *Kalkınma ve Küreselleşme*. Bağlam Yayıncılık.

Dempster, H., & Zimmer, C. (2020). Migrant workers in the tourism industry: How has COVID-19 affected them, and what does the future hold. *CGD Policy Paper, 173*.

EIGE. (2023). *Tourism & Issues of gender inequalities in the policy area*. EIGE. https://eige.europa.eu/gender-mainstreaming/policy-areas/tourism

EUROSTAT. (2018). *Tourism industries – employment*. Eurostat. https://ec.europa.eu/eurostat/statistics-explained/index.php?title=Tourism_industries_-_employment

EUROSTAT. (2023). *Tourism industries – employment*. Eurostat. https://ec.europa.eu/eurostat/statistics-explained/index.php?title=Tourism_industries_-_employment&oldid=445425

Figueroa-Domecq, C., Palomo, J., Flecha-Barrio, M. D., & Segovia-Pérez, M. (2020). Technology double gender gap in tourism business leadership. *Information Technology & Tourism*, *22*(1), 75–106. doi:10.100740558-020-00168-0

Fleming, S. S. (2015). Déjà Vu? An updated analysis of the gender wage gap in the U.S. hospitality sector. *Cornell Hospitality Quarterly*, *56*(2), 180–190. doi:10.1177/1938965514567680

Guimaraes, F. F. C. R., & Silva, J. R. (2016). Pay gap by gender in the tourism industry of Brazil. *Tourism Management*, *52*, 440–450. doi:10.1016/j.tourman.2015.07.003

Hutchings, K., Moyle, C., Chai, A., Garofano, N., & Moore, S. (2020). Segregation of women in tourism employment in the APEC region. *Tourism Management Perspectives*, *34*, 100655. doi:10.1016/j.tmp.2020.100655

ILO. (2011). *Stimulating rural growth, jobs and income through tourism, rural policy brief.* ILO. https://www.ilo.org/wcmsp5/groups/public/---ed_emp/documents/publication/wcms_161153.pdf

ILO. (2017). *ILO guidelines on decent work and socially responsible tourism.* ILO. https://www.ilo.org/wcmsp5/groups/public/---ed_dialogue/---sector/documents/normativeinstrument/wcms_546337.pdf

ILO. (2020a). *ILO Monitor: COVID-19 and the world of work. Third edition, Updated estimates and analysis.* ILO. https://www.ilo.org/wcmsp5/groups/public/@dgreports/@dcomm/documents/briefingnote/wcms_743146.pdf

ILO. (2020b). *COVID-19 crisis and the informal economy, immediate responses and policy challenges.* ILO. https://www.ilo.org/wcmsp5/groups/public/---ed_protect/---protrav/---travail/documents/briefingnote/wcms_743623.pdf

ILO. (2022a). *World employment and social outlook | Trends 2022.* ILO. https://www.ilo.org/wcmsp5/groups/public/---dgreports/---dcomm/---publ/documents/publication/wcms_834081.pdf

ILO. (2022b). *The future of work in the tourism sector: Sustainable and safe recovery and decent work in the context of the COVID-19 pandemic.* ILO. https://www.ilo.org/wcmsp5/groups/public/---ed_dialogue/---sector/documents/meetingdocument/wcms_840403.pdf

ILO. (2023). *Women's entrepreneurship development.* ILO. https://www.ilo.org/empent/areas/womens-entrepreneurship-development-wed/lang--en/index.htm

ILOSTAT. (2018). *Total employed by sex.* ILO. http://ilostat.ilo.org

IPEA. (2008). *Sistema de informações sobre o mercado de trabalhono setor turismo: Caracterização da ocupação formal noturismo, com base nos dados da RAIS de 2002 e 2006, e daocupação informal no turismo, com base nos dados da PNADde 2002 e 2006.* IPEA. http://agencia.ipea.gov.br/images/stories/15_caracterizacao_da_ocupacao_formal_e_informal_2002_2006_setembro_2008.pdf

Irshad, H. (2010). *Rural tourism- An overview.* Government of Alberta, Rural Development Division.

Kimbu, A. N., Afenyo Agbe, E. A., & Adam, I. (2019). *These African women are driving the continent's tourism industry – but they need support.* Research Gate. https://www.researchgate.net/publication/337480264_These_African_women_are_driving_the_continents_tourism_industry_-_but_they_need_support

Kortt, M. A., Sinnewe, E., & Pervan, S. J. (2018). The gender wage gap in the tourism industry: Evidence from Australia. *Tourism Analysis*, *23*(1), 137–149. doi:10.3727/108354217X15143857878697

Monterrubio, C. (2022). The informal tourism economy, COVID-19 and socioeconomic vulnerability in Mexico. *Journal of Policy Research in Tourism, Leisure & Events*, *14*(1), 20–34. doi:10.1080/19407963.2021.2017726

Muñoz-Bullón, F. (2009). The gap between male and female pay in the Spanish tourism industry. *Tourism Management*, *30*(5), 638–649. doi:10.1016/j.tourman.2008.11.007

NAE. (2018). *New American economy, hospitality & tourism, immigrant workers.* NAE. https://www.newamericaneconomy.org/issues/hospitality-tourism/

Nassani, A. A., Aldakhil, A. M., Qazi Abro, M. M., Islam, T., & Zaman, K. (2019). The impact of tourism and finance on women empowerment. *Journal of Policy Modeling*, *41*(2), 234–254. doi:10.1016/j.jpolmod.2018.12.001

Obadic, A. (2016). Gender discrimination and pay gap on tourism labor market. *International Journal of Mechanical and Industrial Engineering*, *10*(3), 823–828.

OECD. (2022). *OECD tourism trends and policies 2022.* OECD Publishing.

Oliver, X., & Sard, M. (2021). Gender wage gap in hospitality. *Journal of Hospitality & Tourism Research (Washington, D.C.)*, *45*(2), 345–372. doi:10.1177/1096348020909423

PUBLICO. (2017). *La discriminación de género en un hotel de Tenerife: ellos 640 euros y ellas 139.* Publico. https://www.publico.es/sociedad/discriminacion-genero-hotel-tenerife-640-euros-139.html

STATCAN. (2019). *Tourism Human Resource Module, 2019.* Statcan. https://www150.statcan.gc.ca/n1/daily-quotidien/201030/dq201030c-eng.htm

Su, M. M., Wall, G., Ma, J., Notarianni, M., & Wang, S. (2020). Empowerment of women through cultural tourism: Perspectives of Hui minority embroiderers in Ningxia, China. *Journal of Sustainable Tourism*, *31*(2), 307–328. doi:10.1080/09669582.2020.1841217

Takenaka, A. K., Villafuerte, J., Gaspar, R., & Narayanan, B. (2020). COVID-19 impact on international migration, remittances, and recipient households in developing Asia. *ADB Briefs*, (August). doi:10.22617/BRF200219-2

The News Minute. (2020). The invisibility of women's labour in tourism and the impact of COVID-19. *The News Minute.* https://www.thenewsminute.com/article/invisibility-women-s-labour-tourism-and-impact-covid-19-123315

Toksöz, G. (2011). *Kalkınmada Kadın Emeği.* Varlık Yayınları.

Travelpulse. (2023). How women are playing a pivotal role in the African travel industry. *Travel Pulse.* https://www.travelpulse.com/News/Destinations/How-Women-are-Playing-a-Pivotal-Role-in-the-African-Travel-Industry

UNCTAD. (2023). *African tourists bring better jobs for women and youth.* UNCTAD. https://unctad.org/news/african-tourists-bring-better-jobs-women-and-youth

UNWOMEN. (2022). *Advancing women's economic empowerment in the tourism sector in covıd-19 response and recovery.* UN Women. https://www.unwomen.org/sites/default/files/2022-05/Advancing-womens-economic-empowerment-in-the-tourism-sector-in-COVID-19-en.pdf

UNWTO. (2020). *2020: A year in review.* UNWTO. https://www.unwto.org/covid-19-and-tourism-2020

Voumik, L. C., Nafi, S. M., Majumder, S. C., & Islam, M. A. (2023). The impact of tourism on the women employment in South American and Caribbean countries. *International Journal of Contemporary Hospitality Management, 35*(9), 3095–3112. doi:10.1108/IJCHM-04-2022-0449

Warren, C. (2014). Reviewing the skilled migration and 400 series visa programme. *Home Affairs.* https://www.homeaffairs.gov.au/reports-and-pubs/files/carlita-warren.pdf

WB-IFC. (2017). *Tourism for development.* WB-IFC. https://documents1.worldbank.org/curated/en/401321508245393514/pdf/120477-WP-PUBLIC-Weds-oct-18-9am-ADD-SERIES-36p-IFCWomenandTourismfinal.pdf

WTO. (2019). *Global report on women in tourism.* World Tourism Organization, UNWTO, Madrid. https://www.e-unwto.org/doi/pdf/10.18111/9789284420384

WTO. (2020). *Regional report on women in tourism in the Middle East.* WTO. https://www.e-unwto.org/doi/epdf/10.18111/9789284422371

WTO. (2022). *Regional report on women in tourism in Asia and the Pacific*. UNWTO, Madrid. https://www.e-unwto.org/doi/epdf/10.18111/9789284423569

WTO-ILO. (2014). Measuring employment in the tourism ındustries – Guide with best practices. World Tourism Organization and International Labour Organization, UNWTO, Madrid.

WTTC. (2019). *Travel & Tourısm: Driving women's success*. WTTC. https://wttc.org/Portals/0/Documents/Reports/2019/Social%20Impact-Driving%20Womens%20Success-Mar%202019.pdf?ver=2021-02-25-182742-097

WTTC. (2022). *Travel & Tourism economic impact 2022, global trends*. WTTC. https://wttc.org/Portals/0/Documents/Reports/2022/EIR2022-Global%20Trends.pdf

Zhang, J., & Zhang, Y. (2020). Tourism and gender equality: An Asian perspective. *Annals of Tourism Research*, *85*, 103067. doi:10.1016/j.annals.2020.103067

KEY TERMS AND DEFINITIONS

Female Employment: Female employment and employment patterns of females globally.

Horizontal Discrimination: Horizontal segregation claims that men and women are concentrated in different kinds of jobs.

Vertical Discrimination: Vertical segregation describes the clustering of men at the top of occupational hierarchies.

Chapter 5
Storytellers' Inside Story:
Exploitation and Domination Practices in Tour Guiding

Burak Düz
Anadolu University, Turkey

ABSTRACT

This chapter draws a conceptual framework for the exploitation and domination practices in tour guiding. Despite the emphasized significance of their role in the tourism industry, tour guides have been reported to be the victims of unfair practices from the stakeholders. Therefore, an extensive literature review was carried out, and the author offered their insider perspective based on their nine-year tour guiding experience. The findings in the body of tour guiding literature were scrutinized, and conceptual discussions were offered to provide an insight into the 'inside story' of tour guides. Eventually, some practical implications were noted, and the need to remediate the conditions of tour guides was confirmed due to their crucial role in the industry.

INTRODUCTION

The body of tour guiding literature has focused on the perceptions of tour guides and tourists (Pearce, 1984; Pizam & Sussmann, 1995; Pizam & Jeong, 1996; Pizam et al., 1997); the role of the guides in tourist satisfaction (Geva & Goldman, 1991; Kuo et al., 2015); the competences of tour guides such as interpretation, communication, and humor (Yu et al., 2001; Ibanez et al., 2003; Leclerc & Martin, 2004; Wynn, 2005; Bryon, 2012; Weiler & Walker, 2014; Lin et al., 2022); or the performance

DOI: 10.4018/978-1-6684-9255-0.ch005

of the labor (Fine & Speer, 1985; Zhang & Chow, 2004; S. Huang et al., 2010; S. Huang, 2010; K. C. Chang, 2014; Sezgin & Düz, 2018). In other words, a large number of studies in the literature inquired about the profession, mainly in the sense of tourists or destinations. Hence tour guiding was regarded as an instrument to assure tourist satisfaction or a favorable destination image. The idealized function of tour guides is self-sacrificing and fulfilling their tasks without question. Hence, their self-expectations are ignored (Zhao & Timothy, 2017). Furthermore, the labor of the tour guides was taken for granted, and the studies (Holloway, 1981; Ap & Wong, 2001; Mak et al., 2011; Nyahunzvi & Njerekai, 2013) examining the inside story is a few. On the other hand, questioning what tour guides are going through to accomplish their roles is crucial for the profession's sustainability. Thus, insights need to be offered into tour guiding since tour guides are held responsible in the industry to the extent of 'making or breaking a tour' (Ap & Wong, 2001).

Labor-based research is ignored the in tour guiding literature. Likewise, the exploitation and domination practices against tourism employees are under-studied (Ong et al., 2014). In tour guiding literature, the studies providing insights into the labor barely defined the phenomenon as 'exploitation', or 'domination'. In this regard, 'issues', 'problems', 'hassles', or 'challenges' are the terms used to define labor-related adverse practices. On the other hand, the exploitation and domination in the tour guiding profession should be defined and clarified since there is an agreement on the importance of tour guiding services in tourism studies. Thus, the efforts to examine and discuss such practices critically will promote the recognition of the tour-guiding profession, which has become widespread in many places in the 21st century. Likewise, decreasing the anxiety and worries of tour guides regarding their job will enhance their joy and reflect on their performance (Güzel et al., 2013). Enhancing the labor conditions of tour guides will yield gains to the circles of tourism, including tour guides themselves, tourists, host communities, tour operators, official authorities, and other service providers.

This chapter aims to draw a framework for the exploitation and domination practices in tour guiding. In this context, the conceptual and theoretical framework asserted by Çıvak (2021) and Çıvak and Besler (2022) was adopted. Thus, exploitation practices noted by Çıvak (2021) and Çıvak and Besler (2022) are as follows: labor-intensive working, legal rights, unjust regulations, manufacturing consent, and the role of the state. Likewise, managerial, customer-driven, and co-workers-driven practices were defined as domination practices. Hence, the aforementioned practices in tour guiding labor were determined through an extensive literature review. Tour guiding studies were reviewed from the seminal studies of Grönroos (1978), Schmidt (1979), Holloway (1981), Pearce (1984), Cohen (1985), and Fine and Speer (1985) to the latest papers. The empirical research findings were formulated based on the adopted conceptual and theoretical framework. Later, each content was evaluated and discussed. Additionally, an insider perspective was carried out by the author, a licensed tour guide with nine-year experience. Eventually, an extensive and depictive framework was offered for exploitation and domination practices in tour guiding regarding critical perspective.

Tour Guiding Service in the Tourism Industry

In tourism studies, the definition of tour guiding has been clarified via the description of the professionals in the field: tour guides. World Federation of Tourist Guide Associations (WFTGA) defined the tour guide as "a person who guides visitors in the language of their choice and interprets the cultural and natural heritage of an area which person normally possesses an area-specific qualification usually issued and/ or recognized by the appropriate authority" (wftga.org, n.d.). Providing knowledge for the tourists regarding the visited sites (Holloway, 1981), making entertaining and exciting interpretations, conducting in a tourist destination, and possessing the competencies for language/s are among the required skills for tour guiding (Ap & Wong, 2001). Besides their knowledge, tour guides' communication skills and service quality determine the experience's quality for tourists (T. Y. Chang et al., 2012). The missions of tour guides have been defined in various aspects due to the roles they adopt in tours, and their roles determine how to handle their duties in different cases. Some of these roles are 'information-giver', 'teacher or instructor', 'motivator', 'ambassador', 'entertainer', 'confidant', and 'disciplinarian' (Holloway, 1981).

The existence of a tour guide is determinative for people to join tours instead of traveling alone (Chan et al., 2015). Because tour guides function in various ways to enhance the experience of tourists. Initially, a guided tour is preferable for tourists because it determines what to see in a limited time, reduces the potential conflicts between the tourists and the host community, and decreases the concerns caused by the 'unknown' (Schmidt, 1979). Moreover, tourists would like to learn about a destination they visit, and this motivation is one of the reasons why they travel (Mancini, 1990). In this respect, a tour guide is a primary source of information about a destination (Nettekoven, 1979). Because the assistance of a tour guide helps to know where to look, when to look, and why to look as well as how to behave (Reisinger & Steiner, 2006). Additionally, tour guides encourage tourists to hear, smell, taste, and feel a foreign culture (Urry, 1990). Hence, tour guides mediate between tourists and destinations to establish an intellectual, emotional, and spiritual bond (Christie & Mason, 2003; Weiler & Ham, 2001). On the other hand, it is not a one-way win for the benefit of tourists. Tour guides help destinations and host communities to yield gains via their translator role, which requires the responsibility to interpret the meaning, values, and significance of the visited environment and culture. Thus, tour guides fulfill their responsibilities towards the destination sources (Hu & Wall, 2012).

Tourists are the leading actors in the industry. It is not possible to mention 'the tourism industry' without the existence of tourists. During a day in the tourism industry, tour guides are the employees who make the most contact with tourists (Grabowski & Geng, 2000; Tang & J. C. Chang, 2006; Zhang & Chow, 2004) and it makes tour guiding service crucial for tourist satisfaction (Geva & Goldman, 1991; Chan et al., 2015). Tour guides' responsibilities towards tourists are due to their leader and mentor roles. Hence tour guides are responsible for meeting the expectations and the needs of tourists. Their answerable attitude toward tourists' questions of where to go, how long to stay, how to visit, which activities to be involved and how to get in touch with the host communities enhance the experience of tourists (Hu & Wall, 2012). Likewise, a tour guide should operate as a psychologist who understands the needs of tourists (Naumov, 2020). In this context, the interaction quality between the tour guides and tour participants is essential since the quality of the interaction could ensure tourist satisfaction (Chiang & Chen, 2014; Kalyoncu & Yuksek, 2017).

The intensifying competition among destinations changed the expectations from tour guides. They are supposed to go beyond simply greeting and informing tourists (Prakash & Chowdhary, 2010). Hence, the role of tour guides has expanded to various assignments, from assisting the tour operation to providing the experience in a destination (Weiler & Black, 2015). In this regard, tourist satisfaction with tour guiding services contributes to the destination (Zhang & Chow, 2004). To exemplify, if the tour guiding service is qualified, tourists are likely to extend their stay, yielding income for the destination (Dahles, 2002). Likewise, guiding services' influence can reach such an extent that the tour guide might sell the next tour (Grönroos, 1978), and it points out the essential role of tour guides for the destinations. Eventually, tour guides are a part of the destination strategies (Prakash & Chowdhary, 2010).

Due to their visibility in the industry and the effect on tourist satisfaction, the role of tour guides in a destination is crucial for the stakeholders. Any improvement in the proficiency and professionalism of tour guides yields the industry higher profits (Mak et al., 2010). For example, tour operators' profit is determined by the tour guiding service because the connection between the service suppliers -such as tour operators- and the tourists is provided through the tour guides. Therefore, tour guiding service is a determinant for tourists to evaluate the tour operator's service quality (Wang et al., 2000). Besides the tourists and tour operators, tour guides interact with other stakeholders, including host communities, public authorities, and other tour guides. The responsibility of tour guides towards the host communities due to their representative role requires being hospitable and understanding with the visitors. Additionally, tour guides fulfill their duty towards the public authorities by yielding gains and enhancing the destination's image. Finally, tour guides are responsible for improving their colleagues' social status (Hu & Wall, 2012).

Labor Exploitation in Tour Guiding

Tour guides face challenges based on their working conditions, employment terms, remuneration, and the actions of the official authorities. Despite the emphasized significance of tour guiding services to the extent of 'making or breaking a tour' (Ap & Wong, 2001), tour guides' labor-based efforts are exploited mainly by tour operators who are the employers in the field. On the other hand, the difficulties in the profession result from the policymakers such as the official authorities. The exploitative practices against tour guides include hard working conditions, job insecurity, uninsured employment, low remuneration, pressure for taking tourists to shops, unfair treatment of the employers, and the lack of monitoring actions. Hence, the nature of tour guiding has been clarified in an extensive framework, and the practices of labor exploitation have been defined.

Labor-Intensive Working Conditions

Tour guiding is an attractive profession for people. It enables individuals to travel, meet new people, learn about new cultures, eat the best of food, and earn a living in the meantime. 'Tour guide' title comes with some privileges, and the charisma offered by the job is alluring for young people in particular. The roles of tour guides such as 'leader' and 'information-giver' are self-satisfying deeds and lead to self-actualization. However, tour guiding requires physical effort to endure traveling for days, walking long-distance routes, and moving around different environments, including various physical conditions such as altitude or weather. Furthermore, tour guides are expected to possess the competencies to perform the job. These competencies include a wide range of qualifications such as knowledge -on almost everything-, organizational abilities, and communication skills. These competencies are certainly not obtained overnight. Eventually, tour guides work very hard and spend years to get to the point where they stand as 'experienced'; yet, the work's nature is hard to sustain.

Tour guides do not work shifts. Any schedule delay or a service chain flaw is a time loss. Tour guides are not paid for overtime, and cannot call it quits on a tour, either. Even after the check-in at a hotel, while tour participants rest in their rooms, the guide might still be working due to the clients' complaints about the hotel. Therefore, long hours of labor are a problem for tour guides and cause them to have to switch jobs or retire as they get old (Köroğlu, 2012). In high seasons, tour guides may not have off days to rest. Once their tour ends, they can be assigned to another tour and switch buses on the route. Likewise, a backup suitcase is ready, and the spare time of a guide is barely enough to stop by the home. Thus, 'back-to-back' tours mean working for long periods and missing important days in their lives. Despite the complications, tour guides cannot say 'no' to the next tour for the fear that someone else will be assigned for it and they will be replaced.

There is no fixed workplace for tour guides, which is a challenge for the profession's members (Li et al., 2020). The nature of the work requires traveling and makes tour guiding an exhausting service. Tour guiding labor includes long and connecting flights, time differences, and jet lag. These conditions harm the health of tour guides in the long term (Canküil & Güzeloğlu, 2021). Hence, health issues cause tour guides to quit or retire. Besides, the labor-intensive structure of tour guiding involves employers serving people. Thus, the human-focused nature of tour guiding creates human-based problems and increases the workload. Even the simple misbehaviors of tourists in group package tours, such as leaving the group without permission, might cause them to get lost. Any disorder leads the whole group to delay the itinerary. Thus, the increase in the guide's workload could be easily triggered by just one member of the tour (Tsaur et al., 2019) and the heavy workload is a hassle for tour guides (Hu & Wall, 2013).

Employment and Remuneration

The tourism industry is vulnerable to local or global affairs. Economic crises, disasters, epidemics, terrorist attacks, military actions, and wars affect the demand for tourism services. For example, the COVID-19 pandemic or the war in Ukraine affected the industry, and employees fell out of work. Tour guides, as service providers in the industry, confront the consequences of such tragic affairs, too. A tour guide might enjoy a year of occupation and make a tremendous income; however, the hopes for the next year might give place to unemployment. Therefore, there is no job security for tour guides (Köroğlu, 2012) as well as the other employees of tourism services. Likewise, due to the seasonal occupation, tour guides confront insecurity, and their commitment to the job weakens over time (Mak et al., 2011). In this respect, job insecurity is a tough challenge for employed tour guides, and prospective guides are also worried about their future career (Yetgin et al., 2018). The concerns of tour-guiding students about their future are a threat to the sustainability of the profession. On the other hand, the lack of social welfare and insurance are the other employment-based challenges in the profession (Hu & Wall, 2013). Thus, tour guides take care of the insurance policies such as health insurance and retirement annuity on their own.

Tour guides can operate as permanent or freelance professionals. Permanent tour guides are paid a monthly salary, and freelance tour guides are paid for a day or hours depending on the itinerary (Zhang & Chow, 2004). In some countries, such as Türkiye (Sezgin & Düz, 2018) and Taiwan (Min, 2012), there is no hour-by-hour remuneration, and a daily-basis payment is practiced. On the other hand, in Hong Kong, tour guides can be hired for hours, too (Zhang & Chow, 2004). Either permanent or freelance, tour guides are hired by tour operators and basic remuneration is crucial for tour guides to earn a living. Nonetheless, low and unfair remuneration has been reported in the profession since the 1980s (Holloway, 1981). Tour operators cut the fees of the guides to recover costs (Mak et al., 2011). Likewise, they take advantage of slack seasons and hire tour guides at a below cost due to the inoccupation (Hu & Wall, 2013). In the tourism industry, inexperienced tour guides are likely to be offered low and unfair wages for tours (İlhan & Soybalı, 2018). Furthermore, in some cases, they procrastinate the payments to tour guides (Karamustafa & Kılıçhan, 2019) or withhold them forever (Yarcan, 2007). Eventually, the remuneration in tour guiding is not perceived as fair compared to the workload of the job (Şenol & Aliyev, 2015). The lack of a reasonable income causes high turnover rates and unqualified service in the tour guiding profession (L. Huang & Kao, 2011).

Some examples prove the existence of other unfair remuneration practices. In Hong Kong and Macau, *Mai tou* -a bribing practice- shifts pressure to tour guides to pay a fee per tourist so that the tour operator can hire them. However, there are other payment issues than the basic salary. Tour operators might cut tour guides' tips to recover costs (Mak et al., 2011). Therefore, tour guides focus on extra gains based on their financial concerns, and it makes way for the tour operators to exploit them even more. Thus, tour operators put pressure on tour guides to make more profits. They set a shopping quota per tourist, and tour guides find themselves in the 'salesperson' role. On the other hand, when tourists are forced to make a substantial purchase, it reduces the quality of the tour and causes tourist dissatisfaction (Mak et al., 2010). Tour operators' practices for tourist consumption result from fierce competition. Hence, they don't pay a wage to the tour guides and expect them to force the tourists to maximize their expenses. Eventually, tour guides have no choice but to lean into commissions and tips (Hu & Wall, 2013). On the other hand, if a tour guide is lucky enough not to receive a complaint because of all the pressure on tourists to purchase, s/he might get an extra income. Because tour guides are paid a bonus determined due to the satisfaction perceived by the tour participants (Geva & Goldman, 1991).

Illegal guiding is one of the exploitation practices in the industry. Tour operators hire unlicensed tour guides to cut costs. However, the employment of unlicensed labor is perceived by tour guides as a neglect of their educational background. Furthermore, licensed tour guides consider that illegal guides steal their tours and hence their profits (Ayaz & Demir, 2019). Illegal guiding practices might lower the labor cost for the tour operators but result in unfair competition. Also, poor service quality damages the destination's brand (Chowdhary & Prakash, 2008). On the other hand, employing unqualified tour guides harms the industry due to incompetency issues (Ap & Wong, 2001). Another practice of lowering the cost for tour operators is to hire fewer tour guides than the requisites. For instance, despite the legal obligation of hiring a tour guide per vehicle in Türkiye, tour operators tend to hire one guide for two vehicles to reduce costs (Türkmen & Deniz, 2022). One tour guide could be hired for even five or six buses in some tours. In most cases, the guide is not informed about the situation and gets shocked after meeting the group(s). In this case, tour guides are asked to switch vehicles during the tour and work as many times as the number of vehicles. Such interferences of tour operators result in poor service quality and customer complaints.

There are some other challenges that tour guides confront due to the unfair practices of the tour operators. Tour operators' business failures make tour guides pay the price (Nyahunzvi & Njerekai, 2013). For example, a last-minute cancellation by the tour operator is an unfair treatment against the tour guides (Karamustafa & Kılıçhan, 2019). Tour operators may even cancel the guides' tours via unreasonable excuses (Türkmen & Deniz, 2022). In this case, it is a challenge for a tour guide to substitute that tour for another one. Hence, tour guides can be the victims of the unprofessionalism demonstrated by the tour operators and get stressed. Also, stress appears when the tour operator does not provide sufficient information for the guide (e.g., itinerary), makes an unequal payment for the same labor, or provides inappropriate equipment (Mackenzie & Kerr, 2013). On the other hand, the unprofessional customs of tour guides reflect on the hiring process. Because personal relationships and acquaintances are determinative for tour operators when assigning tour guides for the tours. Finally, tour operators' ungenerous practices exploit tour guides since it was reported that the tour operators do not pay tour guides' expenses spent during the tour. For example, Turkish tour guides complained about not being paid for the lunch expenses during the tour (Türkmen & Deniz, 2022).

The Role of the Official Authorities

The governments, local administrations, and employee associations are the responsible policy-makers in tour guiding services. The most common problem is the lack of monitoring and inspection efforts. Even though the regulations are adequate and inclusive, defective monitoring actions cause problems (Hu & Wall, 2013). Because monitoring actions are supposed to protect the tour guides, some instances exemplify the challenges that tour guides cannot get support for. For example, the domination by Chinese immigrant tour guides turns out to be a challenge that tour guides face in Hong Kong and Macau (Mak et al., 2011). Unorganized foreign labor causes exploitation in tour guiding. In some cases, tourism employees whose job definitions are unrelated to the 'tour guide title' can be a threat to the profession. In this regard, some tourism attendants assigned to monitor or guard a tourism site exceed their authority and guide illegally. Tour operators tend to hire this illegal and low-cost labor instead of tour guides leading to authorized professionals' unemployment (Büyüktepe et al., 2019; Eser & Şahin, 2020).

Practices such as unfair remuneration and insufficient monitoring reduce the motivation and enthusiasm of tour guides. The lack of regulations to protect tour guides paves the way for tour operators to exploit them (Hu & Wall, 2013). Furthermore, tour guiding is not always sunshine and rainbows and involves danger in some destinations. Therefore, policy-makers must take safety and security precautions, and regular monitoring is essential. For example, tour guides in Zimbabwe were reported to suffer from safety and security issues. Some dangerous activities that tour guides are involved with visitors, such as walking with lions, water rafting, and elephant riding, require insurance and medical aid. These requirements aside, no specific practical-oriented training for dangerous activities was reported (Nyahunzvi & Njerekai, 2013). The laxness of professional organizations -such as tour guide associations- gets in the way of solving the problems the tour guides face (Karamustafa & Kılıçhan, 2019).

Domination Practices in Tour Guiding

The labor-based efforts of tour guides are underrated, and the domination practices are executed to intimidate the professionals and sustain their labor for the interests of the other stakeholders. Tour guides face hardships caused by the tour operators and are put through some unfair practices. The employers and the official authorities use domination practices to intimidate tour guides through unemployment. On the other hand, tourists can be mean to their guides due to different motives. Tourists' complaints can be domination attempts against the guides regardless of the source causing dissatisfaction. Hence, tour guides are considered to be the 'scapegoats' of the industry and are portrayed as the targets to 'solve the problems'. Additionally, tour guides do not face the domination practices of employers and tourists, only. However, their colleagues and co-workers also set challenges for tour guides based on their motives and interests.

Domination of the Employers and Official Authorities

Exploitative practices towards tour guides are normalized and sugar-coated by the tour operators. The heavy workload is baptized as a must-endure test for the passion for the job (Ong et al., 2014). Hence, the employers make way for other unfair practices, and the workload is increased with irrelevant tasks. Full-time tour guides are asked to fulfill inconvenient physical and maintenance duties when there is no tour to be assigned. Fixing the fences or digging trenches are among the irrelevant workload that tour guides are told to do by their employers. Further, in case of resistance against the irrelevant workload, the employers intimidate tour guides with unemployment and remind them of the people who would fill their position (Nyahunzvi & Njerekai, 2013). Thus, the tour guides who do not comply with the irrelevant tasks confront the risk of losing their job, and the exploitative practices of the tour operators turn into domination.

Tour operators evaluate tour guides by the extent of commissions they bring to the enterprise and not by the service quality (Ap & Wong, 2001). Thus, tour operators do not assign tours to the 'unprofitable' guides, and they face barreling towards unemployment. Even a tour guide in China links this pressure to the death of a colleague who had cancer (Xu (Hazel) & McGehee, 2017). Moreover, tour guides might find themselves in a dilemma between providing qualified service and earning their lives. The shops which attract tour guides to receive higher commissions are likely to deceive tourists. Thus, it would be a challenge for tour regarding ethics and conscience when they lead tourists to those shops even though they have to live by the commissions (S. Huang, 2010). Consequently, exploitative practices appear in a sequence in the tourism industry. Tour operators exploit tour guides, and the guides exploit tourists (Ong et al., 2014).

Tour guides are assigned to itineraries they are unprepared for or experts of. However, tour guides hesitate to reject these tours for the fear that they may not be hired again by the same tour operator. However, these practices of the tour operators result in tourist dissatisfaction and complaints. In such cases, tour guides who fail in a destination suffer damage to their reputation. Tour operators' false promises to tourists make the guides pay the price (Chowdhary & Prakash, 2008). Due to the fierce competition, tour operators race to offer the fullest itinerary to tourists regardless of feasibility. Thus, it puts tour guides in trouble at the end of the day (Arslan & Şimşek, 2018). Furthermore, when there is a need for a 'scapegoat' after any failure, the tour guide is the one to point fingers at because tour operators charge tour guides for the consequences of their unprofessional practices (Türkmen & Deniz, 2022).

Domination on tour guides can be executed by the official authorities, too. In some cases, official authorities draw the framework for the narratives of the tour guides. Therefore, the politics in a destination could be determinative of the context disseminated to tourists (Wong, 2013). For instance, the governmental institutions in Indonesia trained the tour guides and set obligations for their narratives. Thus, the tour guides were forced to narrate the official and authorized content (Dahles, 2002). Additionally, tour guides might face unfair treatment from the official authorities and be punished by their license. Such as the tour guides who were forced to work for zero-fare tours in China protested in a strike and asked for changes in the system in 2004. However, they faced sanctions as their certificates were revoked (Xu (Hazel) & McGehee, 2017).

Domination of the Tourists

Tourists' domination practices can be in the form of misbehaviors on tour. Tour members intentionally make difficulties, and it is a challenging effort for the guides to overcome these misbehaviors in the heavy workload. Violating the shared rules of the tour, sexual harassment, or physical violence are among the challenges in guided tours (Tsaur et al., 2019). Some tour participants are mean to their guides as they violate personal boundaries and body shame their guides. Likewise, the practices of tour participants, such as interrupting the narratives, despising the knowledge, and testing the information disseminated by the guides, cause trouble for the professionals. The attempts at grabbing the microphone are another reported misbehavior of tour participants (Tetik Dinç et al., 2022). Tourists' misbehaviors, such as mocking their tour guides, disturbing them for trivial reasons during their resting time (Ceylan, 2021), or attempting to occupy the seat reserved for the guide are the other challenges reported (Efendi & Karakaş Tandoğan, 2019). Moreover, tourists' misbehaviors, such as giving orders to tour guides or bossing them around, are among the hassles (Kurt Dündar & Arslantürk, 2022). Likewise, the doubts expressed by the tour members concerning the tour guide's competencies are a case of crisis for the guide's authority (Cohen, 1985; Tsaur et al., 2019). In some cases, tour participants compare their guide to their colleagues, making them feel worthless, insulting, or walking up to the guides (Tetik Dinç et al., 2022). Also, tour members tell the other tour participants 'not to believe tour guides because they are liars and cheat' (Tsaur et al., 2019).

Tourists were reported to bully their guides when their unreasonable demands could not be met. However, tour guides need to be reasonable while operating the itinerary for the tourists. This dilemma between the tourists' expectations and tour guides' potential service cause further complaints. For example, cancellation of an activity due to the weather conditions -such as a storm- could be the reason to scold the tour guide to tears. Also, asking the tour guide to assign a good room every night on tour can cause quarrels when the demand is not fulfilled. However, tour guides must be fair to each group member, and tourists insisting on personal favors put the guides under pressure (Xu (Hazel) & McGehee, 2017). When the unreasonable demands of tour participants cannot be fulfilled, it leads to conflicts. For example, tourists expect or ask their guide to behave the way they want. In China, tour members complained about their tour guide because she did not tell dirty jokes or sing the songs they wanted to hear (Tsaur & Lin, 2014). Likewise, some tour participants ask their guide to perform a piece of lame entertainment or to serve drinks on the bus. These irrelevant and weird requests result from the lack of recognition for the job and set pressure on tour guides (Arslan & Şimşek, 2018).

Gender is one of the focal points of inequality in tour guiding. Some tourists perceived female tour guides as 'less capable', especially for adventure-themed guiding (Mackenzie et al., 2020). Despite the sexist discrimination against female tour guides reported in the literature (Mackenzie et al., 2020; Köroğlu et al., 2020; Alrawadieh et al., 2022), male tour guides face gender discrimination as well. Due to religious concerns or the characteristics of a group, tour members insist on being guided by a female tour guide (Ateş et al., 2022). Hence, male tour guides are rejected and confront unemployment. On the other hand, gender is not the only basis leading to domination practices in guided tours. Tourists might be provoking against their tour guide. Tourists' proactive comments about critical issues, such as national conflicts, are defined as hostile by the tour guides. Eventually, the misbehaviors of tour participants cause tour guides to feel emotions of depression, anger, and disappointment. Further, these feelings lead tour guides to quit their job (Tetik Dinç et al., 2022). On the other hand, misbehaviors of tour participants harm not only the tour guides but also the mood of the other tour members (Tsaur et al., 2019), threatening the tour's success.

Domination of Colleagues, Co-Workers, and Others

The fierce competition in the profession leads to jealousy among tour guides. The tour guides gaining high commissions can be seen as a threat by their colleagues. Due to jealousy, tour guides' misbehaviors, such as gossiping about each other or defaming their colleagues harm the profession (Düz, 2017). Likewise, tour guides are supposed to cooperate on tours. Disseminating knowledge outside the buildings is one of the ways to cooperate in the field so that they do not disturb the other groups (Kavak, 2020). However, some tour guides might raise their voices during interpretation and disseminate knowledge to disprove a colleague in front of his/her clients. On the other hand, tour guides invite tour operators to sustain labor exploitation and domination practices. Once figuring out that a colleague is engaged to a tour operator for a tour, a tour guide might call that tour operator and offer to be hired for a lower wage (Düz, 2017). Lastly, the support of seniors is valuable in the profession. However, some senior tour guides were reported to harass young and inexperienced colleagues (Chowdhary & Prakash, 2008).

Cooperation among tourism employers is crucial for the success of the whole industry. However, tour guides face domination practices of tourism staff due to their roles and conflicts of interest. In this context, drivers are the tourism employees who tour guides spend most of their working hours with. Nonetheless, the drivers resent that tour guides are the focus of attention because their effort in the tour is overlooked or taken for granted by the tour members. The social distance between the guides and drivers also deepens the friction. Besides, economic reasons such as 'lower tip for the driver' cause conflicts between the two. Eventually, the lack of cooperation ends up with sanctions imposed on the tour guide. Failing to slow down or stop at the visitation spots or defaming the tour guide to the tour operators are the challenges that tour guides might face related to the drivers (Holloway, 1981). Likewise, drivers put the guides under pressure by forcing them to ask the tourists to tip (Karamustafa & Çeşmeci, 2006).

Gender discrimination was noticed among colleagues in tour guiding. Female tour guides face challenges derived from the 'macho' guiding culture built into the profession. For example, female tour guides face gender inequality as their colleagues perceive them as 'less capable' of guiding adventure-themed tours (Mackenzie et al., 2020). As a result, they are interrupted by their male co-workers based on the assumption that 'they cannot achieve the tasks of being a tour guide. On the other hand, male tour guides are not the only co-workers questioning the capability of female tour guides. Even the drivers trouble them because they do not want to be instructed by female tour guides (Köroğlu et al., 2020). The misbehaviors of the drivers can even go beyond passive-aggressive manners. In some cases, female tour guides confront verbal harassment from drivers on tour. Furthermore, a female tour guide was reported to have faced a tragic act of violence from a driver (Arslan & Şimşek, 2018).

Competence-based perceptions are not the only challenge for female tour guides. Disrespectful and sexist comments from co-workers or clients cause pressure and discourage female tour guides from building their careers (Mackenzie et al., 2020). Female tour guides face gender-based challenges in their personal lives as well. Friends, families, or relatives can be inconsiderate or prejudiced. For instance, a female tour guide who posted her tour photographs on her social media account was bullied by her husband's relatives because of the tourists involved in the photograph. She blocked those relatives in her social media accounts (Arslan & Şimşek, 2018). Moreover, sexual harassment harms female tour guides' professional and psychological well-being (Alrawadieh et al., 2022). On the other hand, the problem does not seem to be female-specific, as male tour guides have reported sexual harassment, too (Öter & Sonuç, 2014). Despite the encouragement to expose sexual harassment, tour guides tend to ignore or pass off the improper behaviors in the tours (Yıldırım & Özbek, 2019). In that case, sexual harassment remains a challenge unless more courageous actions are taken.

Tourism employers in different occupations get in the way of tour guides and practice domination for their interests, such as profits. The threats from shopkeepers, local police, hawkers, beggars, local licensed guides, touts, and monument staff make working conditions worse for tour guides. The task of protecting foreign tourists from the disturbance of those is a burden for tour guides. The disturbance sometimes goes even further to the extent of physical harm. Likewise, tour guides complain about not being supported or protected by the official authorities. Civil authorities and managers of the visited sites were reported to treat tour guides 'shabbily' (Chowdhary & Prakash, 2008). Additionally, such practices can be in the form of mobbing which is a challenge for tour guides. Interference of others and problems with service providers were reported as the sources of mobbing that tour guides go through (Erkol Bayram, 2016). Finally, the lack of recognition in society and misperceptions of families upset and discourage the tour guides. They resent that their profession is perceived as an unglamorous job by their parents in particular. The parents do not comprehend the critical role of tour guides and have misconceptions about the job, such as it is nothing more than taking tourists to shops (Ap & Wong, 2001).

DISCUSSIONS AND IMPLICATIONS

Tour guides are hardworking employers in the tourism industry. Due to the workload and challenging nature of the profession, tour guides make an extended effort to accomplish their professional tasks. In this regard, tour guides can barely have off days in the high season, leading to health problems. In this regard, tour guides can barely have off days in high season and it leads to health problems. However, they cannot reject the tours for fear that someone else will replace them, and thus they will not be hired again. Despite the workload and significance of their service, tour guides do not have job security and they cannot receive a recompense for their work. The financial concerns of the employers (tour operators) transform into exploitation practices as they hire tour guides at low remuneration. Likewise, tour guides are even asked to pay the tour operators in advance so that they can 'buy the tour'. Therefore, the tour operators tend to 'squeeze' the guides and pay as little as possible. On the other hand, tour operators hiring illegal labor instead of licensed tour guides victimize the labor of guiding and ignore their educational or intellectual backgrounds. The exploitative perspective of the tour operators forces tour guides to take tourists to shop. Hence, 'the best tour guide' for a tour operator is the one who makes the most profit, and it puts the guides under pressure. Eventually, the lack of monitoring actions of the official authorities makes way for sustaining exploitative practices in the profession.

As a result of the exploitation practices of tour operators, tour guides promote and make way for labor exploitation, too; the competition in the field is used as an excuse by the guides for agreeing to underpayment. Tour guides claim that if they do not submit to low remuneration practices, some other guides will replace them. The tour guides yield to exploitation because of their financial debts and concerns about maintaining their lives (Çakmak, 2020). Hence, establishing a minimum wage rate for tour guides would ensure a stable and reasonable income for the tour guides. Improving the financial conditions of the guides will also help reduce the pressure, such as shopping for commissions (Ap & Wong, 2001), which is a clash in the industry. Because the conflicts between tour guides and tourists regarding shopping-based trips and distorted itineraries had press coverage in China, which caused a crisis in the reliability of the tour guides (Mak et al., 2010). Unrightfully, despite their high-level education, foreign language skills, and expertise in different subjects, tour guides go through stress. As a consequence, they challenge the struggles on their own or end up quitting the profession. Likewise, various incidents affect the productivity of tour guides. Determining these incidents and remediating the labor conditions will enhance the motivations and performance of tour guides (Erkol Bayram & Zengin, 2017).

Tour guides confront domination practices as they are asked to do irrelevant work by their employers. Likewise, tour guides accommodate themselves to the tasks of the tour operators, such as assignation to a tour that does not suit their expertise. The threat of losing their job makes tour guides consent to the domination. The actions of the official authorities put tour guides under pressure as well. Setting a framework for the narratives puts tour guides under restraint. Likewise, imposing sanctions against the tour guides who protest or strike is a domination practice to intimidate them. Furthermore, tourists' domination over tour guides can be in various forms. A wide range of practices, from sexual harassment to violation of boundaries, were reported in tourist-guide interactions. Gender-based discrimination against tour guides comes from tourists, colleagues, or co-workers and gets in the way. Female tour guides suffer from misperceptions such as 'being less capable' and male tour guides may not be preferred by tourists due to religious motives. Co-workers such as drivers practice domination against tour guides and slow down driving service, which puts the guides on the spot. Eventually, the competition among tour guides causes conflicts among colleagues, and tour guides harm each other for their financial interests.

Tour guides experience stress from external factors such as tour operators and tourists. (Mackenzie & Kerr, 2013). The mobbing practices against tour guides can be in various forms such as psychological and sexual harassment, the lack of communication, overtime work, proscription, disinformation on purpose, the humiliation of competencies and knowledge, and, defamation. (Erkol Bayram & Zengin, 2017). Tour participants' disrespectful manners and mobbing practices towards tour guides were reported to reflect tourist domination. In this respect, tourists use complaints as an excuse for domination over tour guides. Some of the complaints raised by tour members are as follows: poor service and food quality of the restaurants, failures on the tour itinerary, unequipped vehicles, unqualified hotels, misbehaviors of the vehicle drivers, tardiness of the other tour members, waiting in line at the attractions and dirty toilets. Likewise, the accord between the tour guide and the tourists determines the course of the tour (Yetgin Akgün et al., 2022). When the guide's expertise does not fit the manners of the group, it is likely to cause complaints (Tsaur & Lin, 2014). Hence, developing proactive strategies for probable complaints might help tour guides to avoid conflicts in the first place (Yetgin Akgün et al., 2022).

Disrespectful manners of the tourists put a burden on the shoulders of the tour guides, and they complain about the vulgarism of tour participants (Efendi & Karakaş Tandoğan, 2019). Additionally, gender is one of the focal points of inequality in tour guiding. Female tour guides stated that they had strategies to tussle the negative perceptions, such as developing skills to compete with male co-workers (Mackenzie et al., 2020). As for sexual harassment, tour guides should be encouraged to speak up and report these acts to their employers. In case of a report, tour operators should take cautions such as reporting a complaint of switching the service provider depending upon the offender. If the offender is a tour member, the tour operator should take action and ensure the hospitable environment is sustained for the rest of the group. Additionally, proactive strategies are crucial to prevent nefarious acts. Hence, codes of ethics could be clarified for the tour members and service providers (Alrawadieh et al., 2022). However, labor exploitation and domination practices in tour guiding reveal the substantial need for mutual understanding and common courtesy in the field in terms of interaction in the first place.

REFERENCES

Alrawadieh, Z., Demirdelen Alrawadieh, D., Olya, H. G. T., Erkol Bayram, G., & Kahraman, O. C. (2022). Sexual harassment, psychological well-being, and job satisfaction of female tour guides: The effects of social and organizational support. *Journal of Sustainable Tourism*, *30*(7), 1639–1657. doi:10.1080/09669582.2021.1879819

Ap, J., & Wong, K. K. (2001). Case study on tour guiding: Professionalism, issues and problems. *Tourism Management*, *22*(5), 551–563. doi:10.1016/S0261-5177(01)00013-9

Arslan, A., & Şimşek, G. (2018). Kadın turist rehberlerinin yaşadıkları mesleki sorunlar: Aydın turist rehberleri örneği. *Journal of Travel and Tourism Research*, (13), 23–49.

Ateş, M., Kazan, İ., & Soybalı, H. H. (2022). Turist rehberliğinde ayrımcılık: Kadın turist rehberlerine yönelik bir uygulama. [TURED]. *Turist Rehberliği Dergisi*, *5*(1), 1–16. doi:10.34090/tured.1064528

Ayaz, N., & Demir, C. (2019). Perceived work stress factors: A study on tour guidesç. *Journal of Tourism and Gastronomy Studies*, *7*(1), 415–427. doi:10.21325/jotags.2019.370

Bryon, J. (2012). Tour guides as storytellers–from selling to sharing. *Scandinavian Journal of Hospitality and Tourism*, *12*(1), 27–43. doi:10.1080/15022250.2012.656922

Büyüktepe, H. S., Gökdemir, S., & Korkmaz, H. (2019). Turist rehberlerinin sorunları üzerine nitel bir araştırma: Çanakkale örneği. *Journal of Travel and Tourism Research*, (14), 94–117.

Çakmak, T. F. (2020). Turist rehberlerini yasa dışı çalışmaya yönlendiren etmenler üzerine bir araştırma. *Journal of Hospitality and Tourism Issues*, *2*(2), 91–107.

Cankül, D., & Güzeloğlu, F. T. (2021). Profesyonel turist rehberlerinin yurtdışı turlarında yaşadıkları sorunlar. *Journal of Humanities and Tourism Research*, *11*(4), 844–862. doi:10.14230/johut1110

Ceylan, U. (2021). Turist rehberlerinin mesleki motivasyonlarında tur üyelerinin yanlış davranışlarının etkisi. *Anemon Muş Alparslan Üniversitesi Sosyal Bilimler Dergisi*, *9*(1), 39–49. doi:10.18506/anemon.683064

Chan, A., Hsu, C. H., & Baum, T. (2015). The impact of tour service performance on tourist satisfaction and behavioral intentions: A study of Chinese tourists in Hong Kong. *Journal of Travel & Tourism Marketing*, *32*(1-2), 18–33. doi:10.1080/10548408.2014.986010

Chang, K. C. (2014). Examining the effect of tour guide performance, tourist trust, tourist satisfaction, and flow experience on tourists' shopping behavior. *Asia Pacific Journal of Tourism Research*, *19*(2), 219–247. doi:10.1080/10941665.2012.739189

Chang, T. Y., Kung, S. F., & Luh, D. B. (2012). The innovative service model: A study on tour guide service design in Taiwan by ICT application. *Academic Research International*, *2*(1), 192–198. doi:10.1300/J162v01n04_06

Chiang, C. Y., & Chen, W. C. (2014). The impression management techniques of tour leaders in group package tour service encounters. *Journal of Travel & Tourism Marketing*, *31*(6), 747–762. doi:10.1080/10548408.2014.889641

Chowdhary, N., & Prakash, M. (2008). Challenges of tourist guiding-An assessment of situation in India. *Tourism Infrastructure. Technology & Operations*, (Part VII), 287–302.

Christie, M. F., & Mason, P. A. (2003). Transformative tour guiding: Training tour guides to be critically reflective practitioners. *Journal of Ecotourism*, *2*(1), 1–16. doi:10.1080/14724040308668130

Çıvak, B. (2021). *Otel işletmelerinde emek sömürüsü, tahakküm ve direniş ilişkilerine dair niteliksel bir araştırma.* [Unpublished doctoral dissertation, Anadolu Üniversitesi, Sosyal Bilimler Enstitüsü, Eskişehir, Türkiye].

Çıvak, B., & Besler, S. (2022). A Critical Analysis of Managerial Control Mechanisms in Hotel Businesses. *SAGE Open*, *12*(1). doi:10.1177/21582440221088850

Cohen, E. (1985). The tourist guide: The origins, structure and dynamics of a role. *Annals of Tourism Research*, *12*(1), 5–29. doi:10.1016/0160-7383(85)90037-4

Dahles, H. (2002). The politics of tour guiding: Image management in Indonesia. *Annals of Tourism Research*, *29*(3), 783–800. doi:10.1016/S0160-7383(01)00083-4

Düz, B. (2017). Turist rehberliğinde meslek etiği ilkelerinin faydacı etik yaklaşımı ile belirlenmesi. *Hitit Üniversitesi Sosyal Bilimler Enstitüsü Dergisi*, *10*(2), 1639–1662. doi:10.17218/hititsosbil.286973

Efendi, M., & Karakaş Tandoğan, G. (2019). Yerli turistlerin paket turlarda yaptıkları etik dışı davranışlar üzerine turist rehberleri gözüyle bir değerlendirme. *Seyahat ve Otel İşletmeciliği Dergisi*, *16*(1), 21–33. doi:10.24010oid.451222

Erkol Bayram, G. (2016). Turist rehberlerinin mobbinge maruz kalma durumunun demografik değişkenler kapsamında incelenmesine yönelik bir araştırma. *Journal of International Social Research*, *9*(47), 1185–1194. doi:10.17719/jisr.2016.1465

Erkol Bayram, G., & Zengin, B. (2017). Turist Rehberlerinin Verimliliğini Etkileyen Psikolojik Faktörler. In Güzel, F. Ö., Altıntaş, V., Şahin. İ. (Eds.), Turist Rehberliği Araştırmaları Öngörüler ve Uygulamalar (pp. 197-218). Detay Yayıncılık.

Eser, S., & Şahin, S. (2020). Turist rehberliği mesleğinde yaşanan güncel sorunlar üzerine bir inceleme. *Türk Turizm Araştırmaları Dergisi*, *4*(2), 1344–1355. doi:10.26677/TR1010.2020.399

Fine, E. C., & Speer, J. H. (1985). Tour guide performances as sight sacralization. *Annals of Tourism Research*, *12*(1), 73–95. doi:10.1016/0160-7383(85)90040-4

Geva, A., & Goldman, A. (1991). Satisfaction measurement in guided tours. *Annals of Tourism Research, 18*(2), 177–185. doi:10.1016/0160-7383(91)90002-S

Grabowski, C. P., & Geng, W. (2000). European silk road tourists' and their tour guides' perceptions of product and service quality. *Journal of Quality Assurance in Hospitality & Tourism, 1*(4), 97–106. doi:10.1300/J162v01n04_06

Grönroos, C. (1978). A Service Oriented Approach to Marketing of Services. *European Journal of Marketing, 12*(3), 588–601. doi:10.1108/EUM0000000004985

Güzel, F. Ö., Atilla Gök, G., & Büyüker İşler, D. (2013). Duygusal emek ve işten ayrılma niyeti ilişkisi: Turist rehberleri üzerinde bir araştırma. *Seyahat ve Otel İşletmeciliği Dergisi, 10*(3), 107–123.

Holloway, J. C. (1981). The guided tour: A sociological approach. *Annals of Tourism Research, 8*(3), 377–402. doi:10.1016/0160-7383(81)90005-0

Hu, W., & Wall, G. (2012). Interpretative guiding and sustainable development: A framework. *Tourism Management Perspectives, 4*, 80–85. doi:10.1016/j.tmp.2012.05.003

Hu, W., & Wall, G. (2013). Tour guide management in Hainan, China: Problems, implications and solutions. *Asia Pacific Journal of Tourism Research, 18*(3), 205–219. doi:10.1080/10941665.2011.640699

Huang, L., & Kao, P. H. (2011). How to tell a good tour guide under different strategic orientations. *African Journal of Business Management, 5*(27), 11146. doi:10.5897/AJBM11.1291

Huang, S. (2010). A revised importance–performance analysis of tour guide performance in China. *Tourism Analysis, 15*(2), 227–241. doi:10.3727/108354210X12724863327803

Huang, S., Hsu, C. H., & Chan, A. (2009). Tour guide performance and tourist satisfaction: A study of the package tours in Shanghai. *Journal of Hospitality & Tourism Research (Washington, D.C.), 34*(1), 3–33. doi:10.1177/1096348009349815

Ibanez, J., Aylett, R., & Ruiz-Rodarte, R. (2003). Storytelling in virtual environments from a virtual guide perspective. *Virtual Reality (Waltham Cross), 7*(1), 30–42. doi:10.100710055-003-0112-y

İlhan, Y., & Soybali, H. H. (2018). Turist rehberlerinin mesleki sorunları üzerine bir araştırma. [TURED]. *Turist Rehberliği Dergisi, 1*(1), 13–23. doi:10.34090/tured.443615

Kalyoncu, M., & Yuksek, G. (2017). The Effect of impression management techniques which tourist guides use in package tours on personal interaction quality. *Journal of Tourism and Hospitality Management, 5*(1), 15–25. doi:10.15640/jthm.v5n1a2

Karamustafa, K., & Çeşmeci, N. (2006). Paket tur operasyonunda turist rehberlerinin karşılaştıkları yönetsel sorunlar üzerine bir araştırma. *Anatolia: Turizm Araştırmaları Dergisi, 17*(1), 70–86.

Karamustafa, K., & Kılıçhan, B. (2019). Delphi yöntemi ile profesyonel turist rehberliği mesleğine ilişkin sorunların tespit edilmesi. *Afyon Kocatepe Üniversitesi Sosyal Bilimler Dergisi, 21*(4), 1348–1365. doi:10.32709/akusosbil.597628

Kavak, M. (2020). Frig kentleri. In T. F. Çakmak (Ed.), *Turist rehberleri için Anadolu'da antik kent ve merkezler* (pp. 123–151). Detay Yayıncılık.

Köroğlu, Ö. (2012). İçsel ve dışsal iş doyum düzeyleri ile genel iş doyum düzeyi arasındaki ilişkinin belirlenmesi: Turist rehberleri üzerinde bir araştırma. *Doğuş Üniversitesi Dergisi, 13*(2), 275–289. doi:10.31671/dogus.2018.131

Köroğlu, Ö., Ulusoy Yıldırım, H., & Balıkoğlu, A. (2020). Kadın turist rehberlerinin mesleki sorunlarının belirlenmesi. *Mehmet Akif Ersoy Üniversitesi Sosyal Bilimler Enstitüsü Dergisi*, (31), 26–40. doi:10.20875/makusobed.586054

Kuo, N. T., Chang, K. C., Cheng, Y. S., & Lin, J. C. (2015). Effects of tour guide interpretation and tourist satisfaction on destination loyalty in Taiwan's Kinmen Battlefield Tourism: Perceived playfulness and perceived flow as moderators. *Journal of Travel & Tourism Marketing, 33*(sup1), 103-122. doi:10.1080/10548408.2015.1008670

Kurt Dündar, A., & Arslantürk, Y. (2022). Paket turlarda etik dışı davranışlar: Yabancı turistler üzerine nitel bir değerlendirme. *Güncel Turizm Araştırmaları Dergisi, 6*(2), 517–535. doi:10.32572/guntad.1101091

Leclerc, D., & Martin, J. N. (2004). Tour guide communication competence: French, German and American tourists' perceptions. *International Journal of Intercultural Relations, 28*(3-4), 181–200. doi:10.1016/j.ijintrel.2004.06.006

Li, Y., Liu, B., Zhang, R., & Huan, T. C. (2020). News information and tour guide occupational stigma: Insights from the stereotype content model. *Tourism Management Perspectives, 35*, 100711. doi:10.1016/j.tmp.2020.100711

Lin, H., Tian, J., Kong, Y., & Gao, J. (2022). Impact of tour guide humor on tourist pro-environmental behavior: Utilizing the conservation of resources theory. *Journal of Destination Marketing & Management, 25*, 100728. doi:10.1016/j.jdmm.2022.100728

Mackenzie, S. H., Boudreau, P., & Raymond, E. (2020). Women's adventure tour guiding experiences: Implications for well-being. *Journal of Hospitality and Tourism Management*, *45*, 410–418. doi:10.1016/j.jhtm.2020.09.006

Mackenzie, S. H., & Kerr, J. H. (2013). Stress and emotions at work: An adventure tourism guide's experiences. *Tourism Management*, *36*, 3–14. doi:10.1016/j.tourman.2012.10.018

Mak, A. H. N., Wong, K. K. F., & Chang, R. C. Y. (2010). Factors affecting the service quality of the tour guiding profession in Macau. *International Journal of Tourism Research*, *12*(3), 205–218. doi:10.1002/jtr.746

Mak, A. H. N., Wong, K. K. F., & Chang, R. C. Y. (2011). Critical issues affecting the service quality and professionalism of the tour guides in Hong Kong and Macau. *Tourism Management*, *32*(6), 1442–1452. doi:10.1016/j.tourman.2011.01.003

Mancini, M. (1990). *Conducting tours: A practical guide*. Southwestern Publishing.

Min, J. C. (2012). A short-form measure for assessment of emotional intelligence for tour guides: Development and evaluation. *Tourism Management*, *33*(1), 155–167. doi:10.1016/j.tourman.2011.02.014

Naumov, K. (2020). Tourist perceptions and satisfaction regarding tour guiding in The Republic of North Macedonia. *European Journal of Multidisciplinary Studies*, *5*(29), 12–26. doi:10.26417/797jhv49a

Nettekoven, L. (1979). Mechanism of cultural interaction. In E. De Kadt (Ed.), *Tourism: Passport to development* (pp. 135–145). Oxford University Press.

Nyahunzvi, D. K., & Njerekai, C. (2013). Tour guiding in Zimbabwe: Key issues and challenges. *Tourism Management Perspectives*, *6*, 3–7. doi:10.1016/j.tmp.2012.11.003

Ong, C. E., Ryan, C., & McIntosh, A. (2014). Power-knowledge and tour-guide training: Capitalistic domination, utopian visions and the creation and negotiation of UNESCO's Homo Turismos in Macao. *Annals of Tourism Research*, *48*, 221–234. doi:10.1016/j.annals.2014.06.010

Öter, Z., & Sonuç, N. (2014). Cultural heritage interpretation in gender context: Case of female tourist guides in Turkey. *In Proceedings of the Heritage, Tourism and Hospitality International Conference (HTHIC)*, İstanbul, Türkiye, 66-77.

Pearce, P. L. (1984). Tourist-guide interaction. *Annals of Tourism Research*, *11*(1), 129–146. doi:10.1016/0160-7383(84)90100-2

Pizam, A., Jansen-Verbeke, M., & Steel, L. (1997). Are all tourists alike, regardless of nationality? The perceptions of Dutch tour-guides. *Journal of International Hospitality, Leisure &. Tourism Management, 1*(1), 19–38. doi:10.1300/J268v01n01_03

Pizam, A., & Jeong, G. H. (1996). Cross-cultural tourist behavior: Perceptions of Korean tour-guides. *Tourism Management, 17*(4), 277–286. doi:10.1016/0261-5177(96)00019-2

Pizam, A., & Sussmann, S. (1995). Does nationality affect tourist behavior? *Annals of Tourism Research, 22*(4), 901–917. doi:10.1016/0160-7383(95)00023-5

Prakash, M., & Chowdhary, N. (2010). Tour guides: Roles, challenges and desired competences a review of literature. *International Journal of Hospitality and Tourism Systems, 3*(1), 1–12.

Reisinger, Y., & Steiner, C. (2006). Reconceptualising interpretation: The role of tour guides in authentic tourism. *Current Issues in Tourism, 9*(6), 481–498. doi:10.2167/cit280.0

Schmidt, C. J. (1979). The guided tour: Insulated adventure. *Urban life, 7*(4), 441-467. https://doi.org/ doi:10.1177/089124167900700402

Şenol, F., & Aliyev, İ. (2015). Türkiye'de Rus turist pazarına hâkim tur acentalarında istihdam edilen turist rehberleri üzerine bir inceleme. *Kastamonu Üniversitesi İktisadi ve İdari Bilimler Fakültesi Dergisi, 8*, 167–181.

Sezgin, E., & Duz, B. (2018). Testing the proposed "GuidePerf" scale for tourism: Performances of tour guides in relation to various tour guiding diplomas. *Asia Pacific Journal of Tourism Research, 23*(2), 170–182. doi:10.1080/10941665.2017.1410196

Tang, C. Y., & Chang, J. C. (2006). *An investigation of service quality between the licensed and un-licensed tour leaders. The 7th Biennial Conference on Tourism in Asia*, Jeonju, Korea.

Tetik Dinç, N., Balıklıoğlu, A., & Ulusoy Mutlu, H. (2022). Turist rehberlerine yönelik turist nezaketsizliği ve rehberlerin nezaketsiz davranışlara ilişkin tepkileri. *Selçuk Üniversitesi Sosyal Bilimler Enstitüsü Dergisi*, (49), 100–117. doi:10.52642usbed.1152188

Tsaur, S. H., Cheng, T. M., & Hong, C. Y. (2019). Exploring tour member misbehavior in group package tours. *Tourism Management, 71*, 34–43. doi:10.1016/j.tourman.2018.09.018

Tsaur, S. H., & Lin, W. R. (2014). Hassles of tour leaders. *Tourism Management, 45*, 28–38. doi:10.1016/j.tourman.2014.03.017

Türkmen, S., & Deniz, G. B. (2022). Seyahat acentalarının turist rehberlerine yönelik sergiledikleri etik dişi davranişlarin belirlenmesi: ÇARO örneği. *Pamukkale Üniversitesi Sosyal Bilimler Enstitüsü Dergisi*, (53), 367–377. doi:10.30794/pausbed.1145202

Urry, J. (1990). The tourist gaze: Leisure and travel in contemporary societies. *Sage (Atlanta, Ga.)*.

Wang, K. C., Hsieh, A. T., & Huan, T. C. (2000). Critical service features in group package tour: An exploratory research. *Tourism Management*, *21*(2), 177–189. doi:10.1016/S0261-5177(99)00047-3

Weiler, B., & Black, R. (2015). *Tour guiding research: Insights, issues, implications*. Channel View.

Weiler, B., & Ham, S. (2001). Tour guides and interpretation. In D. Weaver (Ed.), *Encyclopedia of ecotourism*. CABI. doi:10.1079/9780851993683.0549

Weiler, B., & Walker, K. (2014). Enhancing the visitor experience: Reconceptualising the tour guide's communicative role. *Journal of Hospitality and Tourism Management*, *21*, 90–99. doi:10.1016/j.jhtm.2014.08.001

WFTGA. (n.d.). *What is a professional tourist guide?* WFTGA. https://wftga.org/about/

Wong, C. U. I. (2013). The sanitization of colonial history: Authenticity, heritage interpretation and the case of Macau's tour guides. *Journal of Sustainable Tourism*, *21*(6), 915–931. doi:10.1080/09669582.2013.790390

Wynn, J. R. (2005). Guiding practices: Storytelling tricks for reproducing the urban landscape. *Qualitative Sociology*, *28*(4), 399–417. doi:10.100711133-005-8365-2

Xu (Hazel), Y., & McGehee, N. G. (2017). Tour guides under zero-fare mode: Evidence from China. *Current Issues in Tourism*, *20*(10), 1088–1109. doi:10.1080/13683500.2016.1149155

Yarcan, Ş. (2007). Profesyonel turist rehberliğinde mesleki etik üzerine kavramsal bir değerlendirme. *Anatolia: Turizm Araştırmaları Dergisi*, *18*(1), 33–44.

Yetgin, D., Yılmaz, A., & Çiftci, G. (2018). Krizlerin turist rehberliği öğrencilerinin kariyer planlamasındaki etkisi. *Journal of Tourism and Gastronomy Studies*, *195*(3), 214. doi:10.21325/jotags.2018.278

Yetgin Akgün, D., Ay, E., & Çalık, A. Ö. (2022). Customer complaints in package tours and complaint management of tour guides. *Journal of Gastronomy Hospitality and Travel*, *5*(4), 1482–1494. doi:10.33083/joghat.2022.214

Yıldırım, G., & Özbek, Ö. (2019). Profesyonel turist rehberlerine yönelik cinsel taciz ile ilgili bir araştırma. *Turist Rehberliği Dergisi*, *2*(2), 72–87. doi:10.34090/tured.591586

Yu, X., Weiler, B., & Ham, S. (2001). Intercultural communication and mediation: A framework for analysing the intercultural competence of Chinese tour guides. *Journal of Vacation Marketing*, *8*(1), 75–87. https://journals.sagepub.com/doi/pdf/10.1177/135676670200800108. doi:10.1177/135676670200800108

Zhang, H. Q., & Chow, I. (2004). Application of importance-performance model in tour guides' performance: Evidence from mainland Chinese outbound visitors in Hong Kong. *Tourism Management*, *25*(1), 81–91. doi:10.1016/S0261-5177(03)00064-5

Zhao, S., & Timothy, D. (2017). The dynamics of guiding and interpreting in red tourism. *International Journal of Tourism Cities*, *3*(3), 243–259. doi:10.1108/IJTC-01-2017-0007

ADDITIONAL READING

Pond, K. L. (1993). *The professional guide: Dynamics of tour guiding*. Van Nostrand Reinhold Company.

Weiler, B., & Black, R. (2015). *Tour guiding research: Insights, issues, implications*. Channel View.

Weiler, B., & Ham, S. (2001). Tour guides and interpretation. In D. Weaver (Ed.), *Encyclopedia of ecotourism*. CABI. doi:10.1079/9780851993683.0549

KEY TERMS AND DEFINITIONS

Commission: Tour guides are paid percentages due to the purchases of tourists in shops. In most cases, commissions are shared among the tour operators, tour guides and drivers. Even the hotels hosting the tourists involve the partage.

Domination: It includes any kind of violence towards tour guides and aims to sustain tour guides' labor for the benefit of others. Domination in tour guiding is resulted from the unjust and mean practices of the employers, tour operators and other stakeholders.

Exploitation: It includes the unfair practices that victimize the tour guides such as low remuneration, the lack of job security and legal rights, hiring illegal labor and forcing tour guides to maximize the profits regardless of the probable dissatisfaction on tourists.

Illegal Guiding: This kind of labor appears as unauthorized individuals are employed by the tour operators in an effort to reduce costs. It is unlawful in destinations where the regulations for tour guiding service require licensed or certificated professionals to perform the job.

Remuneration: The regulations establish the payment requirements in tour guiding. The basic salary or daily wage is essential to remuneration and needs to be adequate to maintain one's life.

Stakeholders: It refers to the external factors such as service providers and policy makers in the tourism industry.

Tour Guiding Service: It is performed by the tour guides and includes disseminating knowledge, narrating the values of the visited environment, and taking care of organizational/operational tasks in a destination.

Chapter 6
Commodification of Labor in Tourism:
Human and Non-Human Animals

Nur Kulakoğlu Dilek
Batman University, Turkey

Sebahattin Emre Dilek
Batman University, Turkey

ABSTRACT

Tourism activities contribute to the economic development of countries through the transformation of values such as nature, culture, and history into attractive elements for tourists. However, the continuous growth of the tourism industry brings along numerous environmental and social issues, alongside the economic benefits it generates. To the extent that the tourism industry can perceive a country's culture, natural resources, historical sites, people, and even animals as commodities that can be used for capital gain. In this context, the study primarily evaluates the commodification of human and non-human labor in this industry. Subsequently, all tourism activities that exploit migrants, women, children, and animals, subjecting them to poor working conditions or sexual objectification, are examined through a cause-and-effect relationship. Regarding the concept of labor, a critical assessment is conducted based on the social reality where women, children, and animals are potentially seen as commodities in the tourism sector, and the topic is extensively addressed with examples.

DOI: 10.4018/978-1-6684-9255-0.ch006

INTRODUCTION

The tourism industry is one of the leading sectors contributing to countries' economic development and growth. Moreover, its significant impact on employment opportunities for millions of people worldwide and its influence on national incomes are of great importance. (Mak, 2003). The industry has experienced rapid expansion, particularly in recent years, due to globalization and increased competition. This growth has resulted in the dominance of large-scale enterprises, requiring investments and investors, thereby centralizing control of the sector in the hands of major travel agencies, hotel chains, and other tourism-related businesses. On the other hand, capitalism constitutes an economic system based on private ownership of the means of production and a free-market economy. The relationship between these two concepts is exemplified in how the tourism industry functions within the capitalist economic structure and how labor is commodified in the process. Consequently, the tourism sector is directly associated with capitalism.

Capitalism is a structure that forms the foundation of modern economic systems, characterized by a free-market economy and private ownership (Amable, 2003). One of the most prominent concepts within the capitalist system is commodification. The term commodification refers to the transformation of goods, human labor, or any other value into objects that can be bought and sold (Kosoy& Corbera, 2010; Young & Markham, 2020). In this process, labor or value is perceived and marketed as a tradable commodity. The owner of the labor or value can offer it as a purchasable asset. Commodification involves converting something into tangible or economic value, making it marketable, but it also raises ethical and social considerations.

The commodification of labor implies viewing workers as a service or commodity that can be bought and sold in exchange for wages. This phenomenon is particularly prevalent in capitalist economic systems and can lead to the exploitation of labor and people solely for profit-driven purposes. In this system, capital owners purchase human labor, which plays a significant role among the factors of production. Therefore, human labor is viewed as a commodity within capitalism. This implies that the value of labor is determined by market conditions in a competitive marketplace. As a result, the system creates an environment where the labor of individuals or animals is evaluated and compelled to compete. Furthermore, the commodification of all forms of labor whether human or non-human animal, within the capitalist system, is an important issue to consider from the perspectives of social justice and moral responsibility.

In terms of social justice and moral responsibility, the commodification of human or non-human animal labor is a highly significant issue. The commodification of labor entails the transformation of work into commercial value and is commonly practiced in capitalist economies. However, this practice brings forth certain problems in terms of human rights, justice, and ethical values. When the labor of human or non-human animals is commodified, it creates an environment where an individual's work is limited to the production of goods or services (Thrift, 2006; Coulter, 2016). This situation can lead to the perception of humans or non-human animals solely as factors of production and reduce their value to the economic worth of what they produce. This jeopardizes the respect for the freedom, dignity, and rights of humans and non-human animals.

From a perspective of social justice, the commodification of labor can exacerbate inequalities (Leslie & Catungal, 2012). This is because it can violate the principle of justice that dictates workers should receive compensation proportional to the value they create. In some cases, it can reveal the exploitation of individuals or animals by employers and the presence of unfair working conditions. This generates social injustice and imbalance within society, negatively impacting social harmony. From an ethical responsibility standpoint, the commodification of labor can reduce the worth of individuals or animals to merely economic interests. Humans and non-human animals should not be defined solely by their ability to work or the economic value of what they produce. Everyone should be considered worthy of human rights and respect. The commodification of labor can neglect these ethical values and lead us to perceive individuals or animals as mere means to an end.

In summary, the commodification of human or non-human animal labor is a controversial issue in terms of social justice and moral responsibility. It can increase inequalities, give rise to injustices, and reduce the worth of individuals or animals to solely economic interests. Therefore, policies and practices related to the commodification of labor should embrace an approach that respects human rights, is fair, and sustainable. The labor of humans and non-human animals should be evaluated in alignment with their freedom, dignity, and rights, in addition to their economic value.

The Issue of Human Labor and Commodification in Tourism

Tourism is a large-scale industry that encompasses various service sectors, such as accommodation establishments, travel businesses, recreational facilities, and food and beverage establishments, relying heavily on human labor (Greenwood, 1989; Levy & Lerch, 1991). Globally, the travel and tourism sector accounted for 10.4% of the global gross domestic product (GDP) in 2019 and directly and indirectly employed over 330 million people (1 in 5 of all jobs in the world) (WTTC, 2022). Therefore, the presence of a skilled and educated workforce is crucial for countries' tourism development and success.

In the tourism sector, where the primary aim is service quality, meeting tourist needs, and ensuring guest satisfaction, there exists a significant relationship between employers and employees in the delivery of services. As a result of the transformation in neoliberal labor markets, employers in this system can disregard qualitative factors and issues of insecurity that notably affect the tourism workforce, such as age, gender, race, and ethnic background, by focusing more on the number of tourism jobs to generate higher profits. When considering the sector, it can be observed that a substantial portion of jobs in tourism is performed by low-paid, precarious workers, often belonging to racial minorities, immigrants, and women (Dudley et al., 2021). These low wages, unstable employment, long working hours, or unfavorable working conditions contribute to devaluing the labor in tourism, making it susceptible to exploitation and, in other words, commodification (Aynalem et al., 2016).

The devaluation of labor surplus that arises in the initial cycle of capital, to reduce production costs and increase profit margins, necessitates the commodification of labor, or in other words, subjecting human existence to increasing market relations and profit motives (Overbeek, 2002: 76; Young & Markham, 2020). The theory of the commodification of labor, conceptualized by Marx as the "fetishism of commodities," focuses on the transformation of human labor into a commodity with commercial value, which is then exploited by capital owners (Marx, 2011). According to this theory, as money and capital become the subjects of society, workers devoid of ownership of the means of production are objectified. The commodification of labor leads to the transformation of the relationship between labor and the products of labor into a relationship between commodities, thereby obscuring the unequal relationship between capital and labor (Marx, 2011).

Based on Marx's argument that as the value of material objects increases, the value of the 'human world' diminishes, it can be clearly stated from a moral perspective that commodification is inherently wrong. Therefore, there is a greater need for research that emphasizes the low-paid, exploitative, and precarious nature of the tourism sector and develops solutions (Ladkin, 2011; Baum, 2015; Robinson et al., 2019). In this context, there are significant studies that draw attention to the issue. For example, Ehrenreich's book 'Nickel and Dimed' (2010) explores the struggles of low-wage hotel cleaners, while Jayaraman's 'Behind the Kitchen Door' (2013) discusses prevalent issues of insecure payment, gender discrimination, and racism among workers in US restaurants (Dudley et al., 2021). Both of these studies shed light on the adverse conditions faced by disadvantaged individuals, who form the backbone of the sector and raise the question of how the tourism industry and human labor have become so devalued.

Some various reasons and processes contribute to the commodification of human labor in the tourism industry. An overview will be provided regarding these processes and the resulting issues, followed by an evaluation of disadvantaged groups of human labor in tourism, including migrants, women, children, and animals.

- *Employment Conditions*: The tourism industry has created a distinct labor structure with its seasonality, informality, and unique service delivery, but this has also brought about exploitative relations within the capitalist production process (Aykaç, 2009). In many countries, due to the seasonal nature of the tourism sector, tourism activities are concentrated during certain times of the year (Cannas, 2012). This situation hinders the year-round employment of tourism workers and creates significant problems, particularly for tourism employees in the accommodation sector. When the tourism season ends and demand decreases, employees in accommodation establishments are either laid off or their contracts are suspended. Another characteristic of tourism labor is the prevalence of informal labor (Purkis, 2008). As a result, undeclared employment is widespread in the sector. For instance, out of approximately 3.5 million employees (insured and uninsured total) in the sector, the number of insured workers is around 1.5 million, indicating that one-third of tourism workers are employed informally in Türkiye (Turizm DATABANK, 2022; Yahyaoğlu, 2022). When looking at the global scale, according to calculations by the ILO, approximately two billion people who are employed, on average, around 214 million are unionized. This indicates a unionization rate of 10 percent. However, when self-employed individuals, employers, and family workers are not considered, this rate can reach up to 17 percent (ILO, 2022a). This situation leads to the exploitation of tourism workers' labor and precarious working conditions.
- *Law Wages*: Another issue that arises depending on working conditions is low wages. The problem of low wages in the tourism sector is a widespread issue in many countries and destinations (Choy, 1995). This problem indicates that tourism workers are either working for low wages or lack adequate pay and social benefits. Several key factors contribute to the emergence of this issue. *Competitive market*: The tourism industry is typically a highly competitive sector. Various tourism businesses (hotels, restaurants, tour companies, etc.) strive to attract tourists by keeping their prices low. This means that businesses attempt to cut their costs, leading to the payment of low wages to employees. *Seasonal employment*: Many tourism regions experience seasonal fluctuations, with periods of high tourist activity and limited tourism operations at other times. As a result, businesses only require workers during specific periods and may lay off employees during the off-season. Seasonal employment makes it difficult for workers to have stable incomes and can result in low-paying temporary jobs. *Low-profit margins in the tourism industry*: Many tourism businesses may operate with low-profit margins. High operating costs (staff, rent, suppliers, etc.) and competitive pricing policies can make it challenging for businesses to pay higher wages

to employees. *Legal regulations and weak labor rights*: In some countries, labor rights for workers in the tourism sector may not be adequately protected or may be lacking. Weak labor unions or insufficient regulations concerning labor rights may contribute to workers being forced to work for low wages (Harianto, 2021). *Tourist expectations*: Tourists' demand for low-priced holidays and services may increase the trend of cost-cutting by businesses. In such cases, businesses may try to reduce costs by paying low wages to employees. The issue of low wages can affect tourism workers' living standards, decrease their welfare, and make it challenging to attract and retain a talented workforce within the sector. To address this problem, it is important to adopt fair and sustainable wage policies, for tourism businesses to fulfill their social responsibilities, and for the development of legal regulations that protect workers' rights. Additionally, tourists supporting ethical and sustainable tourism practices can contribute to overcoming the issue of low wages in the tourism sector.

- *Social Security and Job Security*: Precarious employment is defined as working for wages in a job with uncertainty, low income, limited social benefits, and legal rights (Vosko, 2010: 2). Lack of advance notice for termination, uncertainty of working hours, irregular payment of wages, limited or nonexistent social rights and benefits, lack of protection against discrimination and sexual harassment, insufficient support in case of illness, workplace accidents, and unemployment, lack of union representation, and employer-imposed intimidation policies are significant determinants of social insecurity (Kutlu, 2012; Lewchuck et al., 2003). Workers in the tourism sector also typically have limited social security. Particularly, female and child workers, who constitute a significant proportion of the labor force composition, are exposed to high levels of exploitation for the mentioned reasons (Özuğurlu, 2006).
- *Global Labor Mobility*: The increased mobility of capital at the global level has been followed by the geographical mobility of labor, accelerating labor migration towards the global cheap labor market (Purkis, 2008). The tourism sector is also open to global labor mobility. Therefore, another type of labor employed in the sector is migrant labor. Migrants are often depicted as working in adverse conditions at the sharp end of the labor market in sectors such as tourism, construction, sex work, and agriculture (Anderson, 2010). According to the estimates of the International Labor Organization (ILO), in 2019, the number of international migrant workers reached 169 million, accounting for 4.9% of the global labor force in destination countries (ILO, 2019). Sector-specific figures indicate that 66.2% of migrant workers are employed in the service sector. However, there is a significant gender

> difference across sectors. While 79.9% of female migrant workers are employed in the service sector, 56.4% of male migrant workers work in the service sector (ILO, 2019).

The diversification of tourism services and employers' desire to reduce increasing costs in a competitive environment have led to an increased demand for foreign labor as cheap labor. The irregular nature of these migrations and the widespread employment of migrants in informal forms weaken labor organizations on the one hand, and on the other hand, result in the devaluation of labor by reducing wages (Lazaridis, 2007; Schireup et al., 2006). Migrant workers, due to their status as migrants, can more easily find employment in the tourism sector, thanks to their foreign language skills and their desire to benefit from the accommodation, food and beverage, and tipping opportunities provided by the industry. As a result, they may accept low wages and work long hours under precarious conditions (ILO, 2012). However, migrants are not only vulnerable to insecure working and housing conditions but also to various forms of harassment and sexual exploitation (De Zulueta, 2003; McDowell et al., 2009). A study conducted on migrants living in the Antalya region reports that migrant women are forced to do unwanted jobs, their wages are either delayed or not paid at all, their passports are confiscated, and they are constantly subjected to harassment (Kindan, 2022). This situation, which treats labor merely as a production factor and dehumanizes it by stripping away its humane aspects, exacerbates competition in the global labor market and increases the labor exploitation of migrant workers.

The global mobility of labor is not unique to Türkiye but has increased regionally, nationally, and globally since the 1990s (ILO, 2015). The commodification of labor and the migration of unskilled workforce, resulting from neoliberal policies, take place in various forms and speeds in each country depending on different factors. The commodification of labor in tourism can hinder workers' access to decent living conditions and give rise to a range of social and economic issues such as low wages, long working hours, precarious employment, occupational health and safety concerns, and so on.

The Commodification of Women's Labor in Tourism

The relationship between the commodification of women's labor in the tourism sector and capitalism is a complex issue that necessitates a comprehensive discussion and analysis. This relationship should be examined from the perspectives of gender roles, economic structures, and power dynamics. The main points and key connections are presented below:

- Women's Labor and Gender Roles: Long-standing gender roles in societies have traditionally directed women towards domestic care and service sectors. The tourism industry, comprising accommodation, restaurants, entertainment, and services, has thus become an area where women tend to be employed. The employment of women in these sectors reflects gender-based division of labor and perpetuates gender inequality.
- Low-Wage Jobs: Women in the tourism sector are often engaged in low-paid occupations. Especially in developing countries or economies reliant on tourism, women are more likely to be employed in low-wage positions such as cleaning, service, and cooking, while men occupy higher-paying roles. This disparity is a manifestation of gender inequality and can lead to the exploitation of women's labor.
- Temporary and Flexible Employment: The tourism industry can adapt to fluctuating demand on a seasonal or periodic basis, resulting in temporary and flexible employment arrangements. Women may be more prone to accept these flexible job conditions due to their responsibilities in fulfilling family roles. Consequently, this situation may lead to low job security and inadequate social protection for women.
- Gender-Based Discrimination: Women in the tourism sector may face gender-based discrimination and sexual harassment. These challenges hinder their pursuit of workplace equity and justice and create barriers to their career advancement and access to leadership positions.
- Social and Cultural Impacts: The working conditions of women in the tourism sector interact with social and cultural norms. In some societies, women may encounter social pressures against working outside the home and may lack support from their families, limiting their participation in the tourism workforce.

Capitalism is an economic system that treats labor as a commodity and pursues profit. The tourism sector is a part of this capitalist economic system. In this context, women's labor becomes a commodity within the labor market under capitalism. This commodification results in women being perceived merely as a labor force and utilized as a part of economic processes. In summary, the relationship between the commodification of women's labor in the tourism sector and capitalism is intertwined with various factors such as gender-based division of labor, low-paid jobs, flexible employment, gender-based discrimination, and social/cultural influences. These issues can lead to the disempowerment of women economically and socially, contributing to the deepening of gender inequality. Therefore, promoting fair working conditions for women in the tourism industry and addressing gender-based discrimination is crucial for women's rights and gender equality.

The position of women in society and the social roles assigned to them also play a determining role in their placement in the labor market (Sijtsman, 2001). In patriarchal societies, women's participation in labor processes often leads to the reproduction of their exploitation as unpaid domestic workers for a certain remuneration. Women are frequently employed in tasks that resemble traditional domestic roles such as cleaning, cooking, reception, and service, reinforcing their societal roles (Aykaç, 2006; Cave & Kılıç, 2010). As a service-oriented sector, tourism is also characterized as labor-intensive and female-intensive due to its various employment opportunities for women with different qualifications (Akoğlan-Kozak, 1996; Uğuz & Topbaş, 2016). In fact, women account for 54% of the global tourism workforce (UNWTO, 2023). Despite the significant role of the tourism sector in female employment and the substantial presence of women in the sector, the characteristics of tourism employment, such as precariousness and adverse working conditions, can lead to different outcomes for male and female workers. Women, although representing the majority of tourism employees, are often employed in low-paid, insecure, low-career opportunity, low-skill-demanding jobs that are susceptible to exploitation and discrimination (Omay, 2011). Furthermore, they experience significant gender inequality in the sector as they are paid less than their male counterparts for equivalent work. In fact, women in tourism earn 14.7% less than men (UNWTO, 2023). This situation hinders women's economic independence and lowers their social status.

Informal employment and unpaid exploitation of women's labor are also widespread in the tourism sector. The fact that women's labor, particularly in family businesses, is informally employed and exploited without remuneration reflects the long and challenging path that needs to be overcome to create a qualified female workforce (Tucker, 2007; Kaya, 2017). Another issue is that while women constitute the majority of students in formal tourism education, they lack the necessary training in advanced skills and social competencies required for career advancement (Kaya, 2017). As a result, they are underrepresented in higher-level positions within the sector. The intensive and irregular working hours of the industry, uncertainty regarding days off, and the requirement to work during holiday periods create biases against the preference for the female workforce at the managerial level. Additionally, seasonal working periods tend to be shorter for female workers compared to their male counterparts, making it more challenging for them to find employment, and they are unable to benefit from rights such as maternity or childbirth leave (Wigfield, 2001; Magdoff & Magdoff, 2004).

Numerous studies on women's employment in the tourism sector reveal that women face various challenges (Yanardağ & Avcı, 2012; Uğuz & Topbaş, 2016). Women's labor in the tourism sector can be subjected to hazardous working conditions, sexual harassment, gossip, exploitation, and discrimination solely based on their gender (Jucan & Jucan, 2013). Women working in hotels, bars, and nightclubs in tourist areas are often objectified and exposed to sexual harassment and abuse. Furthermore, discriminatory practices such as low wages, poor working conditions, and low social status are common among employers toward female employees.

Moreover, women's labor in the sector is frequently commodified and marketed based on their sexual appeal (Akena, 2022). For example, Asian women are perceived as sexual objects by male tourists and attract visitors within the context of sex tourism (Urry, 2002; Poulin, 2003). Another example is the "Miss Curvy Uganda" beauty pageant organized in 2019 by the Ugandan Minister of Tourism, where Ugandan women showcased it as a way to attract more foreign tourists and generate additional foreign exchange earnings (Akena, 2022). What set this event apart was the targeting of plus-size women based on the claim that curvy and larger-sized women are more sensual. This event received criticism from various groups. As stated by Hall (1997), meaning is ascribed to objects, people, and events within the context of the interpretations imposed on them, and these interpretations can often be objectifying. Women are portrayed as sexual objects in advertisements, brochures, and tourist activities, and this objectification process leads to the commodification of women as a commercial products. In 1987, the Thai government's advertisement stating "Durian, the only fruit more delicious than Thailand's young women" promoted sex tourism and consequently the commodification and exploitation of women's bodies (Hechler, 1995). This situation results in women being perceived as sellable commodities based on their bodies rather than their qualities and skills.

The Commodification of Child Labor in Tourism

The relationship between child labor in the tourism sector and capitalism is a topic that requires careful analysis from various perspectives, including ethical, economic, and social aspects. The presence of child labor in the tourism industry raises significant concerns about the exploitation of vulnerable individuals and their exposure to harmful working conditions. In this context, the key points and connections are presented below:

- Child Labor in the Tourism Sector: Child labor refers to the employment of children under the legal minimum age for work or in jobs that are hazardous to their physical or mental well-being. In some regions, the tourism sector relies on cheap and flexible labor, making children susceptible to exploitation.
- Economic Incentives and Capitalism: Capitalism, as an economic system, prioritizes profit and economic growth. In developing countries or regions heavily dependent on tourism, there might be economic incentives to employ child labor due to its cost-effectiveness and potential to boost profits.

- Poverty and Vulnerability: Poverty is a significant factor contributing to child labor in the tourism sector. In impoverished communities, families may resort to sending their children to work to supplement household income and meet basic needs. Capitalist structures that perpetuate income inequality can exacerbate such vulnerabilities.
- Informal and Unregulated Employment: The informal nature of the tourism sector in some regions can facilitate the engagement of child labor. Informal businesses may not be subject to stringent labor regulations and enforcement, making it easier to employ children without facing legal consequences.
- Social and Cultural Dynamics: Social norms and cultural factors may also play a role in child labor in the tourism sector. In certain societies, children might be expected to contribute to family income from a young age, and such norms can be perpetuated within a capitalist context.
- Impact on Education and Well-being: Child labor in the tourism industry often disrupts children's education, hindering their long-term development and future prospects. Additionally, exposure to hazardous work environments can jeopardize their physical and mental well-being.
- Responsible Tourism Practices: Responsible tourism initiatives can act as a countermeasure to child labor. Ethical and sustainable tourism practices advocate for fair wages, working conditions, and the protection of human rights, including children's rights.
- Role of Government and Regulation: Government policies and regulations play a crucial role in addressing child labor in the tourism sector. Effective enforcement of child labor laws and initiatives to eradicate poverty can mitigate the demand for child labor.

The relationship between child labor in the tourism sector and capitalism is complex and influenced by economic, social, and cultural factors. While capitalism can create economic incentives for child labor in some regions, it is essential to recognize the multidimensional nature of this issue. Addressing child labor in the tourism industry requires a holistic approach, involving responsible tourism practices, government interventions, and societal efforts to protect the rights and well-being of children.

According to the United Nations Convention on the Rights of the Child, every individual below the age of 18 is considered a child (UNICEF, 1989). Any individual below 18 years of age who is engaged in work to contribute to household income is identified as a child laborer. While working in light and non-hazardous jobs is acceptable for children who have reached the age of 15, the minimum age for engaging in hazardous and dangerous work is set at 18. Hazardous and dangerous work includes activities in seasonal agriculture, small and medium-sized enterprises, and street work, which are categorized as the worst forms of child labor and are prohibited.

The International Labor Organization states that the global number of child laborers has risen to 160 million, and approximately 20% (31.4 million) of children are engaged in the service sector (UNICEF, 2021). For instance, The Turkish Statistical Institute (TÜİK, 2023), based on the Household Labor Force Survey 2022 results, indicates that children constitute 22,578,378 of the Turkish population. With 26.5% of the population being children in Türkiye, the labor force participation rate for children aged 15-17 is determined to be 18.7%. When examined by gender, this rate is 27.0% for boys and 10.0% for girls. According to a report published by the Occupational Health and Safety Council, it is stated that there are at least two million child laborers in Türkiye, and this number reaches nearly 5 million during the summer months, including undocumented migrant and refugee children (İSİG Meclisi, 2022). According to the same data, it is indicated that 30.8% of children are involved in agriculture, 23.7% in the industry, and 45.5% in the service sector, while hundreds of thousands of children are exploited on the streets, in small and medium-sized enterprises, and in hazardous jobs.

In the existing literature, there are two different perspectives regarding the use of child labor and child exploitation. Some researchers (Curtin et al., 1997) argue that child labor, regardless of whether it provides economic benefits or not, can contribute to gaining experience, developing skills, socializing, and building self-confidence. Therefore, it is stated that child labor is not exploitative as long as it does not hinder their education and does not involve hazardous conditions. Another perspective suggests that employing children in harmful/dangerous or illegal conditions is morally wrong and prone to exploitation, regardless of the circumstances, as it can harm their education, health, and safety (Silva, 2014). The International Labor Organization defines child labor as an activity that deprives children of education opportunities, forces them to work long hours and under harsh conditions, and causes harm to their mental, physical, and moral well-being (ILO, 2023). The ILO (2023) lists poverty, lack of access to free and quality education, migration, unemployment of adult family members, traditional perspectives and social exclusion, lack of supervision, informality, and demand for cheap labor as the underlying causes of child labor.

Capitalist globalization has brought about unprecedented commodification, especially in the context of child exploitation in the tourism sector, emerging as a global human rights issue (Koops et al., 2017). Child labor, particularly in the tourism sector, can be examined from various factors such as economic, legal, social, psychological, educational, and health perspectives (Graciela et al., 2022). Another classification pertains to hazardous forms of work involving physical, psychological, and sexual abuse. In particular, the rapidly growing phenomenon of sex tourism in the last 30 years has commercialized and turned millions of young women and children into sexual commodities within a global market of sexual exchange (Jeffreys, 1999;

Magablih & Naamneh, 2010). Developments in Internet and mobile technologies can also contribute to the exposure of children to sexual exploitation in tourism through various forms of online sexual exploitation (such as social networks, chat forums, etc.) (United Nations, 2013). Child sex tourism forces approximately two million children worldwide into prostitution and pornography (George & Panko, 2011). The United Nations defines child sex tourism as the sexual exploitation of children by individuals who travel locally or internationally for engaging in sexual activities with children. The profits generated through these exploitations, which can occur in any country or tourist destination, reach significant dimensions (approximately 20 billion dollars annually) (ECPAT, 2021).

Another form of child exploitation in the tourism sector is manifested through various types of tourism or large-scale events, such as volunteer tourism, orphanage tourism, and slum tourism (Reas, 2020). Orphanage tourism refers to visits made by volunteer tourists to interact with and care for disadvantaged children. These visits can be spontaneous or organized as part of packaged tours provided by specialized tour operators or humanitarian organizations. The cost of these tours is typically paid through travel agencies or directly to the destination (Punaks & Feit, 2014). Indeed, volunteer tourism has become a popular product for individuals traveling to underdeveloped or developing countries, turning tourists' sensitivities into an economic value and becoming one of the most profitable sectors within the tourism industry (Van Doore et al., 2016; ECPAT, 2019a; Lyneham & Facchini, 2019). However, this high demand leads to exploitation by stakeholders. In countries like Haiti, Nepal, and Uganda, children are unlawfully gathered by unregulated orphanages and childcare institutions to meet the demand for volunteer tourism experiences (Mulheir & Cavanagh, 2016; ECPAT, 2019b; ECPAT, 2020). For instance, the commodification of children for profit under the guise of "volunteer tourism" in Cambodia is a subject frequently discussed in the literature. The rapidly growing orphanage tourism in Cambodia commodifies vulnerable children as objects of economic value under the umbrella of volunteer tourism (Meintjes & Giese, 2006). Volunteer tourism is considered a contemporary consumer commodity, and the bodies of impoverished children are presented as commodities in the volunteer tourist market (Reas, 2020). For example, in the Bar Street of Siem Reap, a local NGO markets impoverished Cambodian children from an orphanage to tourists for traditional Khmer dance performances every night (Reas, 2013). Orphanage tourism, as an example of the dualistic representation of desperation and touristic allure, unequivocally demonstrates the objectification of children within the realm of tourism, even if done with good intentions. Poverty, lack of education, social exclusion, and weak legal frameworks further enhance the vulnerability of children to exploitation in tourism.

In conclusion, child labor in the tourism sector is often characterized by low wages, hazardous working conditions, and heavy workloads. Children can be employed in hotels, restaurants, street vending, personal services for tourists, or in exploitative areas such as sex tourism. This situation deprives children of their right to education, jeopardizes their physical and psychological well-being, and puts their future at risk. In this regard, the tourism sector should fulfill its social responsibility for both its sustainability and the well-being of children. Considering that today's children will be tomorrow's adults, the existence of a responsible and rights-based tourism sector is crucial for the overall health of society. Such a sector will have a multiplier effect in creating positive impacts, considering the welfare of children.

Animal Labor and Commodification Issues in Tourism

The relationship between animal labor in the tourism sector and capitalism is a topic that necessitates a careful analysis from ethical, economic, and social perspectives. Animal labor refers to the utilization of various animals in the tourism industry for entertainment, transportation, tourist activities, and related purposes. This relationship demands an evaluation of capitalism's effects on animal welfare and natural ecosystems. The main points and connections of this relationship are presented below:

- Animal Labor in the Tourism Sector: The tourism industry involves the travel of individuals for leisure and recreation purposes. During these travels, animals may be employed in various tourist activities, entertainment facilities, and transportation services. For instance, animals such as horses, elephants, camels, and marine mammals might be utilized to attract tourists to different experiences.
- Economic Incentives and Capitalism: Capitalist economies prioritize profit and economic growth. Given that the tourism sector is a rapidly expanding and lucrative industry, animal labor can be commonly used in tourist activities as it may yield economic incentives. The exploitation of animal labor in tourism may emerge as a consequence of capitalist structures aiming to enhance profitability.
- Animal Welfare and Ethical Concerns: Animal labor in the tourism sector raises ethical and animal welfare concerns. Continuous and intensive engagement of animals in tourist activities can lead to physical and psychological health issues and may result in animal abuse. Such practices can also divert animals from their natural behaviors and contribute to the destruction of their natural habitats.

- Natural Ecosystems and Sustainability: The utilization of animal labor in the tourism sector has implications for natural ecosystems. Tourist activities that cause harm to natural habitats can threaten the sustainability of ecosystems and make the conservation of resources challenging for future generations.
- Ethical Standards and Responsibilities: The tourism industry has sparked debates regarding the ethical use of animal labor. Responsible tourism practices are crucial in establishing ethical boundaries for animal labor and safeguarding animal welfare. This requires tourism companies and consumers to be more aware and accountable concerning the use of animal labor.
- Legal Regulations and Oversight: The use of animal labor in the tourism sector is subject to varying legal regulations and oversight across countries. Inadequate regulations may expose animals to exploitation and jeopardize their welfare.

The relationship between animal labor in the tourism sector and capitalism entails significant considerations related to economic gains and animal welfare. The ethical and sustainable use of animal labor necessitates a comprehensive approach that acknowledges the impacts of capitalist economies and the tourism industry on natural ecosystems and animal welfare. Adherence to ethical standards, improvements in legal regulations, and public awareness are essential in promoting more responsible practices regarding animal labor in the tourism sector.

The rise of capitalism has led humans to become increasingly detached from nature, viewing everything around them, both living and inanimate, as commodities existing solely to fulfill their desires and needs. In other words, the historical development of the human-nature relationship has resulted in the adoption of a perspective that sees human domination over nature as a necessity, limiting the relationship in favor of humans at the expense of nature (Dilek, 2017). Based on this misguided viewpoint, a structure has been created in the tourism sector, as in every other sector, where the use of nature/natural life is seen as a right, and the subject is framed solely within the perspective of economic instrumentality. The purely economic and profit-oriented nature of the tourism sector transforms every element, including the natural environment, cultural resources, and animals, into consumer products. Humans have defined and categorized animals based on their usefulness to humans (Carr, 2021). Animals are perceived as workers who exist to satisfy human needs, and they are exploited by attributing human qualities to them in both work and non-work environments. Consequently, animals are held in captivity and their freedom is taken away under the guise of necessity or entertainment in leisure activities (Fennell, 2012). In this context, as Carr and Broom (2018) have questioned, *'Can we consider animals used in the tourism sector as tourism workers since they are not remunerated for their labor?'*

In the concept of sustainability frequently mentioned in the tourism field, nature and animals are considered important and worthy of protection. However, in practice, contrary to this understanding, billions of dollars are generated annually through the exploitation of animals (WSPA, 2021). In other words, animals provide significant economic benefits to both employers and the respective destinations through their work and symbolic value (García-Rosell & Tallberg, 2021). For example, approximately 1.5 million people visit the town of Kaikoura in New Zealand each year solely to observe whales and dolphins (Cloke & Perkins, 2005). This situation turns animals, like humans, into tourism products or tourism workers. Activities focused on animals such as zoos, circuses, safari tours, aquariums, dolphin show centers, horse racing, bullfighting, camel wrestling, bird and wildlife watching, and hunting tourism are in high demand in the tourism sector, but in the process, animal rights and ethical status are disregarded (Hall & Brown, 1996; Regan, 2007; Francione, 2008; Çalık & Çiftçi, 2013). For instance, in Far Eastern countries, elephants are used as tourism workers, being exploited for the enjoyment and entertainment of tourists (Kontogeorgopoulos, 2009). Safari tours that exhibit animals as commodities and trophy hunting that devalues animal life based on their economic value generate approximately $200 million in revenue for Africa, where they are popular (Economists at Large, 2013). In fact, many of the tasks performed by animals in tourism often go unnoticed. While animal-focused tourist activities may come to mind first, animal-themed restaurants, healthcare centers supported by animal therapy, or security dogs used at airports/customs are disregarded in terms of animal labor (Rickly & Kline, 2021).

The most common form of animal labor commodification in the tourism sector is the use of animals as tourism workers to fulfill human needs. Horses, donkeys, camels, elephants, chimpanzees, and many other animals are utilized as transportation means or exhibited as commercial objects for entertainment or photo opportunities. A prominent example illustrating this practice is the use of hundreds of donkeys in Santorini, Greece, where approximately 80,000 tourists visit daily to transport them from the port to the town by making them climb steep stairs under extreme heat (Vasilopoulou, 2021; PETA, 2022). Although the use of animals for transportation purposes dates back to ancient times and has generally been accepted as a role they play, it does not change the fact that they are deprived of their freedom rights.

As a manifestation of corruption in the tourism sector, the recent rise of sex tourism has also extended to the exploitation of animals as sex workers. The sexual abuse of animals not only represents an illegal act but also reflects moral depravity within the context of tourism (Panko & George, 2018). The commodification of animals through violence and sexuality is frequently observed in tourism. For instance, in the orangutan boxing shows held at Safari World in Bangkok, female orangutans are dressed in skirts and bikinis, assuming the role of models, while male orangutans are forced to fight (AFP, 2015). Another form of violent tourism activity is bullfighting. While Spain is traditionally associated with bullfighting, Mexico attracts the highest number of bullfighting tourists with its arenas spread across 15 cities (López-López & Quintero-Venegas, 2021). Bullfighting, presented as a tourist attraction in Mexico, generates income by creating the excitement of violence through the death of bulls. While news reports annually highlight the human fatalities associated with bullfighting for human pleasure and entertainment, little concern is shown for the lives of the animals, thus revealing one of the morally dark aspects of tourism.

Whether exhibited in circuses or zoos, observed in their natural habitats, or subjected to the termination of their lives under the guise of hunting tourism, animals have long been commodified in tourism (Dashper, 2021). Although there has been an increase in academic research focusing on animal welfare in recent years, to prevent such exploitation, it is crucial to grant legal status to animals in government policies and ensure that ethical responsibility is upheld in tourism activities (Carr & Broom, 2018; Von Essen, Lindsjö & Berg, 2020; Fennell, 2022). Both in theory and in practice, tourism stakeholders need to be aware of animal ethics. Putting an end to the cruelty inflicted upon animals requires tourism to cease regarding them as objects of tourist consumption and refrain from showcasing them. Simultaneously, it necessitates that tourists do not demand such activities. Only through these measures can the maltreatment of animals be halted.

CONCLUSION AND RECOMMENDATIONS

Tourism is a sector that has a significant quantitative impact on employment but is characterized by a relatively uneducated and unskilled workforce, high levels of informal employment, long working hours, low wages, limited unionization rates, and ineffective social dialogue. The structure of the sector contributes to the increasing devaluation, exploitation, and commodification of human labor, which is one of its key factors. In this chapter, an assessment is made of the labor issues and labor exploitation in the tourism sector. It is noted that the workforce in the tourism sector is generally low-skilled, low-paid, and employed in informal and unregistered jobs, with limited unionization and ineffective social dialogue. This situation leads to the devaluation and exploitation of human labor.

In tourism regions where seasonality has a significant impact, models should be implemented to provide tourism workers with social and economic rights that ensure their minimum livelihoods, such as unemployment benefits and job guarantees (salary and other rights) during periods of non-employment. These models can be supported by both the government and employers. All countries should sign ILO Convention No. 172 concerning the rights of tourism workers

(ILO, 2022b). Facilitating and promoting unionization should be encouraged, and in small businesses where unionization is not possible, local monitoring mechanisms should be activated to inspect and regulate these establishments. Improving working conditions and professional rights for tourism workers at all levels and in all fields is crucial. Women in the tourism sector should be employed based on gender equality, and their work should be protected from an exploitative labor discipline that leaves them vulnerable to sexual harassment and discriminatory and oppressive practices. Existing legal regulations regarding the employment of foreign workers in the sector should be limited, and the employment of undocumented foreign migrants should be eliminated. Promoting equal pay, combating sexual harassment, and supporting dignified employment opportunities for women in tourism. Effective implementation of gender-sensitive legal and economic policies can also support the economic empowerment of women in the tourism sector. To prevent the commodification of women's labor in the tourism sector, inclusive employment policies and legal regulations should be implemented. Employers should adopt policies that prevent gender-based discrimination and follow a zero-tolerance approach towards behaviors such as sexual harassment. Additionally, it is important to establish legal regulations that meet the demands of women for equal pay and higher status. Priority should be given to education and awareness campaigns to empower women in the tourism sector. Vocational training opportunities should be provided to women, and support programs should be established to encourage their advancement into leadership positions. Furthermore, awareness campaigns and initiatives should be organized to raise awareness among tourism workers about their rights and gender equality. Collaboration and solidarity are crucial to prevent the sexual objectification and commodification of women's labor in the tourism sector. Strong partnerships between trade unions, civil society organizations, employers, and governments should be established to protect the rights of female workers and work towards common goals. Women need to be educated about the sexual objectification and exploitation they may face in the tourism sector. Additionally, tourism businesses should provide training programs to their employees on preventing sexual abuse and promoting equal employment policies. Effective legal regulations and strict monitoring mechanisms are required to prevent sexual objectification and commodification in the tourism sector. Countries should enforce laws that reflect principles of gender equality and impose deterrent penalties to prevent sexual abuse and discrimination. Furthermore, businesses need to comply with these legal regulations and undergo regular inspections. Empowering women and enabling them to rise to leadership roles in the tourism sector is a significant step in preventing sexual objectification and commodification. Providing educational opportunities to women, implementing policies and support programs that encourage their advancement into leadership positions, and fostering an environment of gender equality can help women advocate for their rights and create a more equitable environment in the tourism sector. In addition to that, effective legal regulations and strict monitoring mechanisms are also necessary to prevent the commodification of child labor in the tourism sector. Countries should implement legal regulations that adhere to international standards and human rights norms, specifically prohibiting child labor. It is crucial to effectively monitor these regulations and impose deterrent penalties on violators. Education is a fundamental tool to prevent the commodification of child labor. The right to education for children should be safeguarded, and efforts should be made to ensure that every child has access to schooling. Furthermore, raising awareness among stakeholders in the tourism sector (tourism businesses, governments, and civil society organizations) and organizing education programs on child labor is important. Tourists also need to draw attention to this issue and support ethical tourism practices. Social assistance programs and economic development projects are important in combating poverty and illiteracy, thereby preventing the commodification of child labor. Support should be provided to families, poverty should be addressed, and alternative sources of income should be created. This way, children can focus on their education without the need to work. By implementing these measures, countries can effectively combat the commodification of human labor (especially women and children) in the tourism sector and ensure the protection of human rights. Collaboration between governments, tourism businesses, civil society organizations, and tourists is essential in creating a sustainable and responsible tourism industry that respects the rights and well-being of humans.

In addition to the commodification of human labor, the commodification of non-human animals in the tourism sector is another concerning issue that raises serious concerns for the conservation of natural habitats and the well-being of animals. Below are some proposed solutions to address this problem: *Education and Awareness*: Education and awareness programs on animal rights should be organized for tourists, travel agencies, guides, and other stakeholders in the tourism industry. These programs can provide information to educate people about wildlife conservation and ethical tourism practices. *Legal Regulations*: Strict legal regulations need to be established and implemented to protect animals. Laws should prohibit activities that harm or commodify animals and impose severe penalties for violations. *Conservation of Natural Habitats*: The preservation of natural habitats where wildlife can thrive is of great importance. These areas can be open to tourism activities, but they should be regulated in a way that allows animals to exhibit their natural behaviors and be observed without causing harm. *Ethical Tourism Standards*: Compliance with ethical principles within the tourism industry is a fundamental necessity for protecting animal rights. Zoos, marine parks, and other tourist attractions should adopt and develop best practices to meet the needs of animals. *Promotion of Alternative Tourism Experiences*: Encouraging alternative tourism experiences that allow tourists to observe animals in their natural environments without disturbing them is important. For example, nature-based tours, nature walks, bird-watching activities, and other similar activities enable tourists to observe animals in their natural habitats without causing any disturbance. *Encouragement of Best Practice Examples*: Promotion of good practice examples in animal-friendly tourism should be encouraged. This includes providing incentives for tour operators, hotels, and other tourism stakeholders to adopt and promote animal-friendly practices. *Scientific Research and Monitoring*: Scientific research and monitoring programs should be conducted to track the impact of animals in the tourism sector and conservation efforts. These data can inform policymakers and tourism stakeholders and assist in the development of effective strategies.

These proposed solutions represent some of the steps that can be taken to reduce the commodification of animals in the tourism sector and protect their well-being. Addressing this issue requires further efforts, collaboration, and global awareness. In conclusion, the commodification of the labor of women, children, and animals is a concerning issue from social and ethical perspectives. This situation can lead to exploitation, forced labor, abuse, and unfair treatment. The commodification of the labor of these three groups can result in human rights violations and injustice. Overall, the commodification of the labor of women, children, and animals is a serious problem that leads to social injustice and human rights violations. Addressing this issue requires multidimensional solutions such as education, awareness, legal regulations, ethical standards, and sustainability principles. Public awareness, political will, and international collaboration are crucial for evaluating and protecting the labor of these groups. By doing so, we can contribute to building a more just and sustainable future.

REFERENCES

AFP. (2015). *Boxing orangutans a throwback as Thailand warms to animal rights*. AFP. https://tribune.com.pk/story/948705/boxing-orangutans-a-throwback-as-thailand-warms-to-animal-rights

Akena, F. A. (2022). The "Miss Curvy Uganda" pageant: Representation, commodification, and exploitation of women's bodies. *Social Semiotics*, *32*(2), 224–239. doi:10.1080/10350330.2020.1743524

Akoğlan-Kozak, M. (1996). Konaklama endüstrisinde kadının konumu. *Anatolia Turizm Araştırmaları Dergisi*, *7*(3-4), 16–23.

Amable, B. (2003). *The diversity of modern capitalism*. Oxford University Press. doi:10.1093/019926113X.001.0001

Anderson, B. (2010). Migration, immigration controls, and the fashioning of precarious workers. *Work, Employment and Society*, *24*(2), 300–317. doi:10.1177/0950017010362141

Aykaç, A. (2006). Linking scale of operation and labor transformation on tourism: Comparative analysis on three Turkish cases. *Anatolia*, *17*(2), 257–277. doi:10.1080/13032917.2006.9687189

Aykaç, A. (2009). *Yeni işler, yeni işçiler: Turizm sektöründe emek*. İletişim Yayınları.

Aynalem, S., Birhanu, K., & Tesefay, S. (2016). Employment opportunities and challenges in tourism and hospitality sectors. *Journal of Tourism & Hospitality (Los Angeles, Calif.)*, *5*(6), 1–5. doi:10.4172/2167-0269.1000257

Baum, T. (2015). Human resources in tourism: Still waiting for change? *Tourism Management*, *50*, 204–212. doi:10.1016/j.tourman.2015.02.001

Çalık, A. Ö., & Çiftçi, G. (2013). Animal ethics in tourism. *International Journal of Business and Management Studies*, *1*(5), 43–53.

Cannas, R. (2012). An overview of tourism seasonality: Key concepts and policy. *AlmaTourism*, *3*(5), 40–58.

Carr, N. (2021). Redefining the work of working animals in the tourism industry: An animal-centric reflection. In C. Kline & J. Rickly (Eds.), *Exploring non-human work in tourism: From beasts of burden to animal ambassadors* (pp. 37–50). De Gruyter Oldenbourg. doi:10.1515/9783110664058-003

Carr, N., & Broom, D. M. (2018). *Tourism and animal welfare*. CABI International. doi:10.1079/9781786391858.0000

Cave, P., & Kılıç, S. (2010). The role of women in tourism employment with special reference to Antalya, Turkey. *Journal of Hospitality Marketing & Management*, *19*(3), 280–292. doi:10.1080/19368621003591400

Choy, D. J. (1995). The quality of tourism employment. *Tourism Management*, *16*(2), 129–137. doi:10.1016/0261-5177(94)00023-4

Cloke, P., & Perkins, H. C. (2005). Cetacean performance and tourism in Kaikoura, New Zealand. *Environment and Planning. D, Society & Space*, *23*(6), 903–924. doi:10.1068/d57j

Coulter, K. (2016). *Animals, work, and the promise of interspecies solidarity* (Vol. 26). Palgrave Macmillan. doi:10.1057/9781137558800

Curtin, M., Hossain, T., & Verghese-Choudbury, A. (1997). Children living, learning and working in the city. In J. Beall (Ed.), *A city for all. Valuing difference and working with diversity* (pp. 66–78). Zed Books Ltd.

Dashper, K. (2021). Conceptualizing non-human animals as "workers" within the tourism industry. In C. Kline & J. Rickly (Eds.), *Exploring non-human work in tourism: From beasts of burden to animal ambassadors* (pp. 21–36). De Gruyter Oldenbourg. doi:10.1515/9783110664058-002

De Zulueta, T. (2003). Migrants in irregular employment in the agricultural sector of Southern European countries. Report for the debate in the standing committee. Council of Europe.

Dilek, S. E. (2017). Turizmin ahlaki sorumluluğu ve hayvanlar sorunsalı. In E. Günlü-Küçükaltan & S. E. Dilek (Eds.), *Metalaşan turizm-metalaştırılan hayvanlar* (pp. 9–16). Nobel Akademik Yayıncılık.

Dudley, K. D., Duffy, L. N., Terry, W. C., & Norman, W. C. (2021). The historical structuring of the US tourism workforce: A critical review. *Journal of Sustainable Tourism, 30*(12), 2823–2838. doi:10.1080/09669582.2021.1952417

Economists at Large. (2013). The $200 million question: How much does trophy hunting really contribute to African communities? *A report for the African Lion Coalition*. Economists at Large, Melbourne, Australia. https://iwbond.org/2015/09/03/the-200-million-question/

ECPAT. (2019a). *The link between voluntourism, orphanage tourism and child sexual exploitation*. ECPAT. https://ecpat.org/voluntourism-orphanage-tourism-sexual-abuse/

ECPAT. (2019b). *ECPAT country overview: Uganda.* ECPAT. https://ecpat.org/wp-content/uploads/2021/08/ECPAT-Country-Overview-Report-Uganda-April-2019-2.pdf

ECPAT. (2020). *Nepal – Country Overview*. ECPAT. https://ecpat.org/resource/nepal-country-overview/

ECPAT. (2021). *Sexual Exploitation of Children In Travel And Tourism.* ECPAT. https://ecpat.org/wp-content/uploads/2021/05/ECPAT-Summary-paper-on-Sexual-Exploitation-of-Children-in-Travel-and-Tourism-2020.pdf

Fennell, D. A. (2012). *Tourism and animal ethics*. Routledge.

Fennell, D. A. (2022). An animal welfare literacy framework for tourism. *Annals of Tourism Research*, *96*, 103461. doi:10.1016/j.annals.2022.103461

Francione, G. L. (2008). *Hayvan haklarına giriş: Çocuğunuz mu köpeğiniz mi?* (R. Akman & E. Gen, Trans.). İletişim Yayınları.

García-Rosell, J., & Tallberg, L. (2021). Animals as tourism stakeholders: Huskies, reindeer, and horses working in Lapland. In C. Kline & J. Rickly (Eds.), *Exploring non-human work in tourism: From beasts of burden to animal ambassadors* (pp. 103–122). De Gruyter Oldenbourg. doi:10.1515/9783110664058-007

George, B. P., & Panko, T. R. (2011). Child sex tourism: Facilitating conditions, legal remedies, and other interventions. *Vulnerable Children and Youth Studies*, *6*(2), 134–143. doi:10.1080/17450128.2010.521599

Graciela, C. J. Rocío del C. S. Barquín, Lilia Z. V. & Elva, E. V. M. (2022). Child labor and child work in the touristic sector of Cozumel and Valle de Bravo, Mexico. International Journal of Hospitality & Tourism Administration, 23(3), 599-622.

Greenwood, D. J. (1989). Culture by the pound: An anthropological perspective on tourism as cultural commoditization. In V. L. Smith (Ed.), *Hosts and guests: the anthropology of tourism* (pp. 129–138). University of Pennsylvania Press. doi:10.9783/9780812208016.169

Hall, D., & Brown, F. (1996). Towards a welfare focus for tourism research. *Progress in Tourism and Hospitality Research*, *1*(2), 41–57. doi:10.1002/(SICI)1099-1603(199603)2:1<41::AID-PTH23>3.0.CO;2-6

Hall, S. (1997). The work of representation. In S. Hall (Ed.), *Representation: Cultural representations and signifying practices* (pp. 13–74). Sage.

Harianto, A. (2021, December). Wages Problems of Local Wisdom-based Tourism Development During Covid-19 Pandemic. In *2nd International Conference on Business Law and Local Wisdom in Tourism* (ICBLT 2021) (pp. 31-35). Atlantis Press. https://www.atlantis-press.com/proceedings/icblt-21/125965331

Hechler, D. (1995). *Child sex tourism*. Geocities. http://www.geocities.com/CapitolHilllSenate/893 11 dbt-rep.html.

ILO. (2012). Migrant Workers in the International Hotel Industry. *International Migration Paper No. 112*. ILO. https://www.ilo.org/sector/Resources/publications/WCMS_180596/lang--en/index.htm

ILO. (2015). *The Contribution of Labour Mobility to Economic Growth*. ILO. https://www.ilo.org/wcmsp5/groups/public/---europe/---ro-geneva/---ilo-ankara/documents/legaldocument/wcms_398770.pdf

ILO. (2019). *Küresel yabancı göçmen işçi tahminleri sonuç raporu*. ILO. https://www.ilo.org/wcmsp5/groups/public/---europe/---ro-geneva/---ilo-ankara/documents/genericdocument/wcms_811865.pdf

ILO. (2022a). *The future of work in the tourism sector: Sustainable and safe recovery and decent work in the context of the COVID Report for the Technical Meeting on COVID 19 pandemic*. ILO. https://www.ilo.org/wcmsp5/groups/public/---ed_dialogue/---sector/documents/meetingdocument/wcms_840403.pdf

ILO. (2022b). *Application of international labour standards 2022: Report of the committee of experts on the application of conventions and recommendations*. ILO. ilo.org/wcmsp5/groups/public/---ed_norm/---relconf/documents/meetingdocument/wcms_836653.pdf

ILO. (2023). *What is child labor*. ILO. https://www.ilo.org/ipec/facts/lang--en/index.htm

Jeffreys, S. (1999). Globalizing sexual exploitation: Sex tourism and the traffic in women. *Leisure Studies*, *18*(3), 179–186. doi:10.1080/026143699374916

Jucan, M. S., & Jucan, C. N. (2013). Gender trends in tourism destinations. *Social and Behavioral Sciences*, *92*, 437–444.

Kaya, Ş. (2017). Turizmde kadın emeği. *Karatahta İş Yazıları Dergisi*, *9*, 1–22.

Kindan, S. A. (2022). *Türkiye'de turizm sektöründe göçmen emeği: Antalya ili örneği* [Unpublished master's thesis, Ordu Üniversitesi, Sosyal Bilimler Enstitüsü].

Kontogeorgopoulos, N. (2009). The role of tourism in elephant welfare in Northern Thailand. *Journal of Tourism*, *X*(2), 1–19.

Koops, T., Turner, D., Neutze, J., & Briken, P. (2017). Child sex tourism–prevalence of and risk factors for its use in a German community sample. *BMC Public Health*, *17*(1), 1–8. doi:10.118612889-017-4270-3 PMID:28427370

Kosoy, N., & Corbera, E. (2010). Payments for ecosystem services as commodity fetishism. *Ecological Economics*, *69*(6), 1228–1236. doi:10.1016/j.ecolecon.2009.11.002

Kutlu, D. (2012). Türkiye işgücü piyasasında güncel gelişmeler ve güvencesizleşme örüntüleri. In Ö. Göztepe (Ed.), *Güvencesizleştirme: Süreç, Yanılgı, Olanak* (pp. 61–116). Nota Bene Yayınları.

Ladkin, A. (2011). Exploring tourism labor. *Annals of Tourism Research, 38*(3), 1135–1155. doi:10.1016/j.annals.2011.03.010

Lazaridis, G. (2007). Les infirmières exclusives and migrant Quasi-Nurses in Greece. *European Journal of Women's Studies, 14*(3), 227–245. doi:10.1177/1350506807079012

Leslie, D., & Catungal, J. P. (2012). Social justice and the creative city: Class, gender and racial inequalities. *Geography Compass, 6*(3), 111–122. doi:10.1111/j.1749-8198.2011.00472.x

Levy, D. E., & Lerch, P. B. (1991). Tourism as a factor in development implications for gender and work in Barbados. *Gender & Society, 5*(1), 67–85.

Lewchuk, W., de Wolff, A., King, A., & Polanyi, M. (2003). From job strain to employment strain: Health effects of precarious employment. *Just Labour, 3*. doi:10.25071/1705-1436.165

López-López, Á., & Quintero-Venegas, G. (2021). Animal dark tourism in Mexico: Bulls performing their own slaughter. In C. Kline & J. Rickly (Eds.), *Exploring non-human work in tourism: From beasts of burden to animal ambassadors* (pp. 69–82). De Gruyter Oldenbourg. doi:10.1515/9783110664058-005

Lyneham, S., & Facchini, L. (2019). Benevolent harm: Orphanages, voluntourism and child sexual exploitation in South-East Asia. In A. P. M. Michael Phelan (Ed.), *Trends & issues in crime and criminal justice* (pp. 76–88). Australian Institute of Criminology.

Magablih, K., & Naamneh, M. (2010). Child labor in the tourism industry in Jordan. *Tourism Analysis, 15*(1), 89–97. doi:10.3727/108354210X12724734223711

Magdoff, F., & Magdoff, H. (2004). Disposable workers: Today's reserve army of labor. *Monthly Review (New York, N.Y.), 55*(11), 18. Advance online publication. doi:10.14452/MR-055-11-2004-04_2

Mak, J. (2003). *Tourism and the economy*. University of Hawaii Press.

Marx, K. (2011). Kapital, I. Cilt, (Çev. Mehmet Selik ve Nail Satlıgan). İstanbul: Yordam Kitap.

McDowell, L., Batnitzky, A., & Dyer, S. (2009). Precarious work and economic migration: Emerging immigrant divisions of labour in Greater London's service sector. *International Journal of Urban and Regional Research, 33*(1), 3–25. doi:10.1111/j.1468-2427.2009.00831.x

Meclisi, İ. S. İ. G. (2022). *Çocuk işçiliği.* ISIGMECLISI. https://isigmeclisi.org/20806-cocuk-isciligi-mahmut-boyunegmez

Meintjes, H., & Giese, S. (2006). Spinning the epidemic: The making of mythologies of orphanhood in the context of AIDS. *Childhood, 13*(3), 407–430. doi:10.1177/0907568206066359

Mulheir, G., & Cavanagh, M. (2016). *Orphanage entrepreneurs: The trafficking of Haiti's invisible children.* Lumos. https://lumos.contentfiles.net/media/documents/document/2017/12/Haiti_Trafficking_Report_ENG_WEB_NOV16.pdf

Omay, U. (2011). Yedek işgücü ordusu olarak kadınlar. *Çalışma ve Toplum, 30*, 137-166.

Overbeek, H. (2002). Neo liberalism and the regulation of global labor mobility. *The Annals of the American Academy of Political and Social Science, 581*(May), 74–90. doi:10.1177/000271620258100108

Özuğurlu, M. (2006). Türkiye'den güvencesiz çalışma ve sınıflar mücadelesinin yeni gündemi. In C. Gürkan, Ö. Taştan, & O. Türel (Eds.), *Küreselleşmeye güneyden tepkiler*. Dipnot Kitabevi.

Panko, T. R., & George, B. P. (2018). Animal sexual abuse and the darkness of touristic immorality. In M. Korstanje & B. George (Eds.), Virtual traumascapes and exploring the roots of dark tourism. (pp. 175-189). IGI Global. doi:10.4018/978-1-5225-2750-3.ch009

PETA. (2022). *Santorini donkey rides.* PETA. https://investigations.peta.org/santorini-donkey-rides/

Poulin, R. (2003). Globalization and the sex trade: Trafficking and the commodification of women and children. *Canadian Women's Studies, 22*(3/4), 38–47.

Punaks, M., & Feit, K. (2014). *The paradox of orphanage volunteering: Combating child trafficking through ethical voluntourism.* Next Generation Nepal. http://www.nextgenerationnepal.org/File/The-Paradox-of-Orphanage-Volunteering.pdf

Purkis, S. (2008). Turizmle kalkınmada enformel emek kullanımı: Marmaris örneği. *Ekonomik Yaklaşım Dergisi, 19*(69), 107–126. doi:10.5455/ey.10678

Reas, P. J. (2013). 'Boy, have we got a vacation for you': Orphanage tourism in Cambodia and the commodification and objectification of the orphaned child. *Thammasat Review*, *16*(1), 121–139.

Reas, P. J. (2020). "Children that are cute enough to eat": The commodification of children in volunteering vacations to orphanages and childcare establishments in Siem Reap, Cambodia. *Tourism, Culture & Communication*, *20*(2-3), 83–93. doi:10.3727/109830420X15894802540151

Regan, T. (2007). Kafesler boşalsın: Hayvan haklarıyla yüzleşmek. Çev. Serpil Çağlayan, İstanbul: İletişim Yayınları.

Rickly, J., & Kline, C. (2021). Introduction: Working for the (hu)man in the tourism industry. In C. Kline & J. Rickly (Eds.), *Exploring non-human work in tourism: From beasts of burden to animal ambassadors* (pp. 1–18). De Gruyter Oldenbourg. doi:10.1515/9783110664058-001

Robinson, R. N., Martins, A., Solnet, D., & Baum, T. (2019). Sustaining precarity: Critically examining tourism and employment. *Journal of Sustainable Tourism*, *27*(7), 1008–1025. doi:10.1080/09669582.2018.1538230

Schireup, C. U., Hansen, P., & Castles, S. (2006). *Migration, citizenship and the European welfare state: A European dilemma*. Oxford University Press. doi:10.1093/0198280521.001.0001

Sijtsman, W. (2001). *Tourism training development for women's empowerment: An amendatory methodology* [Unpublished master's dissertation, Netherlands Institute of Tourism and Transport Studies, Netherlands].

Silva, M. (2014). El trabajo infantil rural en la agenda pública del desarrollo sustentable. *Espacio Abierto*, *23*(1), 23–44.

Thrift, N. (2006). Re-inventing invention: New tendencies in capitalist commodification. *Economy and Society*, *35*(2), 279–306. doi:10.1080/03085140600635755

Tucker, H. (2007). Undoing shame: Tourism and women's work in Turkey. *Journal of Tourism and Cultural Change*, *5*(2), 87–105. doi:10.2167/jtcc089.0

TÜİK. (2023). *Çocuk istatistikleri*. https://data.tuik.gov.tr/Bulten/Index?p=Istatistiklerle-Cocuk-2022-49674&dil=1

Turizm Databank. (2022). *Turizm sektörü istihdamı*. https://www.turizmdatabank.com/turizm-istatistikleri/otellerde-toplam-calisan-sayisi/

Uğuz, S. Ç., & Topbaş, F. (2016). Turizmde kadın istihdamı ve ücret ayrımcılığı: Karşılaştırmalı bir analiz. *Anatolia: Turizm Araştırmaları Dergisi*, *27*(1), 62–78.

UNICEF. (1989). *Convention on the rights of the child*. UNICEF. https://www.unicef.org/turkiye/en/convention-rights-child

UNICEF. (2021). *Child labour: Global estimates 2020, trends and the road forward*. UNICEF. https://data.unicef.org/resources/child-labour-2020-global-estimates-trends-and-the-road-forward/

United Nations. (2013). *Combating child sex tourism*. UN. https://www.ohchr.org/en/stories/2013/04/combating-child-sex-tourism

UNWTO. (2023). *Women take 'centre stage' in tourism development*. UN. https://www.unwto.org/news/women-take-centre-stage-in-tourism-development#:~:text=Fifty%2Dfour%20per%20cent%20of,higher%20than%20in%20other%20sectors

Urry, J. (2002). *The Tourist Gaze*. Sage Publications.

Van Doore, K., Martin, F., & McKeon, A. (2016). International volunteering and child sexual abuse. *Global study on sexual exploitation of children in travel and tourism Expert paper*. New York: Better Care Network. https://bettercarenetwork.org/sites/default/files/Expert%20Paper%20-%20International%20Volunteering%20and%20Child%20Sexual%20Abuse.pdf

Vasilopoulou, C. (2021). The donkeys of Santorini: Workers or slaves? In C. Kline & J. Rickly (Eds.), *Exploring non-human work in tourism: From beasts of burden to animal ambassadors* (pp. 159–180). De Gruyter Oldenbourg. doi:10.1515/9783110664058-010

Von Essen, E., Lindsjö, J., & Berg, C. (2020). Instagranimal: Animal welfare and animal ethics challenges of animal-based tourism. *Animals (Basel)*, *10*(10), 1830–1847. doi:10.3390/ani10101830 PMID:33050024

Vosko, L. F. (2010). *Managing the margins: Gender, citizenship, and the international regulation of precarious employment*. Oxford University Press.

Wigfield, A. (2001). Post-Fordism, gender and work. Londra: Ashgate Publishing.

WSPA. (2021). *The contribution of animal welfare and tourism for sustainable development*. WSPA. https://www.responsibletravel.org/wp-content/uploads/sites/213/2021/03/sustainable-tourism-animal-welfare.pdf

WTTC. (2022). *Travel & Tourism could grow to $8.6 trillion in 2022, says WTTC*. WTTC. https://wttc.org/news-article/travel-and-tourism-could-grow-to-8-point-6-trillion-usd-in-2022-say-wttc

Yahyaoğlu, M. (2022). *Turizmin ve turizm işçilerinin sorunları ve çözüm önerilerimiz*. Devturismis. http://www.devturizmis.org.tr/post/turizmin-ve-turizm-iscilerinin-sorunlari-ve-cozum-onerilerimiz

Yanardağ, M. Ö., & Avcı, M. (2012). Turizm sektöründe istihdam sorunları: Marmaris, Fethiye, Bodrum ilçeleri üzerine ampirik bir inceleme. *Ege Stratejik Araştırmalar Dergisi*, *3*(2), 39–62.

Young, M., & Markham, F. (2020). Tourism, capital, and the commodification of place. *Progress in Human Geography*, *44*(2), 276–296. doi:10.1177/0309132519826679

KEY TERMS AND DEFINITIONS

Animal Labor: Animal labor refers to the situation where animals are employed in agriculture, transportation, entertainment, or other human activities. This labor generally aims to utilize the physical strength and abilities of animals. Examples of animal labor include using horses to pull plows in agriculture, employing camels for transportation, or having elephants perform in the entertainment industry.

Capitalism: Capitalism is an economic and social system that is fundamentally based on principles of private ownership, free competition, and a market economy. In capitalism, the means of production, resources, and property are typically controlled by individuals or private enterprises. Economic activities and the production of goods and services often occur under conditions of free market, meaning they are determined by the dynamics of supply and demand.

Child Labor: Child labor refers to the participation of children, usually with inadequate age and developmental levels, in work or job activities. This type of labor can occur under hazardous conditions that may harm children's physical and mental health. Often carried out for low wages or no pay at all, these tasks can restrict children's opportunities for education, play, and normal developmental activities.

Commodification: Commodification refers to the process of turning an intangible asset or concept into something that holds tangible value. This term is commonly used in fields like economics and sociology. In the context of an economy, commodification signifies the transformation of a good or service into a marketable product. A company or individual can accomplish this by producing a product, offering a service, and presenting it for sale or trade.

Non-Human Animals: From a philosophical perspective, the term 'non-human animals' can reflect a notion that humans are different from and superior to other animals. However, modern ethical and philosophical debates have increasingly raised awareness about the potential emotions, consciousness, and rights of non-human animals. This perspective is emphasized by animal rights advocates, while also supporting the idea that humans should engage with their natural environment sustainably and respectfully.

Women Labor: Women's labor refers to the production, services, or work activities carried out by women. This labor can encompass a wide range of areas, from household chores to paid employment sectors. Women often play significant roles both in the paid labor market and within the family, contributing significantly to the functioning of society.

Chapter 7
Authenticity Under the Pressure of Commodification in Tourism

Özgür Devrim Yılmaz
Faculty of Business, Dokuz Eylül University, Turkey

Yavuz Demirci
Institute of Social Sciences, Dokuz Eylül University, Turkey

ABSTRACT

The chapter begins by defining the concept of authenticity and discusses different types of authenticity. The concept of commoditization is then introduced, followed by a review of the literature on commoditization in terms of its positive and negative aspects. As a result, it seems obvious that a broad consensus on topics such as definition, classification, effects, precautions, or management decisions to be made on the idea of authenticity in tourism and the commodification of culture or cultural elements is not yet achievable. While it continues to be discussed in the relevant literature whether tourists today seek authenticity, it can be noted that certain cultural elements have become commodities and started moving away from authenticity in tourism destinations. Some cultural elements that provide a competitive advantage have become ordinary; and this situation is mostly accepted by local people who acquire an income from tourism. Moreover, when the relevant literature is reviewed, there are also studies claiming that commodification positively affects cultural authenticity.

DOI: 10.4018/978-1-6684-9255-0.ch007

INTRODUCTION

Tourism has become a popular field of study due to its economic, environmental, and social impacts, inspiring researchers in the related field as well as other disciplines such as economics, history, geography, business administration, sociology, psychology, philosophy, culture, communication, and language studies. Numerous studies, both within and outside the field, have been conducted to examine the effects of tourism on an event, occurrence, or change, or how the variable in question influences or will affect tourism.

The basic explanation for the increased interest in tourism research is, of course, the sector's favorable economic influence on national economies. Indeed, tourism remains a prominent sector in many developing economies, while it is also under examination in industrialized economies (Mowforth & Munt, 2015). In other words, the positive impact of tourism on national economies cannot be underestimated. In this regard, it is inevitable for tourism studies to expand and grow by including other disciplines. When it comes to studying the environmental implications of tourism, the situation does not appear different. It is worth noting that there are an increasing number of multidimensional studies examining the environmental impacts of tourism, the tourism-environment interaction, environmental protection and sustainability, ecosystem protection, (Weaver, 2006; Buckley, 2012), and management of natural areas. Analyses and assessments of economic and environmental implications, which are largely based on concrete data, may be substituted with more abstract data and evaluations in studies of tourism's impact on society. Tourism, as is well known, has both positive and negative social consequences for both tourists and locals living in the destination (Ap & Crompton, 1998; Mason, 2003; Gursoy, Chi & Lu, 2011). While international tourism mobility has positive social effects such as cultural recognition, the development of awareness of the protection of historical and cultural elements in host societies, the expanding of tourism awareness, and an increase in intercultural communication and interaction, it also has negative effects such as intercultural conflicts, cultural degeneration, local people-tourist conflicts, and tourist-tourist conflicts. For this reason, the authors concentrate on the subjects of authenticity and commodification together, whose negative effects are studied in particular.

Individuals' search for authenticity is closely related to globalization and modernization. In line with globalization and modernization, individuals who are overwhelmed by alienation, mediocrity, and fake or unreal relationships begin to take an interest in places and societies where naturalness and authenticity prevail to get rid of this situation. It is argued that the desire to reach core values and authenticity in increasingly globalized lives is one of the most prominent desire-stimulating factors of modern tourism. In this respect, it is obvious that tourists want to see daily life as it is lived by entering the authentic culture of a destination during their travels (Cohen, 1988; Selwyn, 1996; MacCannell, 1999; Wang, 1999).

The modern individuals' inability to express themselves by following their social position and the pressure to appear different cause people to seek experiences outside of their daily lives and environments are counted among the reasons for people's search for authenticity (Sedmak & Mihalic, 2008). The tourist, who needs to move away from daily life and go to different regions to see and experience differences, is searching for originality (MacCannell, 1973; Turner & Turner, 1978) as a form of satisfaction today, which can somehow bring him face-to-face with false (or fake) experiences (Goulding, 2000). This situation made the concept of authenticity debatable, and the concept started to be discussed in tourism starting in the 1960s when mass tourism destinations increased. By stating that the main problem in determining authenticity is the differences in perspectives and situations, Sedmak and Mihalic (2008) also cited the difficulty of establishing common criteria for different fields in the evaluation of authenticity and the uncertainty of the people who are authorized to determine the criteria (academics, tourists, suppliers, etc.) as a few of the reasons for discussing the concept.

Another concept that does not have a clear and agreed-upon definition has been discussed in terms of its positive and negative consequences and has a cause-and-effect relationship with the concept of authenticity is "commodification". Modern capitalism's transformation of the world into a global home by destroying the concept of distance and deterritorializing everything has also shown itself in the field of tourism. One of the first critical studies on the concept of commodification, whose relationship with tourism has been discussed since the 1980s, was made by Greenwood (1977), who saw the loss of meaning caused by cultural commodification as a problem as important as the inequality in income distribution that emerged with the development of tourism

With the spread of consumer capitalism, diversity becomes standardized and homogenized (Hughes, 1995), while the emerging similarity poses a threat to authenticity. Urry (1992) suggested that both people and places have been lost in global tourism, which has also made tourism a center of globalization. Mass tourism has made tourism activities universal and affected local cultures (Mbaiwa, 2011). In this context, the main purpose of this chapter is to review the concept of authenticity in-depth, to examine the concept of commodification, to reveal how the issues are addressed in tourism research, and to make evaluations on the related studies on the commodification of culture.

AN IN-DEPTH LOOK AT THE CONCEPT OF AUTHENTICITY IN TOURISM

Authenticity remains an intriguing and promising research topic with many unresolved aspects (Heitmann, 2011), and it is a subject that serves as the foundation for many different fields such as sociology, anthropology, archaeology, geography, literature, philosophy, and tourism (Grayson & Martinec, 2004). The range and amount of studies in the field of tourism attract notice when the associated literature is studied. This shows that authenticity is still an essential research topic in tourist studies, and all of these studies on authenticity are important in terms of providing light on many topics. It is nearly hard to construct a definition of authenticity that fits every situation (Frissvoll, 2013), hence researchers' efforts to define the idea continue today in light of emerging new perspectives in various contexts. The word authenticity, which has a Greco-Roman origin (Heitmann, 2011), primarily refers to "genuine", "reliable", and "true" (Arthur, 2006). Since these phrases will not always suffice to define the concept of authenticity (Steiner & Reisenger, 2006), it is clear that diverse definitions of the concept can lead to semantic confusion. Even the field in which a scientific study on authenticity is conducted leads authenticity to be defined differently (Knox, 2008).

The concept of authenticity can be briefly defined as the "essence, pure, sincere, original, real" signs and images of local characteristics that carry the characteristics of the region, country, or region to the present day by preserving its connection with the past and giving people different emotions. Prior studies examining the issues of authenticity in tourism are carried out mainly in areas that have a direct or indirect relationship with tourism, such as touristic local houses (Agyeiwaah, 2013; Wang, 1997; Wang, 2007); souvenirs (Asplet & Cooper, 2000; Chang, Wall & Chang; 2008); museums (Chhabra, 2008; Goulding, 1999; Hede & Thyne, 2010), restaurants (Tsai & Lu, 2012); and heritage sites (Chronis & Hampton, 2008; Cohen, 2011; Goulding, 2000). When these studies carried out by considering different areas of tourism are evaluated together, authenticity in tourism can also be defined as touristic attractions such as architecture, cuisine, intangible heritage, and/or the natural environment that are local and existed before the development of modern tourism. Authenticity in tourism can also be explained as ethnically 'pure', culturally 'traditional," or 'closed' places or things unaffected by Westernization or other global processes. When the first studies on authenticity in tourism are examined, Boorstin's (1964) definition (as stated in Cohen, 1988) is encountered, which regards visitors as passive hedonists who engage in so-called inauthentic activities. Tourists, he claims, ignore reality, enjoy inauthentic attractions, do not seek authenticity in the locations they visit, and strive to fit narrow expectations. According to the author, tourists believe that guided tours allow them to meet locals, and they are satisfied with so-called authentic activities and fall into the misconception that fictional activities are authentic. MacCannell (1973), on the other hand, stated that, unlike Boorstin, modern people are bored with the monotony of daily life, have become alienated from their environment, and their routine life pushes them to seek the authentic; their travels are based on the search for authenticity, and the tourist communicates with the reality in other societies at a higher level. Furthermore, he believed that modern tourists desire authenticity in general. However, MacCannell's assertion elicited a strong reaction, and several authors (Cohen, 1988; Buchmann, Moore & Fisher, 2010; Rickly-Boyd, 2012) argue that not all tourists want authenticity and that many tourists may choose the inauthentic and like the experience expressed.

Authenticity is defined by Sedmak and Mihalic (2008) as "attractions of elements such as cuisine, intangible heritage, and an unspoiled environment that preserve the features that existed prior to the development of tourism, are unique to the region, and are produced without regard for industrial purposes." Authenticity, according to Heitmann (2011), is "an object that is the essence, pure, real, and has not undergone any change or intervention since its creation." Furthermore, Cobb (2014) contends that authenticity is a "created effect" rather than a "real thing." In other words, in some circumstances, the sense of authenticity may differ from person to person (Grayson & Martinec, 2004).

Attempting to define authenticity in various ways has also led studies to try to explain different types of the term. Researchers have used numerous concepts to describe the significance of authenticity in various contexts. MacCannell's (1973) staged authenticity, Selwyn's (1996) hot and cool authenticity, Wang's (1999) object-related and action-related authenticity, Grayson and Martinec's (2004) iconic and indexical authenticity, and Gilmore and Pine's (2007) natural, original, exceptional, referential, and influential authenticity are among these concepts.

According to MacCannell (1973), a number of arrangements are made by product and service providers to guarantee that the experience provided is considered as real by consumers, and this authenticity is referred to as "staged authenticity." The changes made in staged authenticity can range from minor changes to make these elements more prominent to major changes such as the addition of new elements that have nothing to do with the original. Depending on Goffman's (1959) distinction between the front and back areas, MacCannell (1973) split the business areas into six distinct sections depending on their authenticity qualities and labeled each of them as a "stage." It is also believed that tourists seeking authenticity should pay attention to these back areas, as these areas may contain clues to authenticity (Kontogeorgopoulos, 2004). The concept of staged authenticity is similarly related to the concept of "pseudo-events," which Boorstin presented in 1961. According to Boorstin and MacCannell, recognizing inauthenticity is the first step toward comprehending authenticity. Although both academics take an object-oriented approach, Boorstin believes that individuals prefer artificial activities, whereas MacCannell believes that people seek authenticity (Heitmann, 2011). According to Hinch and Higham (2005), when the cultural aspect is included, fake authenticity has negative impacts. These negative effects can be exemplified by tourists meeting the elements that are far from authentic and seeing events or activities as authentic when, in fact, they have been altered and do not reflect reality (Heitman: 2011). Instead of emphasizing what authenticity is, the concept of staged authenticity stresses what it is not. As a result, researchers have developed new viewpoints on authenticity.

Selwyn (1996) explores authenticity through the lens of tourism, dividing it into two categories: hot authenticity and cool authenticity. Hot authenticity is a reflection of the evaluator's feelings, whereas cool authenticity should be founded on different concrete signs and evaluated using an objective manner. The evaluation of experts is seen as the basis of the authenticity decision in the effort to define cold authenticity, but the sentiments and beliefs of the individual play a major part in the determination of hot authenticity. Experts, for example, are contacted to determine whether a prehistoric fossil is authentic or not. This procedure is critical in determining cool authenticity. However, a yoga practitioner's experience of self-existence during the exercise serves as the foundation for determining hot authenticity. As a result, encountering object authenticity results in cool authenticity, while encountering existential authenticity results in hot authenticity.

Wang (1999) divides the concept of authenticity, which he interprets from a tourism perspective, into two primary categories as object-related and action-related. Object-related authenticity is classified as "objective" or "constructive," whereas action-related authenticity is classified as "existential." Authenticity, according to objectivist views, may be universally evaluated, and there is an objective standard used to determine authenticity. It has a controversial and variable structure, according to constructivist approaches (Steiner & Reisenger, 2006), and since personal interpretations are involved in the process of determining the meaning of authenticity, a subjective and pluralistic understanding is dominant (Belhassen, Caton, & Stewart, 2008). In other words, while objective authenticity stresses the originality of things, constructive authenticity emphasizes the symbolic authenticity that evaluators ascribe to the object through aspects such as imagination, expectation, and preference. According to objective authenticity advocates, just as the authenticity of artifacts and objects displayed in museums can be proven, there is a condition of provable authenticity in objective authenticity. What makes anything authentic in constructive authenticity is the evaluator's perspective. Instead of seeking objectivity, the evaluator prefers to imbue the thing with symbolic significance. People's expectations, opinions, ideas, and preconceptions can all have an impact on interpretations. Over time, something inauthentic can become authentic (Wang, 1999). Existential authenticity, sometimes referred to as subjective authenticity, is defined as a unique condition of being. According to Wang (1999), existential authenticity is unrelated to object authenticity. As a result, it is investigated separately from objective and constructive authenticity. Some actions and activities that are outside of the routine of daily life may contribute to the person engaging feel as if they are expressing themselves better and more freely, or they can help them find their genuine self.

Based on Peirce's work on the philosophy of signs, Grayson and Martinec (2004) divide the concept of authenticity into two major categories: iconic and indexical. Iconic authenticity relates to the authenticity of reproductions and is concerned with how people perceive and judge authenticity. The presence of such authenticity is dependent on the copy being made in the same manner as the original. In order to be authentic in the eyes of the consumer, indexical authenticity must be validated by knowledge and expertise. To determine whether a chair is an original Victorian chair, for example, the consumer must have some proof (through certification or in a trustworthy context) that it is, in fact, Victorian. Iconic and indexical authenticity are not mutually exclusive. Despite the fact that perception favors iconicity over indexicality and vice versa, every perceived cue contains both iconic and indexical properties. As a result, certain items are both iconically and indexically authentic. For example, a consumer may believe that a chair was not only built during the Victorian era (indexical authenticity) but also represents Victorian style (iconic authenticity).

Lastly, Gilmore and Pine (2007) examine the economic values that businesses offer to consumers in five categories as raw materials, products, services, experiences, and transformations. Accordingly, raw materials include economic value in the form of natural resources. For example, a necklace made of gold is a product. Services are intangible activities offered or distributed to individual customers. Experiences are staged events that affect individuals personally. Transformations are outputs that lead to changes in customers. Since products and services alone are no longer sufficient to satisfy customers, businesses have endeavored to create individual experiences for consumers. Instead of buying products and services, consumers have started to pay money to spend time with businesses. However, the increasing role of fiction in the process of experience creation leads to a preference for the "real" and "authentic" during the purchase decision by customers.

Based on this distinction Gilmore and Pine (2007) examine authenticity in five genres as natural, original, exceptional, referential, and influential authenticity. Natural authenticity refers to the tendency for people to consider as genuine anything that is present in or of the earth in its natural state, unaltered by human hands—not artificial or fabricated. The incorporation of natural components or properties into an offering constitutes the essence of natural authenticity. Organic food growers appeal to this authenticity genre by avoiding pesticides and fertilizers. When it comes to original authenticity, things that are unique in their design and are the first of their kind—as opposed to copies or imitations—tend to be perceived as genuine. The key to originality is having an offering come before or differ in form from other offerings of the same kind. Leading technology companies are creating items that fit this category and are grabbing consumers' attention. People regard exceptional authenticity as something which is done extremely effectively and with human concern. Care for customizing an offering to match the unique or uncommon preferences of individuals to whom it is presented is a distinguishing trait of exceptional authenticity. Any business that encourages its employees to genuinely care about their clients and respond to their specific requirements demonstrates great authenticity. Referential authenticity refers to the reality that individuals tend to see as authentic something which references to another context, gaining inspiration from human history and reaching into our common memories and longings. To appeal through referential authenticity, a product or service must relate to something that is already viewed as genuine. Finally, influential authenticity refers to the fact that individuals regard as authentic something which generates influence on other entities, thereby urging humans to a higher aim. The key element of influential authenticity is instilling meaning in a product and inspiring people to a higher cause.

These different approaches and perspectives on authenticity indicate that the topic can be examined in almost every facet of tourism, but a precise definition or set of definitions is not likely to be adopted any time soon. Particularly in tourism types that focus on cultural aspects, it is unclear from whose perspective authenticity should be addressed. Additionally, a destination's genuine attributes and attractions contain a lot of components that are mostly utilized in tourism marketplaces. At this point, it is appropriate to assert that cultural elements such as local people's lifestyle, costumes, culture, music, carnivals, festivals, hunting and agricultural traditions, traditional architecture, castles, museums, shopping malls, theme parks, archaeological sites, art galleries, famous places of birth and graves of people, mosques, churches, staged historical events, theaters, furniture, restaurants, ethnic cuisines, dinner tables, weddings, wildlife, public places, rumors, legends, and/or local people-tourist relations have been mostly transformed into tourism products that can be bought and sold today. This necessitates addressing authenticity concerns from the standpoint of commodification.

COMMODIFICATION: WHAT IF CULTURE BECOMES COMMODIFIED THROUGH TOURISM?

Unlike the definition of authenticity, most scholars define the concept of commodification identically, however different perspectives about the concept have emerged as a result of the way the topic is approached. In other words, in the relevant literature, the concept of "commodification" is examined from multiple viewpoints and there is no consensus on the consequences in particular. To quickly review the concept's definition Cohen (1988) defines "commoditization" as the process by which goods (and services) become primarily valued in terms of their exchange value in a context of trade; developed exchange systems in which this exchange value is expressed in terms of prices create a market. Watson and Kopachevsky (1994) assert that commercialization, which involves uniformity of goods, pleasures, and experiences, is a prevalent aspect of contemporary capitalism. Commodification encompasses three primary aspects: the production of products and services for exchange; monetization and market-based exchange; and the pursuit of profit as the driving force behind the sale of commodities and services (Fløysand & Jakobsen, 2007).

It is noticeable that commodification in tourism studies often focuses on the commodification of culture. One of the earliest critical examinations of the concept of commodification and its relationship to tourism may be found in Greenwood's (1977) work. Greenwood states that the loss of meaning engendered by cultural commodification is a concern on par with the rise in income disparity brought on by the expansion of tourism. As consumer capitalism has grown, variety has been standardized and homogenized, compromising authenticity as similarity has increased (Hughes, 1995). The growth of mass tourism has also had an impact on local cultures (Mbaiwa, 2011). Giddens (1990) defined cultural homogeneity as the formation of a common global culture in regions that are not affected by time. The obvious effects of homogeneity include the same architecture found in hotels and shopping malls in far-off cities (Holton, 2000), as well as the standardized tourist experiences provided in plenty of destinations. In other words, even in places with diverse cultures, there are comparable tourist activities due to the increasing similarity of tourist goods and services. Based on the fact that fast-food franchises offer their items on standard menus, apply the same service standards throughout the world, and erode authenticity along with important cultural values, Ritzer (2011) investigates similarity under the term "McDonaldization."

The commercialization of culture is the term used in tourism to describe the packaging of cultural events and artifacts for the tourist industry. Governments are attempting to encourage cultural tourism in a number of developing nations, not least in Africa, to attain economic success (Manwa, 2007; Lenao, Mbaiwa, & Chanda 2015; Mbaiwa & Sakuze, 2009). Cole (2007) and Cohen (2011) stated that several destinations have been in danger of losing their appealing cultural characteristics because of global cultural and architectural homogeneity. In these studies conducted in different fields of tourism, it is possible to see that commodification is addressed with both positive and negative aspects and there is a cause-and-effect relationship between it and the concept of authenticity. The vast majority of the literature on the nature of modern tourism and its effects on host society depends on several paramount presumptions, including that tourism leads to "commoditization," that commoditization destroys the authenticity of local cultural products and human relations, and that "staged authenticity" limits travelers' real desires for authentic experiences (Cohen;1988). Cohen (1988) also states that authenticity and commodification are related to each other and are not two separate ideas. This is because the process of commodification does not always destroy cultural items, for both locals and tourists.

The literature's in-depth discussions of the potential impacts of tourism's commodification concentrate on positive as well as negative consequences (Cole, 2007; Mbaiwa, 2011). According to several researchers (Greenwood, 1977; Swain, 1989; Christiaan, 1990; MacCannell, 1992; Watson & Kopachevsky, 1994; Boynton, 1997; Mowforth & Munt, 1998; Goulding, 2000; Taylor, 2001; Halewood & Hannam, 2001), the effects of commodification include erosion in the authenticity of cultures, the destruction of cultural values, the emergence of cultural disputes, and the loss of local identity. However, a number of scholars are pointing out the benefits associated with growing tourism, involving the ability to save cultural values from extinction and preserve traditions; the emergence of new cultural formations as a result of interactions between tourists and locals; the redefinition of existing values (Cohen, 1988; MacDonald, 1997; Medina, 2003; Xie, 2003; Cole, 2007; Finn, 2009; Su, 2011). Following a thorough examination of previous studies emphasizing the connection between authenticity, commodification of culture, and tourism, it can be summarized as follows.

To start with, Greenwood (1977) conducted one of the earliest critical studies on the idea of commodification, whose relation to tourism has been studied since the 1980s. In his study, Greenwood expresses the following criticism on the topic:

- Culture is packaged, priced, and sold as if some material items (such as buildings, fast food, and room service) as tourism grows. Because cultural commercialization does not require participant consent, local people can be exploited.
- Promises like "if you have money, you have the right to see what you want" act as if natural resources, including cultural traditions, have a cost.
- It causes the local culture, which is promoted as an attraction, to change or vanish.
- Local culture loses its significance for the locals when it is viewed as a commodity on which tourists have a right, and these disregards the cultural rights of the society.

Cohen (1988) does not entirely agree with Greenwood's argument that with time, commodified cultural items lose their significance, excitement, and meaning for the locals and that their formerly "authentic" status has become a staged performance. It is claimed that the commodification of cultural products does not necessarily obliterate their meaning for both tourists and locals; in addition, it is claimed that the constant acquisition of new meanings by tourist-oriented products representing ethnic and cultural identities does not cause the loss of old meanings. In other words, rather than viewing it solely as the destruction or consumption of cultural aspects, it is realized that the commodification process can take on new meanings by introducing a new perspective along with beneficial changes in culture.

Similarly, to Cohen, according to MacDonald (1997) (as cited in Halewood & Hannam, 2001) the use of a commodified format does not imply giving up the concepts of local heritage and identity, on the other hand; rather, it can help to establish the importance of local experience. The commodification process, however, can give the culture new power. Medina (2003) then again, highlights the positive aspects of commodification in the study of Mayan culture, stating that a civilization that has lost its local identity is attempting to reclaim its identity to keep its own culture alive again in response to tourist demand. Su (2011) similarly underlines the positive socio-cultural benefits of commodification by sustaining cultural identities and practices. According to the author, commodification has revitalized neglected ethnic music in China's Lijiang province, which is part of the world's cultural heritage, and has been essential in teaching ethnic music to young people.

Taylor (2001) on the other hand, states that the authenticity of culture is diminished when it is viewed as an event for tourists. Additionally, Swain (1989) and Christiaan (1990) express a similar view by indicating that when the authenticity of culture is diminished; a destination becomes less interesting and unique. Culture becomes commoditized when its resources are developed to become tourist commodities. The value of the product is diminished because of the destination's decreased authenticity. Mowforth and Munt (1998) have criticized the process of cultural commodification that limits cultures to rituals and handicrafts for consumption, claiming that it idolizes cultures and ignores the social relations that contribute to commodity production. According to the researchers, tourists have a dominant position in contrast to locals who sell souvenirs and display their cultural habits.

According to Boynton (1997), traditional communities are forced to display their "primitiveness" for visitors because of the conflicting forces of growth and preservation, even if change is a constant in these societies' everyday lives. One excellent example is the Kagga Kamma Bushmen of the Kalahari, who work during the day at a resort and maintain their ancestral customs in a model village. They actually reside outside of the village in homes with tin roofs, wear baseball caps, and enjoy rap music. The commodification of culture generates tensions and conflicts between local people who may be unwilling to offer their sacred culture to tourists and tourism operators who desire to barter their culture for tourist consumerism. According to Steiner and Reisinger (2006), tensions occur between the use of culture for community expression and economic generation. As an outcome of such tensions, some locals may lose the required drive to continue generating and connecting with cultural characteristics. Finally, Mbaiwa (2011) illustrates the difference in older and younger locals' perspectives on tourism commodification, noting that in Goo-Moremi Village, older people perceive tourism as a destructive threat to local culture, whereas younger people welcome efforts to promote tourism in their communities as an opportunity for improving rural livelihoods. In this regard, while some residents fear that tourism expansion may diminish the sacred value of their culture, others perceive benefits.

CONCLUSION AND DISCUSSION

Different approaches in definition, classification, and handling of the subjects are clearly apparent in the studies that stand out in the literature on authenticity and commodification in tourism, which are the subjects visited in this chapter. At this point, it is obvious that authenticity and commodification in tourism are still feasible for research topics due to variations in how scholars handle the issues and the aspects they identify with the concepts.

To summarize, authenticity in tourism is mainly defined as places or items that have not been influenced by Westernization or other global processes so that they carry their originality. Authenticity can be accepted in a variety of ways, ranging from a product attribute to an experiential outcome. Whilst some scholars hold the view that authenticity has a universal definition, others argue that it displays the characteristics of a subjective approach. Experts, for example, can objectively evaluate the authenticity of a historical item, but individuals recognize and evaluate the authenticity of their own experience from a subjective perspective. In other words, the perception of authenticity can change from person to person. While some studies examine the falsity of staged authenticity, some suggest the fact that the staged authenticity varies from the original is insignificant if it retains some characteristics of the original culture and traditions. This is due to the fact that change is often unavoidable and not all changes result in artificiality and degeneration. However, change brings some consequences with it. While certain methods to personalize tourist products and services integrate with their original structure, others result in major negative changes. The extent to which changes in cultural and natural heritage resources are allowed is also debatable, and modifications of this kind are generally frowned upon in sensitive societies. Moreover, it is highly contentious whether these modifications are ethical or whether they represent authentic cultural components of society. It is not certain from which viewpoint authenticity should be handled, particularly in tourism forms that promote upon cultural elements. Furthermore, a destination's real features and attractions encompass an extensive number of elements that frequently appear in tourism markets. Currently, it is possible to say that most cultural features are now available for purchase and traded as touristic products. This necessitates handling authenticity concerns from a commodification approach.

Commodification, on the other hand, refers to the process by which products (and services) are predominantly valued in terms of their exchange value in a commercial environment; created trading mechanisms in which this exchange value is reflected in terms of prices produce a market. Although there are numerous studies in the relevant literature on the commodification of various concepts (such as women, labor, religion, cuisine, leisure time, emotions, bodies, animals, and so on), the authors focus on studies on the commodification of culture, which is an essential component of tourism, and attempt to address the issue of cultural commodification through discussions on authenticity, which has become a current research area in tourism. It is significant that commodification in tourism studies commonly focuses on the commercialization of culture. The term "commercialization of culture" pertains to the marketing of cultural activities and artifacts for selling to tourists. Commodification is handled in terms of positive and negative consequences in research carried out in different areas of tourism, and there is a causal connection between this phenomenon and the concept of authenticity. The overwhelming majority of works of literature on the dynamics of modern tourism and its effects on host society is predicated on a few key assumptions, which includes that tourism contributes to "commoditization," that commoditization annihilates the authenticity of local arts and crafts and interpersonal interactions, and that "staged authenticity" limits travelers' genuine desires for authentic experiences. Several experts believe that the consequences of commodification include a loss of cultural authenticity, the weakening of cultural values, the creation of cultural conflicts, and an erosion of local culture. Nonetheless, a number of researchers have identified the advantages associated with rising tourism, including having the capacity to preserve cultural values from disappearance and sustain traditions; the formation of new cultural entities as a consequence of interactions among tourists and locals; the redefining of traditional values; and the resulting rise in the level of local adoption of culture. Despite the research carried out with the claim that the tourist is in search of authenticity, it is estimated that the studies, which include the criticism of the commodification of many elements, especially the cultural elements, by the tourism providers, will be increasingly discussed in the related literature. While it is still questioned in the relevant literature whether tourists are in search of authenticity today, it can be said that cultural elements have become commodities in many tourism destinations, they have started to move away from authenticity, and some cultural elements that provide a competitive advantage in terms of tourism have become ordinary, and this situation is accepted by the local people who earn income from tourism.

As a result, it seems reasonable to note that in those studies in which the concepts in issue are being studied, there are first and foremost discussions about whether the tourist is searching for authenticity or not. Though the desire of tourists to escape the routine of daily life and seek novel experiences is known to exist, there is still the possibility that they will not attain what they want in any destination. Furthermore, it is possible that a product or service that a tourist considers authentic may also be staged authenticity. On the other hand, the growing emphasis of research arguing that authenticity is a topic that ought to be addressed with regard to individual tourist evaluations and "authenticity perception" necessitates generalizing individual judgments. Since such an assessment cannot be made for every tourist, authenticity studies are anticipated to be carried out at museums, historical monuments, local food, rural areas, festivals, and events that stand out in the literature. Even with the cultural and historical aspects being displayed in museums, the potential of demonstrating "original-like copies" indicates that debates about objective authenticity will continue to exist. While the standards for determining authenticity in such elements have been established, it maintains the uncertainty by which criterion and by whom the authenticity can be assessed.

On the other hand, it is acceptable to argue that concerns about commodification in tourism are primarily concerned with negative consequences. Most destinations seeking to expand their tourism share are attempting to differentiate their products and services in order to adapt to changing tourist expectations and have begun to employ components that will provide a competitive advantage in tourism. This scenario has expanded the utilization of cultural elements as tourist attractions in particular. The recognition that locals can potentially earn cash through their daily activities, which would otherwise generate no revenue, has further accelerated the process. From this perspective, dealing with the subject of commodification in terms of local people who acquire income (benefit) from tourism and those who do not seem to contribute to the relevant literature.

Last but not least, future studies on authenticity and commodification in tourism will strive to consider and examine in-depth factors such as changing travel motivations, the impacts of current technological developments on tourism, changing consumer behaviors, changes in socio-demographic factors, and tourism awareness.

REFERENCES

Agyeiwaah, E. (2013). International tourists' motivations for choosing homestay in the Kumasi Metropolis of Ghana. *Anatolia: An International Journal of Tourism and Hospitality Research, 24*(3), 405–409.

Ap, J., & Crompton, J. L. (1998). Developing and testing a tourism impact scale. *Journal of Travel Research, 37*(2), 120–130.

Arthur, D. (2006). Authenticity and consumption in the Australian hip hop culture. *Qualitative Market Research, 9*(2), 140–156.

Asplet, M., & Cooper, M. (2000). Cultural designs in New Zealand souvenir clothing: The question of authenticity. *Tourism Management, 21*(3), 307–312.

Belhassen, Y., Caton, K., & Stewart, W. P. (2008). The search for authenticity in the pilgrim experience. *Annals of Tourism Research, 35*(3), 668–689.

Boorstin, D. J. (1961). *The image: A guide to pseudo-events in America*. Harper & Row.

Boynton, G. (1997) 'The search for authenticity: On destroying the village in order to

Buchmann, A., Moore, K., & Fisher, D. (2010). Experiencing film tourism: Authenticity and fellowship. *Annals of Tourism Research, 37*(1), 229–248.

Buckley, R. (2012). Sustainable tourism: Research and reality. *Annals of Tourism Research, 39*(2), 528–546.

Chang, J., Wall, G., & Chang, C. (2008). Perception of the authenticity of Atayal Woven Handicrafts in Wulai, Taiwan. *Journal of Hospitality & Leisure Marketing, 16*(4), 385–409.

Chhabra, D. (2008). Positioning museums on an authenticity continuum. *Annals of Tourism Research, 35*(2), 427–447.

Christiaan, K. P. (1990). Close encounters: "Intimate" tourism in Tibet. *Cultural Survival Quarterly, 14*, 2–5.

Chronis, A., & Hampton, R. D. (2008). Consuming the authentic Gettysburg: How a tourist landscape becomes an authentic experience. *Journal of Consumer Behaviour, 7*(2), 111–126.

Cobb, R. (2014). *The paradox of authenticity in a globalized world*. Palgrave Macmillan.

Cohen, E. (1988). Authenticity and commodification in tourism. *Annals of Tourism Research, 15*(3), 371–386.

Cohen, E. H. (2011). Educational dark tourism at an in popular site: The holocaust museum in Jerusalem. *Annals of Tourism Research, 38*(1), 193–209.

Cole, S. (2007). Beyond authenticity and commodification. *Annals of Tourism Research, 34*(4), 943–960.

Finn, J. (2009). Contesting culture: A case study of commodification in Cuban Music. *GeoJournal, 74*(3), 191–200.

Fløysand, A., & Jakobsen, S. E. (2007). Commodification of rural places: A narrative of social fields, rural development, and football. *Journal of Rural Studies, 23*, 206–221.

Frisvoll, S. (2013). Conceptualising authentication of ruralness. *Annals of Tourism Research, 43*(3), 272–296.

Giddens, A. (1990). *The consequences of modernity*. Stanford University Press.

Gilmore, J. H., & Pine, B. J. (2007). *Authenticity: What customers really want*. Harvard Business Review Press.

Goffman, E. (1959). *The presentation of self in everyday life*. Anchor Doubleday.

Goulding, C. (1999). Contemporary museum culture and consumer behaviour. *Journal of Marketing Management*, *15*(7), 647–671.

Goulding, C. (2000). The commodification of the past, postmodern pastiche, and the search for authentic experiences at contemporary heritage attractions. *European Journal of Marketing*, *34*(7), 835–853.

Grayson, K., & Martinec, R. (2004). Consumer perceptions of iconicity and indexicality and their influence on assessments of authentic market offerings. *The Journal of Consumer Research*, *31*(2), 296–312.

Greenwood, D. J. (1977) "Culture by the pound: An anthropological perspective on tourism as cultural commoditization." from Smith, Valene L. (Eds.), Hosts and Guests: the Anthropology of Tourism (pp. 129-138). Blackwell Publishers.

Gursoy, D., Chi, C. G., & Lu, L. (2011). Antecedents and outcomes of travelers' information-seeking behavior: A study of international visitors to Singapore. *Journal of Travel Research*, *50*(6), 617–631.

Halewood, C., & Hannam, K. (2001). Viking heritage tourism: Authenticity and commodification. *Annals of Tourism Research*, *28*(3), 565–580.

Hede, A.-M., & Thyne, M. (2010). A journey to the authentic: Museum visitors and their negotiation of the inauthentic. *Journal of Marketing Management*, *26*(7-8), 686–705.

Heitmann, S. (2011). Authenticity in tourism. Robinson, P., Heitmann, S. & Dieke, P. (Eds.), Research Themes for Tourism (pp. 45- 58). Wallingford: CABI Publishing.

Hinch, T., & Higham, J. (2005). Sport, tourism and authenticity. *European Sport Management Quarterly*, *5*(3), 243–256.

Holton, R. (2000). Globalization's cultural consequences. *The Annals of the American Academy of Political and Social Science*, *570*(1), 140–152.

Hughes, G. (1995). Authenticity in tourism. *Annals of Tourism Research*, *22*(4), 781–803.

Knox, D. (2008). Spectacular traditional Scottish folksong and authenticity. *Annals of Tourism Research*, *35*(1), 255–273.

Kontogeorgopoulos, N. (2004). Ecotourism and mass tourism in Southern Thailand: Spatial interdependence, structural connections and staged authenticity. *GeoJournal*, *61*(1), 1–11.

Lenao, M., Mbaiwa, J. E., & Chanda, R. (2015). Impacts of cultural tourism on the domestic utility of traditional baskets: Case of Gumare and Etsa 6 villages in the Okavango Delta, Bostana. *Nordic Journal of African Studies*, *24*(3&4), 301–318.

MacCannel, D. (1973). Staged authenticity: Arrangements of social space in tourist settings. *American Journal of Sociology*, *79*(3), 589–603.

MacCannell, D. (1992). *Empty meeting grounds: The tourist papers*. Routledge.

MacCannell, D. (1999). *The tourist: A new theory of the leisure class*. University of California Press.

MacDonald, S. (1997). A people's story: Heritage, identity and authenticity, from C. Rojek & J. Urry, (Eds.), In Touring cultures: Transformations of travel and theory (pp. 155–175) London: Routledge

Manwa, H. A. (2007). Is Zimbabwe ready to venture into the cultural tourism market? *Development Southern Africa*, *24*(3), 465–474.

Mason, P. (2003). *Tourism impacts, planning, and management*. Routledge.

Mbaiwa, J., & Sakuze, L. K. (2009). Cultural tourism and livelihood diversification: The case of Gcwihaba Caves and XaiXai Village in the Okavango Delta, Botswana. *Journal of Tourism and Cultural Change*, *7*(1), 61–75.

Mbaiwa, J. E. (2011). Cultural commodification and tourism: The Goo-moremi Community, Central Botswana. *Tijdschrift voor Economische en Sociale Geografie*, *102*(3), 290–301.

Medina, K. L. (2003). Commoditizing culture: Tourism and Maya identity. *Annals of Tourism Research*, *30*(2), 353–368.

Mowforth, M., & Munt, I. (1998). *Tourism and sustainability: Development and new tourism in the Third World*. Routledge.

Mowforth, M., & Munt, I. (2015). *Tourism and sustainability: Development, Globalisation and New Tourism in the Third World* (4th ed.). Routledge.

Rickly-Boyd, J. M. (2012). Authenticity & aura: A Benjaminian approach to tourism. *Annals of Tourism Research*, *39*(1), 269–289.

Ritzer, G. (2011). *The Mcdonaldization of society 6*. Sage Publication.

(•••). save it'. *Nation (New York, N.Y.)*, *265*(10), 18–21.

Sedmak, G., & Mihalic, T. (2008). Authenticity in mature seaside resort. *Annals of Tourism Research*, *35*(4), 1007–1031.

Selwyn, T. (1996). *The tourist image: Myths and myth making in tourism*. Wiley.

Steiner, C. J., & Reisinger, Y. (2006). Reconceptualizing object auhenticity. *Annals of Tourism Research*, *33*(1), 65–86.

Su, X. (2011). Commodification and the selling of ethnic music tourists. *Geoforum*, *42*(4), 496–505.

Swain, M. (1989). Developing ethnic tourism in Yunnan China: Shalin Sani. *Tourism Recreation Research*, *14*(1), 33–39.

Taylor, J. P. (2001). Authenticity and sincerity in tourism. *Annals of Tourism Research*, *28*(1), 7–26.

Tsai, C., & Lu, P. (2012). Authentic dining experiences in ethnic theme restaurants. *International Journal of Hospitality Management*, *31*(1), 304–306.

Turner, V., & Turner, E. (1978). *Image and pilgrimage in Christian culture*. Columbia Press.

Urry, J. (1992). The tourist gaze "revisited.". *The American Behavioral Scientist*, *36*(2), 172–186.

Wang, N. (1997). Vernacular house as an attraction: Illustration from Hutong Tourism in Beijing. *Tourism Management*, *18*(8), 573–580.

Wang, N. (1999). Rethinking authenticity in tourism experience. *Annals of Tourism Research*, *26*(2), 349–370.

Wang, Y. (2007). Customized authenticity begins at home. *Annals of Tourism Research*, *34*(3), 789–804.

Watson, G. L., & Kopachevsky, J. S. (1994). Interpretations of tourism as commodity. *Annals of Tourism Research*, *21*(3), 643–660.

Weaver, D. (2006). *Sustainable tourism: Theory and practice*. Elsevier Butterworth-Heinemann.

Xie, P. F. (2003). The Bamboo-Beating dance in Hainan, China: Authenticity and commodification. *Journal of Sustainable Tourism*, *11*(1), 5–16.

KEY WORDS AND DEFINITIONS

Authenticity Types: It refers to the different approaches and classifications that emerge when the subject is handled from different aspects of the literature.

Cultural Elements: They are either the direct production of a cultural group or elements that it has borrowed, integrated, and potentially adapted from another cultural group through group-level interactions. Elements of culture include the norms, languages, rituals, holidays, food and diet, art, and architecture of society.

Globalization: It refers to the process by which businesses or other organizations develop international influence or start operating on an international scale.

Local Culture: It includes the set of behaviors, beliefs, and customs that exist in a country as everything from how people eat and dress to the way they work and communicate. It is the life of a nation in the form of attitudes, values, traditions, and customs that arise over time.

Residents: It refers to people who live at a particular place for a prolonged period or who were born there.

Staged Authenticity: It refers to a process where tourism providers put their culture and/or cultural elements on display to attract travelers/tourists who seek genuine experiences.

Tourism: It refers to a social, cultural, and economic phenomenon and relationships arising from the travel and stay of non-residents, insofar as they do not lead to permanent residence and are not connected with any earning activity.

Tourist: It refers to people traveling to and staying in places outside their usual environment for at least 24 hours in the country visited and for not more than one consecutive year for leisure, business, and other purposes.

Chapter 8
Food Porn:
A Review From Social, Cultural, and Economic Perspectives

Pınar Şenel
Anadolu University, Turkey

ABSTRACT

Slow, valuable, and need-oriented production and consumption have been replaced by profit and purpose-oriented overproduction and consumption today. Traditional and healthy production has turned into a fast, unhealthy, and profit purposed production. Food has become a hedonic pleasure and status tool for consumers. Commodified food is explained by the concept of food porn in the literature. In this chapter, food porn history and the socio-cultural and economic reasons that cause the commodification of the food phenomenon were explained first, and then the concept of food porn was explained with its dimensions.

INTRODUCTION

In the 21st century, the effects of capitalism have increased with rapid production and living conditions, urbanization, modernization, and globalization (Novak, 1982). Slow, valuable, and need-oriented production and consumption have been replaced by profit and purpose-oriented overproduction and consumption (Preteceille & Terrail, 1986; Ritzer & Jurgenson, 2010). Food is one of the most affected phenomena by sociocultural and economic developments related to society, has been exposed to this wild production and consumption (Hasan & Yu, 2020). Traditional and healthy food production has turned into a fast, unhealthy, and untraditional production (Wells, 2012;

DOI: 10.4018/978-1-6684-9255-0.ch008

Rose&Lourival, 2019). On the other hand, food has become a hedonic pleasure tool beyond meeting the need. All eating requests that are not intended to satisfy the need for nutrition and hunger are termed hedonic eating (Schneider-Kamp, 2021). Hedonic eating is defined as the consumption of delicious and appetizing food for pleasure. In this context, food turns into an emotional satisfaction tool (Lowe & Butryn, 2007). This food that provides hedonic pleasure is generally unhealthy and non-traditional food contained intense calories, fat, and sugar (Paul, McCarthy & McCarthy, 2015). Also, food consumption has turned into a purposeful consumption activity in the last decade. Individuals give certain messages to the social environment through the economic values and symbolic meanings of the food they consume (Little, Newton & Beveridge, 2016). Although the consumption of food has changed according to different groups of society over the past centuries, it is not possible to explain with natural consumption distinction that case. The consumption that enables individuals to show themselves as rich and to be attributed as wealthy in society is explained by Veblen (Veblen, 1988) with the concept of conspicuous consumption. It can be stated that food has become a pretentious product consumed to create an artificial status distinction in the last two decades. Considering Bourdieu's study (Bourdieu, 1994), food consumption is related to a distinction and the opportunity to resemble an upper segment, as in every consumption phenomenon. In this context, individuals who the majority of society, live in the city away from traditional production and values, and are socio-culturally and economically in the middle class perform conspicuous consumption.

Mass media are the most important tools that have supported this production and consumption. It is an important public relations tool for individuals and businesses and is widely used by all segments of society (McBride, 2010). Especially on social media, remarkable images and videos related to food have been actively shared by individuals, businesses, and dominant groups. Social media is where any post can spread and extremely difficult to control ethically urges sharing that have destroyed the traditional, healthy, and valuable things about food. While the food media on social media has been used by individuals for ostentation purposes, it has been used by business and capital groups for customer acquisition and sales increase purposes. So, it has become commoditized and a hedonic tool (Mc Donnell, 2016).

The commodification of food and it has become a tool of hedonism and status, can be explained by the concept of food porn in the literature (Dejmanee, 2016). Although food porn was first used in 1977 in the sense of similarities between food, its preparation and presentation, and sexuality (Dechmanee, 2016), food that is untraditional and unhealthy, becoming a status tool are issues that should be examined within the scope of food porn. In this section, food porn history and the socio-cultural and economic reasons that cause the commodification of the food phenomenon were explained first. Then the concept of food porn was explained with its dimensions.

Food Porn History

Food is closely related to pleasure and emotions in the sense that it is one of the most basic needs (Peters et al., 2002). Eating emerges as a result of instinct and provides satisfaction after it is realized. In the context of the relationship between food, nutrition, and health (Bays, 2009) it has been also seen as the most important determinant of being strong and strong throughout history (Patel, 2012). Considering these features, it is possible to encounter applications of food porn dating back to the first ages.

The practices of humanity to obtain and cook food in the hunter-gatherer period reveal the relationship between motivation, power, and satisfaction with food, dating back to the early times of humanity (Gremillion, 2011; Aarseth, & Olsen, 2008; Sobal, 2005). In the period, the task of hunting animals, collecting food, and cooking in the fire with primitive methods was attributed to the powerful and as a result of meeting this need, authority in the society was achieved (Gvion, 2011; Nath, 2011). Although the changing practices of obtaining and preparing food in the Neolithic period alleviated the relationship of food with feelings of motivation, power, and satisfaction, the meaning of power and authority of food increased again with the kingdoms and empires (Bryce, 2005. In this period, which was dominated by the land regime, military power was determined as the most important power element. Kingdoms and empires had given special importance to the food they consume to gain superiority in military power and to show this superiority (Işın, 2018). Throughout history, dinner parties had been given to manage both domestic and foreign political relations in kingdoms and empires and the meals served in these invitations have been tried to be shown military and political. As the variety of dishes and rare food on the tables increased, it was understood that the political and military power of the kingdom or empire was greater (Agoston, 1999).

Food has not only explained the superiority in military and political relations but throughout history has turned into a means of showing off superiority in different classes of society. In Ancient Greece, the people ate in earthenware decorated with black figures while the noble class in earthenware decorated with rare red figures. In Rome, the people ate basic food at Thermopolium but the noble class ate special 13 different food at one meal (Hollander, 2018; Civitello, 2004). Tablecloths, plates, and cutlery were especially used on the tables of the noble class in empires such as the Byzantine and Ottoman Empires (Anagnostakis, 2013).

Food has been a representative of the superiority established by the different sexes over each other, except for the classes in society. Since the primitive society period, male superiority in obtaining and cooking food has continued this gender superiority. In the banquets in Ancient Greece, androgynous dining rooms, which were only for men, symposiums that were identified with food in Ancient Rome (Swancutt, 2007), in which only men attended, and with the Industrial Revolution, male dominance in industrial kitchens can be listed in this context (Allen & Sachs, 2007).

Food was also used as a means of pleasure by the noble class of society. There are meals accompanied by dance and music in Ancient Egypt, feasts where rare meats and special dishes are consumed in Ancient Greece (Zhou, 2010; Dalby, 2013), and meal rituals that are eaten in rooms called triclinium in Ancient Rome and end with vomiting (Sokolov, 2007).

Apart from the first applications of food porn in history, it is also necessary to mention the socio-cultural and economic developments that paved the way for these applications. As a result of the Industrial Revolution and male dominance in industrial kitchens (Gorman-Murray, 2008), there have been negative developments regarding the content, preparation, presentation, and consumption of food (McDowell, 2002). The emergence of the phenomenon of eating out with fast living conditions and urbanization, the meals prepared and served quickly in industrial kitchens, and the consumption of this food, which has very low health and nutritional values, can be evaluated in this context. Fast food practices, dating back to the 18th century, have destroyed the labor and value in the process of serving food and in the consumption phase. Fast meals have created a new and unhealthy consumption concept by damaging the nutritional, traditional, and local values of the food (De Graaf & Kok, 2010).

Globalization, which emerged interactively with the capitalist order, has been another factor that negatively affects the locality and traditionality value of food (Inglis & Gimlin, 2015; Phillips, 2006). As a result of globalization, fast food restaurants have become increasingly widespread, and even eclectic cuisines have emerged that offer food, cooking, and presentation techniques from different cuisines on one plate (Oddy, 2003). Globalization has also led to the commodification of food as a marketing tool and its use by countries for superiority (Rockower, 2012). Coffee houses that served the American pop culture and spread all over the world, fast restaurants such as meat restaurants, dishes, and tea houses to impose the Far East culture, desserts such as Magnolia, San Sabastian, and Tiramisu, which are marketed all over the world by France, Spain, and Italy can be evaluated in this sense.

The most effective factor that supports the commodification of food as a marketing tool and causes the emergence of food porn has been the mass media (Görgenyi et al., 2017). The mass media, with its guiding functions for society and good effects on bringing the gastronomy and food phenomenon to the fore, have created extremely negative effects on the content, value, and meaning of the food, as they are managed by the dominant capital groups (Frost et al., 2016 Michael & Fuste-Forne, 2022). It has become a marketing tool used by chefs, influencers, food businesses, and even food photographers to attract attention, and customers and gain popularity (Onorati & Giardullo, (2020). Acting solely for self-interest in these marketing practices has revealed food pornography.

Food Porn

Porn is defined as media that are used contrary to moral values and to activate sexual urges. Porn has the effect of creating strong emotions and desires in individuals. In this context, there are examples of applications in many areas. There are different types and contents of media such as death, suicide, disaster, and real estate porn Segal, 1990; Bergen & Bogle, 2000).

On the other hand, it is not possible to evaluate the concept of food porn only as sexually related media. Food porn was first considered as a phenomenon in 1957. In 1977, the French researcher Roland Barthes used the term in the sense of similarities between the preparation of food and sexuality, but in 1979 the term was explained with the relationship between food and health. Michael Jacobson, the founder of the "Center for Science in the Public Interest", which looks at food and consumer welfare, used food porn to describe the decoupling between healthy and unhealthy food (Cockburn 1977).

Since food porn is a fairly new and complex phenomenon, experts have also tried to explain it in various aspects. Kishnendu Ray, who has important academic studies in the field of nutrition, stated that food porn is a part of mass culture and defined the concept with unqualified preparation of food and exaggerated presentation. Author Alan Madison stated that media containing light nudity and female images are interesting in mass culture and mentioned that food porn is also the media created about these images. American chemist Frederick Kaufman emphasized that food is prepared with labor and effort while preparing, but the food is advertised in food porn media as a phenomenon that is devalued instead of this labor. Also, Kaufman criticized this situation but stated that we like to watch this media. Famous American chef Chris Cosentino said that the concept is not new, even in ancient Greece, food was eaten for pleasure (McBride, 2010).

Food porn, taking into account the explanations of experts, can be expressed as a food trend that emerged under the influence of mass culture. Mass culture can be explained by moving away from the traditional and cultural values of societies with increasing urbanization and directing them to a culture that was created later (Donohue, 1973). The orientation of societies to this culture, which was created later, takes place through mass media (Wilson, 1993; MacDonald, 1953). Mass media have functions in society such as attracting attention to a product, service, or phenomenon, arousing curiosity, creating demand, and creating a desire to buy

(Adoni, 1979). Although food is a theme for increasing demand for fast food after the Industrial Revolution, in 20. century, it has been a theme in the context of addressing health problems that arise as a result of eating fast food. Publications have been made in the mass media aimed at healthy nutrition, organic food, traditional dishes, and even gastronomic tourism to consume traditional dishes at the destination (Verbeke, 2005; Kim, 2019; Taşpınar & Temeloğlu, 2018). With the spread of social media in the last 20 years, the supply and demand for food have started to be driven by governments, sovereign capital groups, as well as phenomena in social media, chefs, and food and beverage business owners (Holmberg et al., 2016; Rutsaert et al., 2013). In this context, it can be stated that the government and capital groups serving the capitalist order carry out publications aimed at faster and more food consumption, phenomena, chefs, and food and beverage business owners to attract interest and gain profit (Stevens et al., 2018; Filieri, 2013).

Food porn is a trend that has emerged as a result of all these marketing-oriented applications and where the dimensions of health, nutrition, traditionality, sanctity, privacy, and worthiness of food have all been ignored and become a marketing tool or even a commodity. The need to characterize food porn as a trend, although it has similar applications in history, can be explained by the fact that it affects the whole world both in terms of food supply and demand. after these explanations about food porn, it is necessary to examine this phenomenon or trend with its dimensions.

Food Porn Related to the Content and Traditionality of Food

Food is a cultural product whose recipes were created with the experience given over centuries and passed down from generation to generation (Edelstein, 2010). Recipes have been tried and finalized with various food, cooking methods, and equipment from history to the present. In this context, it can be said that they reached the most ideal versions of food in the geography to which they belong in terms of health, nutrition, and sensory characteristics (Cang, 2018). In addition to transferring food as a cultural product from history to the present, it is necessary to mention the geography in which it is produced and consumed. Food is shaped according to the living practices in the geography and values such as religion, tradition, and custom. Therefore, it contains an identity specific to a country, region, or nation (Ferguson, 2010). All these cultural values related to the content of food have been changing with the mass media, especially social media (Simeone & Scarpato, 2020).

Nowadays, recipes and various media have been shared through various websites, blogs, YouTube channels, and mass media and social media tools such as Facebook, and Instagram (Vaterlaus et al., 2015). Because it is quite challenging to control the mass media, misinformation, and false redirects have been encountered regarding food recipes (Suarez-Lledo, V., & Alvarez-Galvez, 2021). In particular, websites, blogs, and influencers that provide recipes for further monitoring or tracking, without any expertise in the relevant culinary culture or food, have committed violations related to the recipe and cultural value of food (Bryne et al.,2017).

Ethical violations related to food have been made for profit purposes and have reached a different dimension in the last 10 years. With the increasing demand for experience and different searches related to food, food, and beverage business owners and chefs have started to apply recipes that are quite different from the contextual and cultural value of food that has been going on since history (Veber et al., 2021). These videos, which have been shared to take advantage of the interaction power of social media, have created a new digital culture (Lewis, 2018). It can be stated that this immoral digital culture is fed by the rise of fast and multi-consumption in food and globalization (Ritzer & Jurgenson, 2010).

It has been observed that the dishes prepared on social media with much more content and calories than their traditional content are also shared by influencers and food photographers to attract attention and customers. It can also be said that these dishes destroy a cultural identity, and are eclectic, and unhealthy (Mejova et al., 2016). Studies in the literature term these dishes, which are prepared unhealthily with excess content and calories, with food porn. Mejova et al. (2015), examined the Foursquare data in their study, which deals with food porn in the context of fast-food restaurants, and stated that while small and local restaurants offer healthy meals, fast food restaurants lead to obesity with products. Petit et al. (2016), stated that consumers are influenced by the images of unhealthy and fast meals instead of healthy meals on social media, and as a result of this interaction, they acquired new and negative eating habits. Seal and others (2022), emphasized that fast and Junky food is shared more in these media and it caused the spread of food porn by affecting consumers in a cognitive sense. When considering the studies, it is possible to state that these media shared on social media and related to food porn consist of fast, and pleasure-oriented meals.

Within the scope of fast and pleasure-oriented meals, food porn may also be associated with hedonic eating (Harun et al., 2020). Hedonism advocates activities that give pleasure to people and the maintenance of these activities. On the other hand, hedonic eating explains people's eating behaviors to get pleasure and be happy, despite their feelings of satiety (Saper et al., 2002). It is observed that high-calorie food such as bread, meat, and sweets, which are indicated as pleasure-oriented food, are often included in the media (Pohjanheimo et al., 2010). In this context, it can be stated that the media related to food porn aims to arouse interest, curiosity, and impulse despite the feeling of satiety in people. It is stated that frequent consumption of delicious but high-calorie foods due to hedonic eating brings many diseases such as obesity, and diabetes mellitus (Berthoud et al., 2010). These images on social media, which are used more frequently by young people, may affect new generations and cause the extremely unhealthy food porn trend to become widespread. Uçar (2020), found that the consumption tendency towards hedonic food is higher in high school students than in adults. Ngqangashe et al. (2021), stated that media related to hedonic food and food porn was found to be interesting by young people and affected food consumption. Chauhan et al. (2022), emphasized that the watched media related to food porn affects food preference in young people. Considering all these studies, it can be stated that food porn regarding the content of food ignores individual and public health, is prepared and presented for hedonic eating and pleasure, and creates a new digital food culture. It is changing the contents of both traditional and fast meals.

Food Porn Related to the Meaning and Value of Food

Food has been one of the items attributed the most significant importance by societies throughout human history with its physiological and socio-cultural value (Griffith et al., 2010). The culinary culture and the food related to the destination have been formed by various life practices, beliefs, rituals, and traditions throughout history (Mintz & Du Bois, 2002; Sterckx, 2011). In this context, food has gained different meanings beyond just meeting nutrition and energy requirements. Meanings such as sanctity, privacy, and healing the soul along with the body have been attributed to food (Benson, 2001; Kesten, 2011; Arjona, 2017; Aronsson & Gottzen, 2011).

All these sacred, intimate, and precious meanings of food have been changing with the social media tools that are spreading rapidly today (Abril et al., 2022; Bigo, 2019). It is seen that food photos and videos occupy a large space among the media shared on social media tools (Soonsan, & Somkai, 2023). With the increasing importance given to gastronomy, the mass media and social media coverage of food quite often have increased the meaning and value of food, as well as destroyed this meaning and value with incorrect applications. The shares made on social media by individuals, customers, influencers, food and beverage owners, chefs, and food photographers to gain followers, profit, and status have turned the food into a marketing tool. (Magee, 2007; Ehrmann et al., 2009).

The use of food as a marketing tool on social media is related to the visual and aesthetic presentation of food (Narvanen, 2018). The fact that food presentation based on visual aesthetics attracts the attention of consumers and ensures that food is seen as a special experience (Gambetti & Han, 2022) explains the food preparation and presentation videos often shared by food and beverage owners, chefs, and food photographers on social media. These preparation and presentation videos published through various Photoshop applications, filters, and effects (Afshari-Jouybari & Farahnaky, 2011) about to attract more attention, have damaged the labor-based, qualitative importance and reality of food (Northup, 2014). In addition, in the media where the food is prepared and presented by food and beverage owners and chefs with a visual show, various treats such as throwing, slapping, and liking the food lead to the destruction of the sacred meaning of food. In the literature, all these shares that harm the laborious and quality-based value of food and its sacred meaning with pornographic images and videos are called food porn (Çetin, 2022; Bound Alberti, 2021).

It is observed that food has been shared on social media also by individuals, customers, and influencers to gain status and followers (Dietler, 1996; Alruwaily et al., 2020). On platforms such as Instagram, Facebook, Twitter, TikTok, and Foursquare, users have shared the food they eat and the media of the restaurants they visit and aim to gain likes, followers, or status with these shares. In this context, expensive food and luxury restaurants give status to individuals and customers, while returning to influencers as likes and followers (Salleh et al., 2015; Mhlanga & Tichaawa, 2017). These shares, in which the intimate value of food is destroyed have turned food into a commodity shared on social media (Rousseau, 2012). Ibrahim (2015), stated that food porn emerged through food sharing on social media and food has become a digital commodity with these shares. Piluso (2021), emphasized that the reality and value of food have been destroyed by sharing on social media and stated that the food, which has become a commodity.

Food porn and the sharing of food media for status purposes are also associated with conspicuous consumption (Ciobanu, & Florescu, 2019). Conspicuous consumption is explained by the purchase of luxury products and services by wealthy individuals to achieve status (Bagwell & Bernheim, 1996). Individuals buy luxury products and services to differentiate themselves from lower-status individuals (Bourdieu, 1994; Costa & Murphy, 2015). Although wealthy individuals tend to consume luxury foods as a taste according to their socio-cultural background (Oygard, 2020; De Morais Sato et al., 2016), sharing these foods and the restaurants where they are consumed on social media is criticizable. Food becoming a distinction tool in society may be criticized from a humanitarian point of view, taking into account the hunger rates in the world (Adebayo & Ojo, 2012; Chichaibelu et al., 2021). On the other hand, from a sectoral point of view, it may affect local, boutique, and qualified food service businesses negatively.

CONCLUSION

Food porn which is a nutrition trend, has emerged with marketing-oriented applications in food&beverage sector. It has spread, especially with the social media sharing of restaurant owners, food photographers, customers, and influencers. Considering literature studies and food&beverage sector applications on food porn, it is clear that food porn has changed eating habits and food's meaning. Food porn has created unhealthy eating habits by changing traditional food recipes and creating digital food. Also, it has damaged food's sacred meaning and value by turning food into a marketing and status tool.

Although food porn applications cannot be prevented because control of social media is very difficult, they should be reduced by strict applications (Dewald, 2022; Griswold, 2013). These applications should be carried out by the government and nutrition authorities. The sharings of restaurant owners, food photographers, customers, and influencers on social media should be examined continuously. The sharings related to food porn should be punished. İt can be used different key terms for controlling the sharings and accounts which are determined related to food porn sharing can be restricted.

Additionally, food porn should be investigated more thoroughly in the literature. Food porn applications in the food & beverage sector and sharings on the media should be investigated with different research methods. The studies examining food porn in terms of moral, ethical, and legal views should be made.

REFERENCES

Aarseth, H., & Olsen, B. M. (2008). Food and masculinity in dual-career couples. *Journal of Gender Studies*, *17*(4), 277–287. doi:10.1080/09589230802419922

Abril, E. P., Tyson, K., & Morefield, K. (2022). SnapChat this, Instagram that The interplay of motives and privacy affordances in college students' sharing of food porn. *Telematics and Informatics*, *74*, 1–12. doi:10.1016/j.tele.2022.101889

Adebayo, P. F., & Ojo, E. O. (2012). Food security in Nigeria: An overview. *European Journal of Sustainable Development*, *1*(2), 199–199. doi:10.14207/ejsd.2012.v1n2p199

Adoni, H. (1979). The functions of mass media in the political socialization of adolescents. *Communication Research*, *6*(1), 84–106. doi:10.1177/009365027900600106

Afshari-Jouybari, H., & Farahnaky, A. (2011). Evaluation of Photoshop software potential for food colorimetry. *Journal of Food Engineering*, *106*(2), 170–175. doi:10.1016/j.jfoodeng.2011.02.034

Agoston, G. (1999). *European warfare 1453–1815*. Springer.

Allen, P., & Sachs, C. (2007). Women and food chains: The gendered politics of food. *International Journal of Sociology of Agriculture and Food*, *15*(1), 1–23. doi:10.48416/ijsaf.v15i1.424

Alruwaily, A., Mangold, C., Greene, T., Arshonsky, J., Cassidy, O., Pomeranz, J. L., & Bragg, M. (2020). Child social media influencers and unhealthy food product placement. *Pediatrics*, *146*(5), 1–10. doi:10.1542/peds.2019-4057 PMID:33106342

Anagnostakis, I. (2013). Byzantine diet and cuisine. In between ancient and modern gastronomy. In I. Anagnostakis (Ed.), *Flavors and delights tastes and pleasures of ancient and Byzantine cuisine* (pp. 43–64).

Arjona, R. (2017). John Calvin on the Lord's Supper: Food, rest, and healing for shivering souls. *Pastoral Psychology*, *66*(2), 177–190. doi:10.100711089-016-0740-2

Aronsson, K., & Gottzén, L. (2011). Generational positions at a family dinner: Food morality and social order. *Language in Society*, *40*(4), 405–426. doi:10.1017/S0047404511000455

Bagwell, L. S., & Bernheim, B. D. (1996). Veblen effects in a theory of conspicuous consumption. *The American Economic Review*, *86*(3), 349–373. https://www.jstor.org/stable/2118201

Bays, J. C. (2009). *Mindful Eating: A guide to rediscovering a healthy and joyful relationship with food--includes CD*. Shambhala Publications.

Benson, E. P. (2001). Why sacrifice? In E. P. Benson & A. G. Cook (Eds.), *Ritual sacrifice in ancient Peru* (pp. 1–20). University of Texas Press. doi:10.7560/708938-002

Bergen, R. K., & Bogle, K. A. (2000). Exploring the connection between pornography and sexual violence. *Violence and Victims*, *15*(3), 227–234. doi:10.1891/0886-6708.15.3.227 PMID:11200099

Berthoud, H. R., Zheng, H., & Shin, A. C. (2012). Food reward in the obese and after weight loss induced by calorie restriction and bariatric surgery. *Annals of the New York Academy of Sciences*, *1264*(1), 36–48. doi:10.1111/j.1749-6632.2012.06573.x PMID:22616827

Bigo, D. (2019). Shared secrecy in a digital age and a transnational world. *Intelligence and National Security*, *34*(3), 379–394. doi:10.1080/02684527.2019.1553703

Bound Alberti, F. (2021). Fat shaming, feminism and Facebook: What 'women who eat on tubes' reveal about social media and the boundaries of women's bodies. *European Journal of Cultural Studies*, *24*(6), 1304–1318. doi:10.1177/13675494211055499

Bourdieu, P. (1994). *Distinction: A social critique of the judgment of taste* (R. Niece, Trans.). Harvard University Press.

Bryce, T. (2005). *The kingdom of the Hittites*. OUP Oxford. doi:10.1093/acprof:oso/9780199281329.001.0001

Byrne, E., Kearney, J., & MacEvilly, C. (2017). The role of influencer marketing and social influencers in public health. *Proceedings of the Nutrition Society, 76*(OCE3), E103, 21-23. 10.1017/S0029665117001768

Cang, V. (2018). Japan's Washoku as intangible heritage: The role of national food traditions in UNESCO's cultural heritage scheme. *International Journal of Cultural Property*, *25*(4), 491–513. doi:10.1017/S0940739118000267

Çetin, A. (2022). The extraordinary show of food: A sociological evaluation on some Instagram content. *Turkish Studies-Social Sciences*, *17*(4), 517–532.

Chauhan, J., Hedaoo, R., & Patil, M. (2022). Gastroporn on Social Media and Its Association with Food Choices and Body Mass Index Among Youth. *Advanced Studies in Biology*, *14*(1), 137–151. doi:10.12988/asb.2022.91588

Chichaibelu, B. B., Bekchanov, M., Von Braun, J., & Torero, M. (2021). The global cost of reaching a world without hunger: Investment costs and policy action opportunities. *Food Policy*, *104*(12), 102151. doi:10.1016/j.foodpol.2021.102151

Ciobanu, E., & Florescu, C. M. (2019). Food Porn in, and (). *East-West Cultural Passage, 19*(2), 96-120. . https://doi.org/ doi:10.2478/ewcp-2019-0014

Civitello, L. (2004). *Cuisine and Culture: A History of food and people*. John Wiley&Sons.

Costa, C., & Murphy, M. (2015). *Bourdieu, habitus and social research: The art of application*. Springer.

Dalby, A. (2013). *Siren feasts: a history of food and gastronomy in Greece*. Routledge.

De Graaf, C., & Kok, F. J. (2010). Slow food, fast food and the control of food intake. *Nature Reviews. Endocrinology*, *6*(5), 290–293. https://www.nature.com/doifinder/10.1038/nrendo.2010.41. doi:10.1038/nrendo.2010.41 PMID:20351697

De Morais Sato, P., Gittelsohn, J., Unsain, R. F., Roble, O. J., & Scagliusi, F. B. (2016). The use of Pierre Bourdieu's distinction concepts in scientific articles studying food and eating: A Narrative review. *Appetite*, *96*, 174–186. doi:10.1016/j.appet.2015.09.010 PMID:26368578

Dejmanee, T. (2016). "Food porn" as postfeminist play: Digital femininity and the female body on food blogs. *Television & New Media*, *17*(5), 429–448. doi:10.1177/1527476415615944

Dewald, S. (2022). Detectives and technological frames: Integrating technology and social media into everyday work. *Policing and Society, 33*(1), 111,118. doi:10.1080/10439463.2022.2064858

Dietler, M. (1996). Feasts and commensal politics in the political economy: food, power, and status in prehistoric Europe. In P. Wiessner & W. Schiefenhövel (Eds.), *Food and the status quest: An interdisciplinary perspective* (pp. 87–125). Berghahn Books.

Donohue, G. A., Tichenor, P. J., & Olien, C. N. (1973). Mass media functions, knowledge and social control. *The Journalism Quarterly*, *50*(4), 652–659. doi:10.1177/107769907305000404

Edelstein, S. (2010). *Food, cuisine, and cultural competency for culinary, hospitality, and nutrition professionals*. Jones & Bartlett Publishers.

Ehrmann, T., Meiseberg, B., & Ritz, C. (2009). Superstar effects in deluxe gastronomy-An empirical analysis of value creation in German quality restaurants. *Kyklos*, *62*(4), 526–541. doi:10.1111/j.1467-6435.2009.00449.x

Ferguson, P. P. (2010). Culinary nationalism. *Gastronomica*, *10*(1), 102–109. doi:10.1525/gfc.2010.10.1.102 PMID:21539054

Filieri, R. (2013). Consumer co-creation and new product development: A case study in the food industry. *Marketing Intelligence & Planning*, *31*(1), 40–53. doi:10.1108/02634501311292911

Frost, W., Laing, J., Best, G., Williams, K., Strickland, P., & Lade, C. (2016). *Gastronomy, tourism and the media* (Vol. 74). Channel View Publications.

Gambetti, A., & Han, Q. (2022). Camera eats first: Exploring food aesthetics portrayed on social media using deep learning. *International Journal of Contemporary Hospitality Management*, *34*(9), 3300–3331. doi:10.1108/IJCHM-09-2021-1206

Garber, A. K., & Lustig, H. R. (2011). Is fast food addictive? *Current Drug Abuse Reviews*, *4*(3), 146–162. doi:10.2174/1874473711104030146 PMID:21999689

Görgenyi, E. H., Mate, B., Vafaei, S. A., & Fekete, M. F. (2017). The role of social media in gastronomy industry. *Apstract, Applied Studies in Agribusiness and Commerce*, *11*(3-4), 95–103. doi:10.19041/APSTRACT/2017/3-4/14

Gorman-Murray, A. (2008). Masculinity and the home: A critical review and conceptual framework. *The Australian Geographer*, *39*(3), 367–379. doi:10.1080/00049180802270556

Gremillion, K. J. (2011). *Ancestral appetites: Food in prehistory*. Cambridge University Press. doi:10.1017/CBO9780511976353

Griffith, C. J., Livesey, K. M., & Clayton, D. (2010). The assessment of food safety culture. *British Food Journal*, *112*(4), 439–456. http://www.emeraldinsight.com/0007-070X.htm. doi:10.1108/00070701011034448

Griswold, A. (2013). Digital detectives and virtual volunteers: Integrating emergent online communities into disaster response operations. [Digital detectives and virtual volunteers: Integrating emergent o...: Ingenta Connect]. *Journal of Business Continuity & Emergency Planning*, *7*(1), 13–25. PMID:24113633

Gvion, L. (2011). Cooking, food, and masculinity: Palestinian men in Israeli society. *Men and Masculinities*, *14*(4), 408–429. doi:10.1177/1097184X11411269

Harun, A. F., Ruslan, N., Adnan, W. A. W., Suliman, S. I., Ismail, J., & Baharin, H. (2020). Digitizing food experience: Food taste perception on digital image and true form using hashtags. *Bulletin of Electrical Engineering and Informatics*, *9*(5), 2141–2153. doi:10.11591/eei.v9i5.2252

Hasan, R., & Yu, Z. (2020). Capitalism leads to unhealthy food consumption. *Strategic Change*, *29*(6), 633–643. doi:10.1002/jsc.2379

Hollander, D. B. (2018). *Farmers and agriculture in the Roman economy*. Routledge Publish. doi:10.4324/9781315103884

Holmberg, C., Chaplin, J. E., Hillman, T., & Berg, C. (2016). Adolescents' presentation of food in social media: An explorative study. *Appetite*, *99*, 121–129. doi:10.1016/j.appet.2016.01.009 PMID:26792765

Ibrahim, Y. (2015). Food porn and the invitation to gaze: Ephemeral consumption and the digital spectacle. [IJEP]. *International Journal of E-Politics*, *6*(3), 1–12. doi:10.4018/IJEP.2015070101

Inglis, D., & Gimlin, D. (2015). The globalization of food. Nova York.

Işın, P. M. (2018). *Bountiful empire: A History of ottoman cuisine*. Reaktion Books.

Kesten, D. (2011). *The healing secrets of food: A practical guide for nourishing body, mind, and soul*. New World Library.

Kim, Y. H. (2019). Organic shoppers' involvement in organic foods: Self and identity. *British Food Journal*, *121*(1), 139–156. doi:10.1108/BFJ-03-2018-0202

Lerma-Cabrera, J. M., Carvajal, F., & Lopez-Legarrea, P. (2015). Food addiction as a new piece of the obesity framework. *Nutrition Journal*, *15*(1), 1–5. doi:10.118612937-016-0124-6 PMID:26758504

Lewis, T. (2018). Digital food: From paddock to platform. *Communication Research and Practice*, *4*(3), 212–228. doi:10.1080/22041451.2018.1476795

Little, D. C., Newton, R. W., & Beveridge, M. C. M. (2016). Aquaculture: A rapidly growing and significant source of sustainable food? Status, transitions, and potential. *The Proceedings of the Nutrition Society*, *75*(3), 274–286. doi:10.1017/S0029665116000665 PMID:27476856

Lowe, M. R., & Butryn, M. L. (2007). Hedonic hunger: A new dimension of appetite? *Physiology & Behavior*, *91*(4), 432–439. doi:10.1016/j.physbeh.2007.04.006 PMID:17531274

Macdonald, D. (1953). A theory of mass culture. *Diogenes*, *1*(3), 1–17. doi:10.1177/039219215300100301

Magee, R. M. (2007). Food puritanism and food pornography: The gourmet semiotics of Martha and Nigella. *Americana: The Journal of American Popular Culture, 6*(2).

McDonnell, E. M. (2016). *Food porn: The conspicuous consumption of food in the age of digital reproduction. Food, media and contemporary culture*. Palgrave Macmillan. doi:10.1057/9781137463234_14

McDowell, L. (2002). Masculine discourses and dissonances: Strutting 'lads', protest masculinity, and domestic respectability. *Environment and Planning. D, Society & Space*, *20*(1), 97–119. doi:10.1068/d43j

Mejova, Y., Abbar, S., & Haddadi, H. (2016). Fetishizing food in digital age: Fodporn around the world. In *Proceedings of the International AAAI Conference on Web and Social Media* (Vol. 10, No. 1, pp. 250-258). AAAI. 10.1609/icwsm.v10i1.14710

Mejova, Y., Haddadi, H., Noulas, A., & Weber, I. (2015). Foodporn: Obesity patterns in culinary interactions. In *Proceedings of the 5th International Conference on Digital Health 2015* (pp. 51-58). ACM. 10.1145/2750511.2750524

Mhlanga, O., & Tichaawa, T. M. (2017). Influence of social media on customer experiences in restaurants: A South African study. *Tourism: An International Interdisciplinary Journal, 65*(1), 45-60. https://hrcak.srce.hr/178622

Michael, N., & Fuste-Forne, F. (2022). Marketing of luxurious gastronomic experiences on social media: The visual storytelling of luxury hotels. *International Journal of Tourism Research*, *24*(6), 827–838. doi:10.1002/jtr.2548

Mintz, S. W., & Du Bois, C. M. (2002). The anthropology of food and eating. *Annual Review of Anthropology*, *31*(1), 99–119. https://www.jstor.org/stable/4132873. doi:10.1146/annurev.anthro.32.032702.131011

Narvanen, E., Mesiranta, N., Sutinen, U. M., & Mattila, M. (2018). Creativity, aesthetics and ethics of food waste in social media campaigns. *Journal of Cleaner Production*, *195*, 102–110. doi:10.1016/j.jclepro.2018.05.202

Nath, J. (2011). Gendered fare? A qualitative investigation of alternative food and masculinities. *Journal of Sociology (Melbourne, Vic.)*, *47*(3), 261–278. doi:10.1177/1440783310386828

Naughton, P., McCarthy, M., & McCarthy, S. (2015). Acting to self-regulate unhealthy eating habits. An investigation into the effects of habit, hedonic hunger, and self-regulation on sugar consumption from confectionery foods. *Food Quality and Preference*, *46*, 173–183. doi:10.1016/j.foodqual.2015.08.001

Ngqangashe, Y., Maldoy, K., De Backer, C. J., & Vandebosch, H. (2022). Exploring adolescents' motives for food media consumption using the theory of uses and gratifications. *Communications*, *47*(1), 73–92. doi:10.1515/commun-2019-0164

Northup, T. (2014). Truth, lies, and packaging: How food marketing creates a false sense of health. *Food Studies: An Interdisciplinary Journal, 3*(1), 9-18. ww.food-studies.com.

Novak, M. (1982). *The Spirit of Capitalism*. Madison Books.

Oddy, D. J. (2003). *From plain fare to fusion food: British diet from the 1890s to the 1990s*. Boydell Press.

Onorati, M. G., & Giardullo, P. (2020). Social media as taste re-mediators: Emerging patterns of food taste on TripAdvisor. *Food, Culture, & Society*, *23*(3), 347–365. doi:10.1080/15528014.2020.1715074

Oygard, L. (2000). Studying food tastes among young adults using Bourdieu's theory. *Journal of Consumer Studies & Home Economics, 24*(3), 160-169.

Patel, R. C. (2012). Food sovereignty: Power, gender, and the right to food. *PLoS Medicine*, *9*(6), 1–5. doi:10.1371/journal.pmed.1001223 PMID:22745604

Peters, J. C., Wyatt, H. R., Donahoo, W. T., & Hill, J. O. (2002). From instinct to intellect: The challenge of maintaining healthy weight in the modern world. *Obesity Reviews*, *3*(2), 69–74. doi:10.1046/j.1467-789X.2002.00059.x PMID:12120422

Petit, O., Cheok, A. D., & Oullier, O. (2016). Can food porn make us slim? How brains of consumers react to food in digital environments. *Integrative Food, Nutrition and Metabolism*, *3*(1), 251–255. doi:10.15761/IFNM.1000138

Phillips, L. (2006). Food and globalization. *Annual Review of Anthropology*, *35*(1), 37–57. doi:10.1146/annurev.anthro.35.081705.123214

Piluso, F. (2021). # Foodporn: Fetishized sharing of food and its images. *Digital Age in Semiotics & Communication*, *4*(1), 31–41. doi:10.33919/dasc.20-21.3-4.3

Pohjanheimo, T., Paasovaara, R., Luomala, H., & Sandell, M. (2010). Food choice motives and bread liking of consumers embracing hedonistic and traditional values. *Appetite*, *54*(1), 170–180. doi:10.1016/j.appet.2009.10.004 PMID:19835923

Preteceille, E., & Terrail, J. P. (1986). Capitalism, consumption, and needs. *Capital and Class*, *10*(2), 226–229. doi:10.1177/030981688602900115

Ritzer, G., & Jurgenson, N. (2010). Production, consumption, presumption: The nature of capitalism in the age of the digital 'prosumer'. *Journal of Consumer Culture*, *10*(1), 13–36. doi:10.1177/1469540509354673

Rockower, P. S. (2012). Recipes for gastrodiplomacy. *Place Branding and Public Diplomacy*, *8*(3), 235–246. doi:10.1057/pb.2012.17

Rose, N., & Lourival, I. (2019). Hegemony, counter-hegemony, and food systems literacy: Transforming the global industrial food system. *Australian Journal of Environmental Education*, *35*(2), 110–122. doi:10.1017/aee.2019.9

Rousseau, S. (2012). *Food and social media: You are what you tweet.* Rowman Altamira. doi:10.5040/9781350042193

Rutsaert, P., Regan, A., Pieniak, Z., McConnon, Á., Moss, A., Wall, P., & Verbeke, W. (2013). The use of social media in food risk and benefit communication. *Trends in Food Science & Technology*, *30*(1), 84–91. doi:10.1016/j.tifs.2012.10.006

Salleh, S., Hashima, N. H., & Murphy, J. (2015). Instagram marketing: a content analysis of top Malaysian restaurant brands. *E-Review of Tourism Research, 6*, 1-5.

Saper, C. B., Chou, T. C., & Elmquist, J. K. (2002). The need to feed: Homeostatic and hedonic control of eating. *Neuron*, *36*(2), 199–211. doi:10.1016/S0896-6273(02)00969-8 PMID:12383777

Schneider-Kamp, A. (2021). Inclusion of the excluded: Consumers' quest for hedonism in food consumption. *International Journal of Consumer Studies*, *45*(3), 320–334. doi:10.1111/ijcs.12612

Seal, A., Gavaravarapu, S. M., & Konapur, A. (2022). Can foodporn prime healthy eating? Thinking beyond digital gazing and satiety. *European Journal of Clinical Nutrition, 1493-1496*(11), 1493–1496. doi:10.103841430-022-01139-w PMID:35411027

Segal, L. (1990). Pornography and violence: What the 'experts' really say. *Feminist Review, 36*(1), 29–41. doi:10.1057/fr.1990.43

Simeone, M., & Scarpato, D. (2020). Sustainable consumption: How does social media affect food choices? *Journal of Cleaner Production, 277*, 1–6. doi:10.1016/j.jclepro.2020.124036

Sobal, J. (2005). Men, meat, and marriage: Models of masculinity. *Food & Foodways, 13*(1-2), 135–158. doi:10.1080/07409710590915409

Sokolov, R. (2007). Piscinae and the myth of Roman decadence at table. In S. R. Friedland (Ed.), *Food & morality: Proceedings of the Oxford Symposium on Food And Cookery* (pp. 263-68). Prospect Books.

Soonsan, N., & Somkai, U. (2023). Dimensions of gastronomic experience affecting on sharing experience: Place attachment as a mediator and length of stay as a moderator. *Journal of Hospitality and Tourism Insights, 6*(1), 344–361. doi:10.1108/JHTI-07-2021-0171

Sterckx, R. (2011). *Food, sacrifice, and sagehood in early China*. Cambridge University Press. doi:10.1017/CBO9780511736247

Stevens, T. M., Aarts, N., Termeer, C. J. A. M., & Dewulf, A. (2016). Social media as a new playing field for the governance of agro-food sustainability. *Current Opinion in Environmental Sustainability, 18*, 99–106. doi:10.1016/j.cosust.2015.11.010

Suarez-Lledo, V., & Alvarez-Galvez, J. (2021). Prevalence of health misinformation on social media: Systematic review. *Journal of Medical Internet Research, 23*(1), e17187. Advance online publication. doi:10.2196/17187 PMID:33470931

Swancutt, D. M. (2007). Still before sexuality:" Greek" androgyny, the roman imperial politics of masculinity and the roman invention of the Tribas. In D. M. Swancutt (Ed.), *Mapping gender in ancient religious discourses* (pp. 11–61). Brill., doi:10.1163/ej.9789004154476.i-582.12

Taşpınar, O., & Temeloğlu, E. (2018). Influence of Tv cooking shows on the behavioral intention of participating in gastronomic tourism. *Journal of Tourism & Gastronomy Studies, 6*(2), 24–40. doi:10.21325/jotags.2018.200

Uçar, A. (2020). *Determination of hedonic hunger and mindful eating in high school students and its relationship with nutritional status* [Unpublished Masters' Thesis, Baskent University].

Vaterlaus, J. M., Patten, E. V., Roche, C., & Young, J. A. (2015). Gettinghealthy: The perceived influence of social media on young adult health behaviors. *Computers in Human Behavior*, *45*, 151–157. doi:10.1016/j.chb.2014.12.013

Verbeke, W. (2005). Agriculture and the food industry in the information age. *European Review of Agriculture Economics*, *32*(3), 347–368. doi:10.1093/eurrag/jbi017

Weber, P., Ludwig, T., Brodesser, S., & Grönewald, L. (2021, May). "It's a kind of art!": Understanding food influencers as influential content creators. In *Proceedings of the 2021 CHI Conference on Human Factors in Computing Systems* (pp. 1-14). 10.1145/3411764.3445607

Wells, J. C. (2012). Obesity as malnutrition: The role of capitalism in the obesity global epidemic. *American Journal of Human Biology*, *24*(3), 261–276. doi:10.1002/ajhb.22253 PMID:22383142

Wilson, S. L. R. (1993). *Mass media/mass culture: an introduction*. McGraw-Hill, Inc.

Zhou, Y. (2010). Festivals, feasts, and gender relations in ancient China and Greece. Cambridge University Press. doi:10.1017/CBO9780511762468

ADDITIONAL READING

Çetin, A. (2022). The extraordinary show of food: A sociological evaluation on some Instagram content. *Turkish Studies-Social Sciences*, *17*(4), 517–532.

Civitello, L. (2004). *Cuisine and Culture: A History of food and people*. John Wiley&Sons.

Costa, C., & Murphy, M. (2015). *Bourdieu, habitus and social research: The art of application*. Springer.

Görgenyi, E. H., Mate, B., Vafaei, S. A., & Fekete, M. F. (2017). The role of social media in gastronomy industry. *Apstract, Applied Studies in Agribusiness and Commerce*, *11*(3-4), 95–103. doi:10.19041/APSTRACT/2017/3-4/14

Petit, O., Cheok, A. D., & Oullier, O. (2016). Can food porn make us slim? How brains of consumers react to food in digital environments. *Integrative Food, Nutrition and Metabolism*, *3*(1), 251–255. doi:10.15761/IFNM.1000138

Saper, C. B., Chou, T. C., & Elmquist, J. K. (2002). The need to feed: Homeostatic and hedonic control of eating. *Neuron*, *36*(2), 199–211. doi:10.1016/S0896-6273(02)00969-8 PMID:12383777

Schneider-Kamp, A. (2021). Inclusion of the excluded: Consumers' quest for hedonism in food consumption. *International Journal of Consumer Studies*, *45*(3), 320–334. doi:10.1111/ijcs.12612

Seal, A., Gavaravarapu, S. M., & Konapur, A. (2022). Can foodporn prime healthy eating? Thinking beyond digital gazing and satiety. *European Journal of Clinical Nutrition*, *1493-1496*(11), 1493–1496. doi:10.103841430-022-01139-w PMID:35411027

Segal, L. (1990). Pornography and violence: What the 'experts' really say. *Feminist Review*, *36*(1), 29–41. doi:10.1057/fr.1990.43

Simeone, M., & Scarpato, D. (2020). Sustainable consumption: How does social media affect food choices? *Journal of Cleaner Production*, *277*, 1–6. doi:10.1016/j.jclepro.2020.124036

Vaterlaus, J. M., Patten, E. V., Roche, C., & Young, J. A. (2015). Gettinghealthy: The perceived influence of social media on young adult health behaviors. *Computers in Human Behavior*, *45*, 151–157. doi:10.1016/j.chb.2014.12.013

KEY TERMS AND DEFINITIONS

Conspicuous Consumption: Consumption of luxury products and services aimed at obtaining and maintaining status.

Food Porn: The fact that food becomes a digital commodity and a marketing and status tool by moving away from its content, conventionality, meaning, and value.

Hedonic Eating: Eating for pleasure and satisfaction purposes other than to meet the nutritional needs of the body.

Hedonism: An approach that argues that people should perform activities that they enjoy.

Mass Communication Tool: Highly effective mean of communication usually managed by sovereign capital groups, through which all kinds of information can be transmitted to the masses such as radio, television, internet, and social media.

Social Media: Social media is a mass communication tool where people can interact with each other on digital platforms and share content.

Thermopolium: Food and beverage establishments where people ate fast food during the Ancient Roman Period, which are like artisan restaurants today

Traditional Food: Food that belongs to the culinary culture of some destination with its content, preparation or processing type, cooking method and presentation.

Chapter 9
Drugs and Psychedelics in Tourism

Feyza Alyu Altınok
https://orcid.org/0000-0001-9146-3841
Anadolu University, Turkey

ABSTRACT

Psychedelics are a class of substances that induce profound alterations in perception, cognition, and consciousness. They include substances such as psilocybin (found in certain mushrooms), LSD, and DMT. Drug tourism, on the other hand, refers to the practice of traveling to destinations specifically to engage in the consumption or exploration of drugs, both legal and illegal. It encompasses a range of experiences, from seeking out prescription medications at lower costs to participating in psychedelic retreats or engaging in recreational drug use. Drug tourism, encompassing legal and illegal drugs, is a significant part of the tourism industry, generating substantial revenue and attracting tourists to various destinations. This chapter delves into the fascinating and complex world of drugs and psychedelics in tourism, examining the resurgence of interest in psychedelic substances, the implications for tourist destinations, the therapeutic potential of psychedelics, and the associated health risks and ethical considerations.

INTRODUCTION

The term “psychedelic” originates from the Greek words “psyche” (referring to the mind or soul) and “delos” (meaning to show), and it was first coined by psychiatrist Humphry Osmond in 1956. Psychedelic substances have a long history of human use, dating back hundreds, if not thousands, of years, for various purposes including

DOI: 10.4018/978-1-6684-9255-0.ch009

recreation, spirituality, and healing. These substances have the remarkable ability to induce profound shifts in consciousness and cognition, leading to transient changes in emotions, perceptual processing, self-perception, and feelings of connectedness. Although scientific interest in psychedelics waxed and waned throughout the 20th century, experiencing a surge of research during the 1950s and 60s, regulatory restrictions limited further investigation until the early 1990s, when psychedelic science began to regain momentum (Aday et al., 2020; Johnson & Griffiths, 2017).

The incorporation of drugs and psychedelics in the realm of tourism has given rise to a fascinating and complex phenomenon known as psychedelic tourism. This chapter delves into the multifaceted dynamics, implications, and trends surrounding the intersection of drugs and psychedelics with the tourism industry.

The resurgence of psychedelic substances has played a significant role in driving the exploration of psychedelic tourism. Recent scientific research has shed light on the potential therapeutic benefits of psychedelics in treating mental health conditions, challenging traditional perspectives, and igniting renewed interest in these substances. Studies by Carhart-Harris et al. (2016, 2018) and Griffiths et al. (2016) have demonstrated their efficacy in addressing conditions such as treatment-resistant depression or anxiety. These findings have sparked a shift in public perception and a reevaluation of the role of psychedelics in promoting healing, personal growth, and spiritual exploration.

Understanding drug tourism, encompassing both legal and illegal drugs in tourist destinations is crucial for comprehending the motivations, challenges, and consequences associated with this phenomenon. It involves examining the factors driving tourists to seek out drug-related experiences, the complexities faced by destinations dealing with drug legalization, or the implications of the illicit drug trade. This exploration provides insights into the diverse motivations of tourists, the economic and social impacts on destinations, and the legal and ethical considerations surrounding drug use in tourism.

The impact of psychedelic tourism on indigenous communities and their cultural practices is another significant aspect to be explored. Indigenous communities have a long-standing relationship with psychedelic substances, often within the context of traditional and religious practices. The incorporation of psychedelic tourism raises questions about cultural appropriation, the preservation of indigenous knowledge, and the potential effects on indigenous communities as they navigate their involvement in the tourism industry.

This chapter explores the resurgence of psychedelic substances, the understanding of drug tourism, the impact on indigenous communities, the therapeutic potential and shifting public perception, responsible development, health risks, and safety considerations, ethical implications, the role of education and decision-making, as well as current trends and future directions in psychedelic tourism research.

The Resurgence of Psychedelic Substances and the Emergence of Psychedelic Tourism

In recent years, there has been a notable resurgence of interest in psychedelic substances, accompanied by the emergence of a new trend known as psychedelic tourism. This phenomenon has captured the attention of researchers, policymakers, and the public alike, as it represents a significant cultural shift and raises intriguing questions about the intersection of drugs and tourism.

Historical Context: From Stigma to Resurgence

In the past, psychedelics were predominantly associated with the illicit recreational drug culture, overlooking their extensive use in psychiatry prior to 1967. The spread of this culture on a global scale resulted in the classification of psychedelic drugs into the Schedule I class of the United Nations Convention on Drugs in 1967, effectively leading to their prohibition. Despite the diminishing prominence of this negative stigma, remnants of it persisted. However, a shift in the status quo has been initiated by the compelling scientific evidence that confirms the therapeutic potential of these drugs in treating various disorders. As a result, the stigma is gradually diminishing, paving the way for a more informed and open perspective toward these substances. During the 1960s, the global proliferation of psychedelic drugs beyond their traditional cultural contexts resulted in the implementation of stringent drug control laws in many Western nations. Even today, psychedelics remain classified as Schedule I substances, leading to negative stigma, vilification, and their criminalization. This ongoing controversy continues to impede scientific research and hinder full acceptance within the medical community (Lowe et al., 2022).

For a long time, the spiritual, religious, and medicinal value of these drugs could not be explored through scientific investigation. However, a recent resurgence of psychedelic research, known as the second wave, is now focusing on their potential as neuropharmaceuticals for the treatment of alcohol and tobacco addiction, general mood and anxiety disorders, and depression related to cancer. A wealth of promising evidence-based data has emerged, affirming the medicinal value of psychedelics, and validating the long-standing anecdotal evidence. These natural therapeutic alternatives, such as psychedelic drugs, offer a safe and effective alternative to conventional medications used for mood and anxiety disorders. Particularly in Western societies, there is a growing interest in psychedelic drugs as therapeutic agents for mood and anxiety disorders. This interest is fueled by increasing global rates of these disorders, shifting social attitudes, government regulations, and expanding investment opportunities that facilitate rigorous scientific studies and the generation of evidence-based data. Alternative psychotherapeutic interventions are also gaining attention due to their low physiological toxicity, relatively low abuse potential, safe psychological effects, and the absence of persistent adverse physiological or psychological effects during and after use. In contrast, conventional psychiatric drugs and antidepressants are becoming less favored due to their undesirable side effects (Lowe et al., 2022; Rucker et al., 2018).

The Second Wave of Psychedelic Research

The resurgence of psychedelics has been strongly influenced by scientific research, which has played a pivotal role in shedding light on their therapeutic potential. In particular, landmark studies conducted by Carhart-Harris et al. (2016, 2018) and Griffiths et al. (2016) have demonstrated the remarkable effects of substances like psilocybin, a naturally occurring psychedelic compound found in certain species of mushrooms commonly known as "magic mushrooms." These studies have provided compelling evidence for the ability of psilocybin to alleviate conditions such as depression and anxiety, challenging conventional perceptions and paving the way for further exploration of psychedelics in therapeutic contexts. Through these groundbreaking investigations, the scientific community has set the stage for a resurgence in the understanding and application of psychedelics for mental health and well-being.

In indigenous cultures, substances like psilocybin and other classic psychedelics have long been used as sacred sacraments. In the mid-twentieth century, these substances garnered attention within psychiatry as tools for understanding brain function and exploring experimental therapeutics. However, their popularity declined in the late 1960s and early 1970s when they became associated with the counterculture movement. In recent years, there has been a resurgence of scientific interest in classic psychedelics. Promising studies have reignited exploration into their potential therapeutic benefits, validating earlier research findings. These substances have shown promise in the treatment of mental health conditions such as alcohol and tobacco addiction, general mood and anxiety disorders, and cancer-related depression and post-traumatic stress disorder (PTSD) (Fonseka & Woo, 2023; Lowe et al., 2022).

The resurgence of psychedelics in modern psychiatry can be attributed to a pivotal period in 2017-2018. During this time, the Food and Drug Administration (FDA) granted breakthrough therapy status to psilocybin for the treatment of depression that does not respond to conventional therapies, as well as to 3,4-methylenedioxymethamphetamine (MDMA) for PTSD. This recognition by the FDA marked a significant milestone in acknowledging the therapeutic potential of these substances and paved the way for further research and clinical trials exploring their efficacy and safety in addressing these challenging mental health conditions.

(Fonseka & Woo, 2023; Lowe et al., 2022). This recognition by the FDA not only validated the therapeutic potential of psychedelics but also sparked a wave of interest and investment in further research and clinical trials.

Shifting Social Attitudes and Societal Acceptance

Alongside scientific research, there has been a notable shift in societal attitudes toward psychedelic substances. The stigma and legal restrictions that have long surrounded these substances are gradually giving way to a more open-minded approach. This is exemplified by the decriminalization efforts in cities like Denver and Oakland, as well as the recognition by the FDA of 3,4-methylenedioxymethamphetamine-assisted therapy (MDMA-assisted therapy) as a "breakthrough therapy" (Danforth, 2016; Inouye and Wolfgang, 2022). These shifts in public opinion and policy have paved the way for increased acceptance and exploration of psychedelic experiences.

The movement to decriminalize psilocybin in the United States gained momentum in 2019 when Denver, Colorado became the first city to decriminalize the substance in May of that year. This was followed by Oakland and Santa Cruz, California, which decriminalized psilocybin in June 2019 and January 2020, respectively. The trend continued with Washington, D.C. decriminalizing psilocybin in November 2020, and Somerville, Massachusetts in January 2021. The cities of Cambridge and Northampton in Massachusetts also joined the list in February and March 2021, respectively. In October 2021, Seattle, Washington became the largest U.S. city to decriminalize psilocybin, followed by Detroit, Michigan in November 2021. Supporters of the decriminalization movement have pointed to emerging research that suggests potential medical uses for psilocybin. Oregon took a significant step by passing a 2020 ballot measure that not only decriminalized psilocybin but also legalized its supervised use, making it the first state to do so. Colorado followed suit with a similar measure in 2022. However, it is important to note that the use, sale, and possession of psilocybin remain illegal under federal law in the United States (Collins & McCorkle 2023; Marks, 2022).

The designation FDA provided regarding MDMA for PTSD has not only contributed to the growing interest in psychedelic-assisted therapies but has also catalyzed a shift in social attitudes and societal acceptance towards these novel treatment approaches. The FDA's recognition of MDMA-AT as a breakthrough therapy has fostered greater openness and support from the medical community, funding agencies, and the general public, paving the way for expanded research and clinical exploration in this field (Inouye & Wolfgang, 2022).

Understanding Drug Tourism: Legal and Illegal Drugs in Tourist Destinations

Drug tourism is a complex and multifaceted phenomenon that involves individuals traveling to specific destinations with the intention of using and experiencing drugs that may be legally or illegally available in those locations. It may be defined as '…the phenomenon by which persons become attracted to a particular location because of the accessibility of licit or illicit drugs and related services' (Valdez & Sifaneck, 1997).

Over the past decades, tourists have embarked on journeys within their own countries and across borders, driven by the desire to acquire and consume psychoactive drugs for medicinal or recreational purposes. This phenomenon can be traced back to the 1950s, when a notable route known as the 'hippie trail' emerged, attracting alternative tourists who embarked on extensive overland voyages between Europe and South Asia. Segev et al. (2005) conducted a study involving predominantly Israeli, British, Swedish, German, and Australian travelers, with an average age of 25.3 years, and found that 54.3% of them engaged in drug use during their travels in Southeast Asia. The study revealed a higher prevalence of drug use among male travelers. Cannabis was the most consumed drug, although substances such as lysergic acid diethylamide (LSD) were also reported. It was noted that drug use rates were significantly higher in India compared to other Asian countries.

In tourist destinations around the world, the availability and legality of drugs can vary significantly. It's important to note that drug laws are subject to change, and what may be legal or illegal today could be different in the future.

Alcohol is legal in most countries and is widely available in tourist destinations. However, the drinking age is typically ranging from 18 to 21 years old. Alcohol consumption is legal in most countries. In Afghanistan, Saudi Arabia, Libya, Mauritania, Brunei, and Yemen it is prohibited and illegal while in Kuwait, Sudan, and Iran, alcohol is illegal for Muslims, but there are some exceptions for non-Muslim minorities. In Maldives, alcohol is prohibited for locals but allowed in tourist resorts. In Tunisia, sales are only banned on Fridays and during Ramadan. And for Qatar, it is legal for tourists at hotels and bars as well as expatriates with permits. It should be noted that the regulations can change over time.

Regarding cannabis, legality varies widely. Some countries and states have decriminalized or legalized its recreational or medicinal use, while others strictly prohibit it. Examples of countries where recreational use is legal include Uruguay (the first country in the world to fully legalize cannabis in 2013), Canada, the Netherlands (specifically in coffee shops), and certain states in the United States (such as Colorado and California). In countries where cannabis is legal, there are often regulations in place regarding age restrictions, possession limits, and places where it can be consumed. Canada legalized recreational cannabis nationwide in October 2018. Adults can purchase and consume cannabis from licensed retailers, and individuals can also grow a limited number of cannabis plants at home for personal use. In the Netherlands, the sale and consumption of cannabis in small quantities (up to 5 grams) are tolerated in licensed coffee shops. However, it is important to note that cannabis cultivation and supply to these coffee shops are technically illegal but tolerated under certain conditions. In addition to Uruguay, several other countries have decriminalized or legalized cannabis for medicinal use, including Argentina, Australia, Brazil, Chile, Colombia, Germany, Israel, Mexico, Portugal, South Africa, and Thailand. Each country has its specific regulations and limitations regarding cultivation, possession, and consumption. Cannabis laws in the United States vary by state. As of September 2021, recreational cannabis is legal for adults aged 21 and older in several states, including Alaska, California, Colorado, Illinois, Massachusetts, Oregon, Washington, and others. Medicinal cannabis is legal in many more states with varying regulations (Bonny-Noach & Sagiv-Alayoff, 2020).

Prescription drugs are legal when used with a valid prescription from a licensed healthcare professional. However, it's important to carry the necessary documentation and ensure that the medications comply with the regulations of the destination country. Prescription drug availability and regulations can vary significantly from one country to another. This variation can attract individuals who are seeking specific medications or looking to acquire drugs without a valid prescription. In some cases, individuals may engage in prescription drug tourism to obtain medications that are not approved or available in their home country. "Pharma tourism," essentially the reimportation of prescription drugs from offshore countries at considerably lower prices sets an example.

Regarding the matter of illegal drugs, their use, possession, and trafficking are strictly prohibited in most countries. Examples of illegal drugs include cocaine, heroin, methamphetamine, LSD, MDMA (ecstasy), and many others. Penalties for drug-related offenses can be severe and may include fines, imprisonment, or even the death penalty in some countries. While the Netherlands has a reputation for its tolerant approach to drug policy, LSD is categorized as a Schedule I substance and is technically illegal. However, the enforcement of LSD-related offenses is generally low, and the focus is more on targeting large-scale production and distribution. In the Czech Republic, the possession and use of small quantities of LSD are considered a misdemeanor rather than a criminal offense. The country has adopted a more liberal stance on drug policy, emphasizing harm reduction and treatment instead of punishment.

Legalization and Decriminalization of Drugs in Tourist Destinations

Certain destinations have implemented policies that have led to the legalization or decriminalization of specific drugs, creating a unique market for drug tourism. For instance, Amsterdam in the Netherlands and certain states in the United States have legalized or decriminalized marijuana, attracting tourists interested in exploring the cultural and recreational aspects of drug use. The famous "coffee shops" in Amsterdam, where marijuana is available for sale and consumption, have become popular attractions for both locals and tourists. Furthermore, it is important to recognize that drug tourism destinations exhibit considerable variation worldwide, with distinct countries and regions offering diverse experiences for individuals interested in drug-related activities. Table 1 represents a comprehensive overview of popular destinations renowned for drug tourism.

Illegal Drug Markets as Tourist Destinations

In addition to destinations with legalized or decriminalized drugs, some tourists seek out locations known for their illegal drug markets. These destinations are often associated with the production, distribution, and consumption of illicit substances. For example, cities like Medellin in Colombia, once known for the operations of drug lord Pablo Escobar, or the Golden Triangle region in Southeast Asia, recognized for its opium production and trafficking, attract individuals seeking adventurous experiences or access to prohibited substances, despite the city's efforts to transform its image (Naef, 2018). Similarly, the Golden Triangle region in Southeast Asia, encompassing parts of Thailand, Laos, and Myanmar, has long been associated with opium production and trafficking (Chin, 2017). Tourists may be enticed by the allure of exploring the underground world of drug cultivation and trade, albeit with inherent risks and legal consequences.

Goa, a popular beach destination in India, has gained notoriety for its illicit drug trade, particularly involving substances like cannabis and MDMA (ecstasy). Despite efforts by law enforcement to curb drug-related activities, some tourists are drawn to the clandestine drug markets that have emerged within the region. Similarly, the Rave Party scene in Goa has been associated with drug use and attracts individuals seeking a vibrant nightlife intertwined with substance experimentation (Gupta, 2014).

Another example is the Jamaican city of Negril, known for its marijuana culture and the availability of locally grown cannabis (Boxill, 2000). While marijuana remains illegal in Jamaica, its consumption and trade have become deeply embedded in the local culture, attracting tourists who are interested in exploring the reggae scene and partaking in the island's cannabis culture.

Additionally, certain regions in Southeast Asia, such as the Vang Vieng region in Laos and the Gili Islands in Indonesia, have gained a reputation for drug-friendly environments, particularly catering to backpackers and partygoers. These destinations have been associated with the availability of drugs like marijuana, MDMA (ecstasy), and magic mushrooms, often facilitated through underground channels.

Prescription Drugs and Medical Tourism: Seeking Affordable Medications Abroad

The rising costs of prescription medications in many countries have led to a growing phenomenon known as medical tourism, where individuals travel to foreign destinations to seek affordable medications and medical treatments. This practice has gained popularity among those facing high healthcare costs, lack of insurance coverage, or limited access to certain medications in their home countries. One of the primary motivations for medical tourists is the significant cost savings associated with purchasing prescription drugs abroad. Many countries, particularly in Asia and Latin America, offer medications at a fraction of the cost compared to industrialized nations. This price disparity is often due to variations in healthcare systems, manufacturing processes, and government regulations that affect drug pricing (Gaines & Lee, 2019; Horowitz et al., 2007). Popular destinations for prescription drug shopping include India, Thailand, Mexico, Turkey, and several Eastern European countries (Horowitz et al., 2007).

Among Organization for Economic Co-operation and Development (OECD) countries, Germany stands out as the leading importer of medical care, with imports totaling US$1.5 billion. Following closely behind are the United States and the Netherlands, with imports amounting to US$660 million and US$618 million, respectively. On the other hand, the United States takes the lead as the largest exporter of health care, reporting exports worth approximately US$2.3 billion. The Czech Republic and Turkey follow with exports amounting to US$418 million and US$409 million, respectively (Runnels & Carrera, 2012).

India is one such destination known for its well-established medical tourism industry. The country offers a wide range of medications at competitive prices. With a vast network of hospitals and clinics catering to international patients, India provides access to high-quality healthcare services and a diverse array of prescription drugs.

Thailand has also gained recognition as a hub for medical tourism, attracting individuals seeking affordable medications. The country boasts modern medical facilities and a reputation for excellence in healthcare. Its reputable pharmacies offer a range of prescription drugs, often at significantly lower prices compared to many Western countries.

Mexico is another popular destination for medical tourists seeking cost-effective medications. The country's proximity to the United States and its well-developed healthcare infrastructure makes it an accessible option for individuals looking to purchase prescription drugs at more affordable prices.

Turkey has become a sought-after destination for medical tourists seeking affordable medications. The country has made significant strides in developing a robust medical tourism industry, offering modern healthcare facilities and reputable pharmacies. One of the primary reasons medical tourists choose Turkey for prescription drug shopping is the substantial cost savings compared to many other countries. Medications are often available at significantly lower prices, making them more accessible to individuals facing high healthcare expenses or limited insurance coverage. Turkey's well-established healthcare infrastructure, skilled medical professionals, and stringent quality control standards contribute to the trust and confidence that medical tourists have in the country's pharmacies. Additionally, Turkey offers a wide range of prescription drugs, both brand-name and generic, ensuring that patients have various options to meet their specific healthcare needs. With its combination of affordable medications and high-quality healthcare services, Turkey has emerged as a preferred destination for medical tourists in search of affordable and reliable prescription drugs.

Several Eastern European countries, including Hungary, Poland, and the Czech Republic, have become popular destinations for medical tourists seeking affordable medications. These countries have developed well-established medical tourism industries, offering a wide range of prescription drugs at competitive prices. Medical tourists are drawn to these destinations due to the significant cost savings compared to their home countries. Moreover, Eastern European countries boast reputable pharmacies and healthcare facilities that cater to international patients, ensuring high standards of quality and safety.

The Impact of Psychedelic Tourism on Indigenous Communities and Cultural Practices

Indigenous communities often have a deep spiritual and cultural connection to psychedelic plants and substances, considering them sacred and integral to their traditional rituals and ceremonies. Psychedelic substances, such as ayahuasca, peyote, and psilocybin mushrooms, have long been used in indigenous rituals and spiritual ceremonies, integral to their cultural heritage. However, the growing interest in psychedelic tourism has led to increased interaction between outsiders and indigenous communities, posing potential risks to their cultural heritage. With appropriate attention from a multidisciplinary scientific approach, psychedelic tourism holds the potential to serve as a safeguard for indigenous communities and their resources against exploitative industries such as mining, logging, and large-scale agriculture. By incorporating responsible practices and oversight, this emerging form of tourism can contribute to the protection and preservation of host communities and their rich cultural heritage. However, if left unchecked and lacking proper supervision, the consequences arising from this delicate form of tourism can be disastrous not only for the indigenous communities and their culture but also for the well-being of the visitors themselves. Therefore, it is crucial to approach psychedelic tourism with careful consideration and implement measures that ensure its sustainable and respectful development (Vidriales & Ovies, 2018).

The influx of tourists engaging in psychedelic experiences can disrupt the social fabric and dynamics of indigenous communities. These communities may face challenges in managing the increased demand for psychedelic ceremonies or accommodating the cultural expectations and behaviors of visitors. The sacred spaces and natural environments where these ceremonies take place may also be negatively impacted by the growing tourism activities, leading to ecological degradation or loss of cultural significance.

The historical presence of exploitative and appropriative psychedelic tourism, which has regrettably neglected to prioritize respect for indigenous communities and their religious practices, is of utmost relevance. Over the years, the phenomenon of psychedelic tourism has often been marred by instances where individuals from outside these communities have engaged in practices that fail to acknowledge and honor the rich cultural heritage and spiritual significance associated with indigenous traditions (Feinberg, 2018; Lutkajtis, 2020; Smith & Appelbaum, 2022).

It is crucial to engage in responsible and respectful psychedelic tourism practices that prioritize the well-being and cultural integrity of indigenous communities. This involves establishing meaningful partnerships and collaborations with indigenous leaders and organizations, ensuring their active participation and decision-making in the development of tourism initiatives. Additionally, education and awareness programs for tourists can play a vital role in promoting cultural sensitivity and respect. Regulatory frameworks can also play a crucial role in mitigating the negative impacts of psychedelic tourism on indigenous communities. Governments and local authorities can implement policies that protect indigenous rights, regulate tourism activities, and ensure that economic benefits from psychedelic tourism are shared equitably with the communities involved.

The Potential Therapeutic Benefits of Psychedelics and the Shift in Public Perception

In recent years, there has been a growing interest in exploring the potential therapeutic benefits of psychedelics, marking a significant shift in public perception, and understanding of these substances. This shift has sparked renewed scientific research and a reevaluation of their therapeutic value in various mental health conditions.

Psychedelics, such as psilocybin (found in magic mushrooms), LSD, and MDMA, have been traditionally associated with counterculture movements and recreational use. However, emerging studies have demonstrated their potential to facilitate profound psychological and emotional experiences, which may be beneficial for individuals struggling with mental health disorders.

One area of exploration is the use of psychedelics in treating depression and anxiety. Clinical trials have shown promising results, with some patients experiencing significant improvements in their symptoms after psychedelic-assisted therapy sessions. These substances have the potential to enhance introspection, promote emotional breakthroughs, and facilitate a sense of connectedness and meaning, which can contribute to long-lasting therapeutic effects.

Depression, a prevalent mental health condition affecting a substantial number of individuals worldwide, poses a significant burden with high mortality rates and disability rates. There is growing interest in exploring the potential of psychopharmacological therapies involving psychedelics, particularly psilocybin, for the treatment of depression and other related conditions. These therapies have shown promising outcomes, including the rapid and substantial improvement of depressive symptoms and a prolonged sense of well-being that can extend for several months following treatment. Furthermore, the potential of psilocybin therapy extends beyond the treatment of depression, as researchers have begun to explore its efficacy in addressing other disorders (Johnson & Griffiths, 2017). Specifically, studies have investigated the effects of psilocybin on anxiety, an area of interest highlighted in several articles included in this review. Additionally, there have been investigations into the use of psilocybin for treating various addictions, including alcoholism and smoking. Furthermore, speculation has arisen regarding its potential effectiveness in addressing obsessive-compulsive disorder and PTSD, among other conditions. These emerging research avenues shed light on the broader applications of psilocybin therapy beyond depression (Dawood Hristova & Pérez-Jover, 2023).

The study conducted by Carhart-Harris et al. (2016) aimed to assess the feasibility, safety, and efficacy of psilocybin in patients with unipolar treatment-resistant depression through an open-label feasibility trial involving twelve patients (six men, six women) with moderate-to-severe unipolar treatment-resistant major depression. Psychological support was provided throughout the sessions. The primary outcome measure for feasibility was the self-reported intensity of psilocybin's effects. Adverse reactions were monitored during and after the dosing sessions. Depressive symptoms were evaluated using standardized assessments from one week to three months post-treatment, with the 16-item Quick Inventory of Depressive Symptoms (QIDS) serving as the primary efficacy measure. The acute psychedelic effects of psilocybin typically manifested within 30 to 60 minutes after dosing, peaked at 2 to 3 hours, and diminished to negligible levels after at least 6 hours. Psilocybin was well-tolerated by all patients, with no serious or unexpected adverse events. Transient anxiety during drug onset, transient confusion or thought disorder, mild and transient nausea, and transient headache were the observed adverse reactions. Compared to baseline, depressive symptoms showed marked reductions at 1 week and 3 months after high-dose treatment. Significant and sustained improvements in anxiety and anhedonia were also observed. In the six-month follow-up study, the findings suggest that psilocybin treatment demonstrated good tolerability, significant effect sizes, and rapid symptom improvements. These positive effects were observed after only two psilocybin treatment sessions and remained significant six months after treatment in a cohort of individuals with treatment-resistant depression (Carhart-Harris et al, 2016,2018). This indicates that psilocybin holds promise as a potential paradigm for addressing depression that is unresponsive to conventional treatments.

In the study conducted by Griffiths et al. (2016), involving 51 cancer patients with symptoms of depression and anxiety, the effects of psilocybin were investigated. The study employed a randomized, double-blind, cross-over design, comparing a very low dose (1 or 3 mg/70 kg) to a high dose (22 or 30 mg/70 kg) of psilocybin. The participants received the doses in a counterbalanced sequence, with a five-week interval between sessions and a six-month follow-up. The study minimized expectancy effects through instructions given to participants and staff. Throughout the study, participant moods, attitudes, and behaviors were rated by participants, staff, and community observers. The results demonstrated that the administration of a high dose of psilocybin led to significant reductions in both clinician- and self-rated measures of depressed mood and anxiety. Additionally, participants reported increases in quality of life, life meaning, and optimism, as well as decreases in death anxiety. These positive changes were sustained at the six-month follow-up, with approximately 80% of participants continuing to experience clinically significant decreases in depressed mood and anxiety. Participants attributed improvements in various aspects of their lives, including attitudes about life/self, mood, relationships, and spirituality, to the high-dose psilocybin experience. More than 80% of participants endorsed moderately or significantly increased well-being and life satisfaction. The ratings provided by community observers aligned with these reported changes. It was found that the mystical-type psilocybin experience on the session day mediated the therapeutic outcomes associated with the dose of psilocybin administered.

Furthermore, psychedelics have shown promise in addressing addiction and substance abuse disorders. Studies have indicated that substances like psilocybin and ayahuasca may help individuals gain insights into the root causes of their addictive behaviors, increase their motivation to change, and facilitate personal growth and spiritual experiences. For instance, MDMA-AT for severe PTSD was found to induce subclinical improvements in alcohol use (Nicholas et al., 2022). The effects of psychedelics, coupled with appropriate therapeutic support, have the potential to transform the treatment landscape for addiction (Van Der Meer ve ark., 2023; Zafar et al., 2023).

Numerous studies in the literature have extensively explored the positive effects of psilocybin in various conditions, including depression, trauma-related disorders in AIDS patients, complex trauma, and neurodegenerative disorders. These studies have shed light on the therapeutic potential of psilocybin in these contexts. Similarly, research on MDMA has demonstrated its positive effects in addressing conditions such as PTSD, depression, substance use disorders, sleep disorders, eating disorders, and end-of-life anxiety associated with life-threatening illnesses. These findings contribute to the growing body of evidence supporting the efficacy of psychedelics in diverse therapeutic applications (Fonseka & Woo, 2023; Perkins et al., 2023). Ongoing investigations are being conducted to explore the therapeutic potential of both classic and non-classic psychedelic drugs in the treatment of trauma, depressive disorders, and other psychopathologies. Classic psychedelic drugs, such as psilocybin, LSD, and ayahuasca, along with non-classic drugs like MDMA and ketamine, have garnered significant interest for their potential in addressing these conditions.

Existing studies have begun to monitor sustained changes. The long-term efficacy and safety profile of the psychedelics should be determined, and there is a need for more rigorous large-scale randomized clinical trials, larger and more diverse samples, employing longer-term longitudinal designs, using stronger control conditions, and standardizing dosages (Aday et al., 2020).

As public perception continues to evolve, regulatory frameworks are also adapting to accommodate this changing landscape. Some jurisdictions have started to explore alternative approaches to drug policy, allowing for scientific research, clinical trials, and the controlled use of psychedelics in therapeutic settings (Marks, 2022). These progressive steps reflect a recognition of the potential benefits of psychedelics and a departure from the historically punitive and prohibitionist approaches.

Responsible Development of Psychedelic Tourism: Regulations and Safeguards

One of the primary concerns in psychedelic tourism is the risk of unregulated and unscrupulous practices. Without appropriate regulations in place, there is a potential for exploitation, misinformation, and unsafe experiences. Therefore, it is essential to develop comprehensive regulatory frameworks that govern the use of psychedelics in tourism settings (Celidwen et al., 2023; Haden et al., 2016). These regulations should address multiple aspects of psychedelic tourism, including training and qualifications for facilitators and guides, adherence to ethical standards, safety protocols, and informed consent processes. Clear guidelines must be established to ensure that individuals seeking psychedelic experiences are provided with accurate information, understand the potential risks involved, and are adequately supported throughout their journey.

The European Drug Report 2023: Trends and Developments offers the most recent analysis conducted by the EMCDDA (European Monitoring Centre for Drugs and Drug Addiction) on the drug landscape in Europe. The report provides a detailed examination of illicit drug use, associated harms, and drug supply. It presents an extensive collection of national data covering these topics, along with highlighting key harm-reduction interventions. For more detailed information, the report can be consulted as a valuable resource.

Education and awareness campaigns are crucial components of responsible psychedelic tourism development. These initiatives should target both tourists and local communities, providing accurate information about the risks, benefits, and responsible use of psychedelics.

Lastly, monitoring and evaluation mechanisms should be implemented to assess the impact of psychedelic tourism and ensure compliance with established regulations. This includes ongoing research, data collection, and analysis to understand the social, cultural, economic, and environmental effects of psychedelic tourism on local communities and ecosystems.

Psychedelic Retreats and the Integration of Wellness Experiences

Psychedelic retreat settings, which include collective psychedelic use, have emerged as a rapidly growing phenomenon that offers a rich opportunity to investigate the psychosocial effects of psychedelics (Kettner et al., 2021; Orozco & Harris, 2023). In countries where certain psychedelic substances remain legal, there is a global demand for structured and safe psychedelic experiences. This demand has given rise to an industry that provides immersive psychedelic retreat programs, often spanning multiple days and featuring guided group sessions known as "ceremonies." By integrating various wellness modalities, such as meditation, yoga, breathwork, and therapeutic practices, psychedelic retreats aim to enhance the overall well-being of participants.

These sessions are typically led by experienced individuals who have undergone formal training or have extensive experience in caring for others during psychedelic journeys. In the context of ayahuasca ceremonies, these individuals are referred to as "shamans," "ayahuasceros," or "curanderos," while Western settings may use terms like "experience facilitators," "sitters," or "guides." These practitioners undertake preparatory activities before the sessions to help participants cultivate a mindset conducive to the psychedelic experience. They also prioritize creating a safe and comfortable environment, employing various elements such as decorative or ritual objects, special lighting, incense, or bonfires. Furthermore, they offer emotional support, provide educational information, and help participants navigate their inner landscapes who are under the influence of psychedelics. During the sessions, facilitators or shamans may engage with participants directly or indirectly, often utilizing music as a medium of interaction. Rattling, drumming, singing, and whistling, sometimes involving the active participation of the group, are commonly observed elements in Meso and South American settings where mescaline (such as peyote or San Pedro) and ayahuasca are used.

One of the key aspects of psychedelic retreats is the emphasis on preparation and integration. Prior to the psychedelic experience, participants engage in preparation activities that may include intention setting, education about the substance, and emotional and psychological readiness. This preparation helps individuals approach the psychedelic journey with intention and clarity, increasing the likelihood of a meaningful and transformative experience.

Integration is equally important, as it allows participants to reflect on and integrate their psychedelic experiences into their daily lives. Integration practices may involve individual or group therapy sessions, support networks, follow-up consultations, and the incorporation of mindfulness practices. By providing a structured framework for integration, psychedelic retreats aim to help participants integrate their insights and experiences into their personal growth and well-being.

Furthermore, the integration of wellness experiences in psychedelic retreats complements the psychedelic journey by supporting participants' physical, mental, and emotional well-being. Practices like yoga and meditation can help individuals cultivate presence, develop self-awareness, and cultivate a deeper connection with themselves. These wellness practices provide a foundation for personal growth and can enhance the therapeutic potential of the psychedelic experience.

Psychedelic retreats also offer an opportunity for individuals to connect with like-minded individuals and foster a sense of community. Group activities, sharing circles, and communal meals create a supportive and inclusive environment, where participants can feel understood and supported in their psychedelic journeys. The sense of community can contribute to a collective healing experience and facilitate the sharing of insights, challenges, and reflections.

Shamanic Practices and Indigenous Religious Contexts: A Longstanding Tradition of Psychedelic Tourism

Shamanic practices and indigenous religious contexts have a deep-rooted history intertwined with the use of psychedelics. Across various indigenous cultures, the ceremonial and therapeutic use of psychedelic substances has played a pivotal role in spiritual journeys and healing rituals for centuries. This longstanding tradition has garnered interest from individuals worldwide, giving rise to a form of psychedelic tourism that aims to immerse oneself in these ancient practices and gain a profound understanding of indigenous spiritual traditions.

Indigenous cultures, such as those found in the Amazon rainforest, the Native American tribes of North America, and the indigenous communities of Africa and Australia, have long embraced psychedelics as sacred tools for communion with the divine and exploring the depths of consciousness. For instance, the consumption of ayahuasca, a potent plant medicine containing the psychoactive compound DMT, is integral to shamanic ceremonies in Amazonian regions of Peru, Brazil, and Ecuador (Petersen et al., 2022). Starting from the 1970s, Ayahuasca has garnered significant attention from non-Indigenous individuals, both as a substance and as a shamanic practice. This heightened interest has manifested in various forms, including tourism and medicinal experimentation. Consequently, Ayahuasca has created numerous economic opportunities for both locals and foreigners, like the economic booms generated by plants like rubber, timber, or vanilla in the past. However, what sets Ayahuasca apart from these other booms is its direct involvement of Indigenous shamans, which attracts a larger influx of tourists to the region compared to the exportation of the plants themselves. Similar to ayahuasca, peyote, a hallucinogenic cactus, has been used for centuries by Native American tribes, particularly in the southwestern United States and Mexico, as a means of connecting with spiritual realms and seeking healing and guidance (Dasgupta, 2019).

Psychedelic tourists are drawn to these shamanic practices and indigenous religious contexts for a variety of reasons. Some seek profound spiritual experiences, aiming to deepen their understanding of the self, the universe, and their place within it. Others seek healing and transformation, hoping to address emotional traumas, overcome addictions, or gain clarity on life's purpose. The ancient wisdom, cultural immersion, and connection to nature that these practices offer are also appealing factors for many.

Apart from the Amazon and North America, other regions have become popular destinations for psychedelic tourism in indigenous contexts. For example, the San Pedro cactus ceremonies in the Andean region, including Peru and Bolivia, attract visitors seeking spiritual insights and personal growth (Carod-Artal & Vazquez-Cabrera, 2006). In Africa, the consumption of iboga, a powerful psychedelic derived from the root bark of the Tabernanthe iboga plant, is central to the Bwiti tradition in Gabon, drawing individuals interested in its unique spiritual and therapeutic properties (Mash, 2023). Similarly, in Australia, indigenous communities utilize various plant medicines, such as the pituri plant, for ceremonial purposes and connection with ancestral spirits (Ratsch et al., 2010).

Health Risks and Safety Considerations in Psychedelic Tourism

Engaging in drug use while abroad introduces additional risks linked to unfamiliar sources of supplies, limited knowledge of local health services, and potential isolation from the support of friends and family. The surge in drug-related activities during international leisure travel has been associated with detrimental consequences such as unprotected sexual encounters, drug-impaired driving accidents, and increased susceptibility to mental health issues (Flaherty et al., 2017).

One of the primary health risks in psychedelic tourism is the improper use or dosage of psychedelic substances. Psychedelics can induce intense altered states of consciousness, which may be overwhelming or psychologically challenging for some individuals. Without proper guidance and supervision, there is an increased risk of adverse psychological reactions, including panic, anxiety, and disorientation. In extreme cases, these experiences can lead to hallucinogen-persisting perception disorder (HPPD) or trigger underlying mental health conditions.

To address these risks, reputable psychedelic retreats and facilitators prioritize participant safety by employing experienced and trained guides who are knowledgeable about the substances, their effects, and harm-reduction strategies. These guides provide support, create a safe environment, and help participants navigate their psychedelic experiences. Additionally, comprehensive screening processes, including medical evaluations and psychological assessments, can help identify individuals who may be at higher risk of adverse reactions and determine their suitability for participation.

Another health concern in psychedelic tourism is the potential for physical harm or accidents. Nonetheless, the use of psychedelics has been found to be irrelevant to suicidality (Zeifman et al., 2021), psychedelics can alter perception, impair coordination, and distort judgment, increasing the risk of accidents and injuries. Participants need to be in safe environments with adequate safeguards to prevent harm. This includes creating spaces that are free from hazards, ensuring proper lighting, and implementing guidelines for safe movement and interaction. Moreover, responsible psychedelic tourism providers prioritize the presence of trained staff who can respond swiftly and effectively in case of emergencies.

In addition to the immediate health risks, there are also broader public health considerations associated with psychedelic tourism. The increased substance abuse that can occur within this context may strain local healthcare systems, particularly in regions with limited resources. The consumption and sharing of drugs can contribute to the spread of infectious diseases, such as HIV and hepatitis, underscoring the importance of implementing harm reduction measures. These measures may include providing access to clean needles and syringes, promoting safe sexual practices, and offering educational resources on substance use and harm reduction.

The Complexities of Ethical Implications in Psychedelic Tourism

Engaging in psychedelic tourism raises a range of ethical considerations that must be carefully examined and addressed (Pilecki et al., 2021). As the popularity of psychedelic substances and their therapeutic potential grow, it is crucial to navigate these complexities with a strong ethical framework that prioritizes the well-being of individuals, communities, and the environment.

One of the primary ethical concerns in psychedelic tourism revolves around informed consent and participant autonomy. Participants should be provided with comprehensive information about the potential risks and benefits associated with psychedelic experiences, ensuring they have a clear understanding of what they are consenting to. Informed consent should also encompass the context in which the psychedelic journey takes place, including the cultural, spiritual, and historical significance of the substances used. Respecting participant autonomy means acknowledging their right to make informed decisions regarding their own mental and physical well-being.

Cultural appropriation is another significant ethical consideration in psychedelic tourism. Psychedelic substances have long been integral to indigenous and traditional practices, often deeply rooted in specific cultural, spiritual, and healing contexts. When these substances are commodified and appropriated in a commercialized tourism setting, it raises concerns about respect for cultural heritage, intellectual property rights, and the potential for exploitation. Engaging with indigenous communities in a mutually respectful and collaborative manner, seeking their guidance and permission, and sharing the benefits of psychedelic tourism with these communities are essential steps toward ethical engagement.

Furthermore, sustainable, and responsible practices in psychedelic tourism are crucial to minimize environmental impacts and preserve natural resources. Sustainable sourcing of psychedelic substances, minimizing waste and pollution, and supporting local economies and conservation efforts are essential considerations. Responsible use of psychedelics also includes promoting harm reduction measures, ensuring participant safety, and providing adequate support and integration services after the psychedelic experience.

Addressing these ethical complexities requires ongoing dialogue, collaboration, and the involvement of various stakeholders, including researchers, facilitators, participants, indigenous communities, and regulatory bodies. Establishing ethical guidelines, codes of conduct, and industry standards can help shape the responsible and ethical development of psychedelic tourism, fostering a sustainable and respectful approach that supports individual growth, cultural preservation, and environmental stewardship.

Current Trends and Future Directions in Psychedelic Tourism Research

Since the 1960s, young countercultural travelers and spiritual seekers have been drawn to Latin American indigenous communities, seeking drug-induced expansion of consciousness and shamanic-assisted spiritual healing. However, in recent years, the Millennial generation has assumed a crucial role in revitalizing indigenous spiritual practices on a global scale. Through the production and dissemination of information via digital media platforms, Millennials have become instrumental in raising awareness about the therapeutic potential of psychedelic plant medicines. As a result, there has been a noticeable increase in the popularity of shamanistic health services among spiritual Millennial travelers. Additionally, psychedelic-plant ceremonies held in indigenous communities have gained significant value as sought-after commodities within the global marketplace of spiritual experiences (Ovies, 2021).

As the popularity of psychedelic substances for therapeutic and recreational purposes grows, research efforts have begun to focus on understanding the effects, benefits, risks, and socio-cultural aspects of psychedelic tourism.

One current trend in psychedelic tourism research is the investigation of the therapeutic potential of psychedelics in a controlled and supportive setting. Studies are examining the efficacy of psychedelic-assisted therapy in treating mental health conditions such as depression, anxiety, PTSD, and addiction. Researchers are exploring the mechanisms of action underlying the therapeutic effects and investigating optimal protocols for integrating psychedelics into therapeutic settings.

Another important area of research in psychedelic tourism is the examination of the socio-cultural impacts on local communities and indigenous practices. Researchers are exploring the ways in which psychedelic tourism affects local cultures, traditions, and sacred rituals. They are investigating the potential benefits and challenges of integrating psychedelic experiences into indigenous religious contexts and exploring the ethical implications of such interactions.

The field of psychedelic tourism research is also embracing interdisciplinary approaches and collaborations. Scholars from various disciplines, including psychology, anthropology, sociology, neuroscience, and public health, are joining forces to gain a comprehensive understanding of the complex phenomena associated with psychedelic tourism. This interdisciplinary approach allows for a more holistic examination of the psychological, sociocultural, and physiological aspects of psychedelic experiences.

Furthermore, there is a growing interest in the study of harm reduction strategies and safety considerations in psychedelic tourism. Researchers are exploring ways to minimize potential risks, ensure participant safety, and promote responsible use of psychedelics. This includes investigating the development and implementation of harm reduction guidelines, training programs for facilitators, and the integration of safety measures in retreat settings.

Looking to the future, psychedelic tourism research will likely continue to expand and diversify. There is a need for longitudinal studies that assess the long-term effects and outcomes of psychedelic experiences. Research exploring the impact of different cultural contexts on psychedelic tourism, as well as the influence of diverse participant characteristics, is also essential.

Additionally, as psychedelic substances gain more recognition and acceptance, there is a need for research on policy implications and regulatory frameworks. This includes studying the legal and ethical aspects of psychedelic tourism, assessing the potential for misuse and abuse, and exploring models for responsible regulation that prioritize safety and accessibility while respecting cultural sensitivities.

In conclusion, current trends in psychedelic tourism research encompass therapeutic applications, socio-cultural impacts, interdisciplinary collaborations, and harm reduction strategies. The field is continuously evolving, and future research will likely focus on long-term outcomes, cultural contexts, policy implications, and ethical considerations. By expanding our knowledge in these areas, we can better understand and navigate the complexities of psychedelic tourism, ultimately contributing to its safe and responsible development.

REFERENCES

Aday, J. S., Mitzkovitz, C. M., Bloesch, E. K., Davoli, C. C., & Davis, A. K. (2020). Long-term effects of psychedelic drugs: A systematic review. *Neuroscience and Biobehavioral Reviews*, *113*, 179–189. doi:10.1016/j.neubiorev.2020.03.017 PMID:32194129

Bonny-Noach, H., & Sagiv-Alayoff, M. (2020). Cannabis tourist destinations: Risk for vulnerable travellers with pre-existing mental disorders. *Journal of Travel Medicine*, *27*(4), taz098. doi:10.1093/jtm/taz098 PMID:31821507

Boxill, I. (2000). Overcoming social problems in the Jamaican tourism industry. *Tourism in the Caribbean*, 17-40.

Carhart-Harris, R. L., Bolstridge, M., Day, C. M., Rucker, J., Watts, R., Erritzoe, D. E., Kaelen, M., Giribaldi, B., Bloomfield, M., Pilling, S., Rickard, J. A., Forbes, B., Feilding, A., Taylor, D., Curran, H. V., & Nutt, D. J. (2018). Psilocybin with psychological support for treatment-resistant depression: Six-month follow-up. *Psychopharmacology*, *235*(2), 399–408. doi:10.100700213-017-4771-x PMID:29119217

Carhart-Harris, R. L., Bolstridge, M., Rucker, J., Day, C. M., Erritzoe, D., Kaelen, M., Bloomfield, M., Rickard, J. A., Forbes, B., Feilding, A., Taylor, D., Pilling, S., Curran, V. H., & Nutt, D. J. (2016). Psilocybin with psychological support for treatment-resistant depression: An open-label feasibility study. *The Lancet. Psychiatry*, *3*(7), 619–627. doi:10.1016/S2215-0366(16)30065-7 PMID:27210031

Carod-Artal, F. J., & Vazquez-Cabrera, C. B. (2006). Mescaline and the San Pedro cactus ritual: Archaeological and ethnographic evidence in northern Peru. *Revista de Neurología*, *42*(8), 489–498. doi:10.33588/rn.4208.2006040 PMID:16625512

Celidwen, Y., Redvers, N., Githaiga, C., Calambás, J., Añaños, K., Chindoy, M. E., Vitale, R., Rojas, J. N., Mondragón, D., Rosalío, Y. V., & Sacbajá, A. (2023). Ethical principles of traditional Indigenous medicine to guide western psychedelic research and practice. *Lancet Regional Health. Americas*, *18*, 100410. doi:10.1016/j.lana.2022.100410 PMID:36844020

Chin, K. L. (2017). *The golden triangle: Inside southeast Asia's drug trade*. Cornell University Press.

Collins, C., & McCorkle, W. (2023). Decriminalization of Drugs and the Support for Free Markets in the United States. *Journal of Advanced Research in Social Sciences*, *6*(1), 34–53. doi:10.33422/jarss.v6i1.870

Danforth, A. L., Struble, C. M., Yazar-Klosinski, B., & Grob, C. S. (2016). MDMA-assisted therapy: A new treatment model for social anxiety in autistic adults. *Progress in Neuro-Psychopharmacology & Biological Psychiatry*, *64*, 237–249. doi:10.1016/j.pnpbp.2015.03.011 PMID:25818246

Dasgupta, A. (2019). Abuse of magic mushroom, peyote cactus, LSD, khat, and volatiles. In *Critical issues in alcohol and drugs of abuse testing* (pp. 477–494). Academic Press. doi:10.1016/B978-0-12-815607-0.00033-2

Dawood Hristova, J. J., & Pérez-Jover, V. (2023). Psychotherapy with Psilocybin for Depression: Systematic Review. *Behavioral Sciences (Basel, Switzerland)*, *13*(4), 297. doi:10.3390/bs13040297 PMID:37102811

Feinberg, B. (2018). Undiscovering the pueblo magico: Lessons from Huautla for the psychedelic renaissance. *Plant Medicines, Healing and Psychedelic Science: Cultural Perspectives*, 37-54.

Flaherty, G. T., Maxemous, K. K., Nossier, R. E., & Bui, Y. G. (2017). The highs and lows of drug tourism: A travel medicine perspective. *Journal of Travel Medicine*, *24*(6). Advance online publication. doi:10.1093/jtm/tax068 PMID:29088481

Gaines, J., & Lee, C. V. (2019). Medical tourism. In *Travel medicine* (pp. 371–375). Elsevier. doi:10.1016/B978-0-323-54696-6.00039-2

Griffiths, R. R., Johnson, M. W., Carducci, M. A., Umbricht, A., Richards, W. A., Richards, B. D., Cosimano, M. P., & Klinedinst, M. A. (2016). Psilocybin produces substantial and sustained decreases in depression and anxiety in patients with life-threatening cancer: A randomized double-blind trial. *Journal of Psychopharmacology (Oxford, England)*, *30*(12), 1181–1197. doi:10.1177/0269881116675513 PMID:27909165

Gupta, P. (2014). Frozen vodka and white skin in tourist Goa. *Tourism and the Power of Otherness: Seductions of Difference*, *34*, 95–109. doi:10.21832/9781845414177-007

Haden, M., Emerson, B., & Tupper, K. W. (2016). A public-health-based vision for the management and regulation of psychedelics. *Journal of Psychoactive Drugs*, *48*(4), 243–252. doi:10.1080/02791072.2016.1202459 PMID:27430375

Horowitz, M. D., Rosensweig, J. A., & Jones, C. A. (2007). Medical tourism: Globalization of the healthcare marketplace. *Medscape General Medicine*, *9*(4), 33. PMID:18311383

Inouye, A., & Wolfgang, A. (2022). 3, 4-Methylenedioxymethamphetamine (MDMA)-Assisted Therapy in Hawaii: A Brief Review. *Cureus*, *14*(6). doi:10.7759/cureus.26402 PMID:35915689

Johnson, M. W., & Griffiths, R. R. (2017). Potential therapeutic effects of psilocybin. *Neurotherapeutics; the Journal of the American Society for Experimental NeuroTherapeutics*, *14*(3), 734–740. doi:10.100713311-017-0542-y PMID:28585222

Lowe, H., Toyang, N., Steele, B., Grant, J., Ali, A., Gordon, L., & Ngwa, W. (2022). Psychedelics: Alternative and potential therapeutic options for treating mood and anxiety disorders. *Molecules (Basel, Switzerland)*, *27*(8), 2520. doi:10.3390/molecules27082520 PMID:35458717

Lutkajtis, A. (2020). Lost saints: Desacralization, spiritual abuse and magic mushrooms. *Fieldwork in Religion*, *14*(2), 118–139. doi:10.1558/firn.40554

Marks, M. (2022). The varieties of psychedelic law. *Neuropharmacology*, 109399. PMID:36565855

Mash, D. C. (2023). Ibogaine: Ibogaine therapy as a method of care treatment for opioid dependence. *Psychedelics as Psychiatric Medications*, 75.

Naef, P. (2018). "Narco-heritage" and the Touristification of the Drug Lord Pablo Escobar in Medellin, Colombia. *Journal of Anthropological Research*, *74*(4), 485–502. doi:10.1086/699941

Nicholas, C. R., Wang, J. B., Coker, A., Mitchell, J. M., Klaire, S. S., Yazar-Klosinski, B., Emerson, A., Brown, R. T., & Doblin, R. (2022). The effects of MDMA-assisted therapy on alcohol and substance use in a phase 3 trial for treatment of severe PTSD. *Drug and Alcohol Dependence*, *233*, 109356. doi:10.1016/j.drugalcdep.2022.109356 PMID:35286849

Orozco, M., & Harris, S. (2023). Psilocybin and the Meaning Response: Exploring the Healing Process in a Retreat Setting in Jamaica. *Anthropology of Consciousness*, *34*(1), 130–160. doi:10.1111/anoc.12162

Ovies, D. H. (2021). Millennials in the search for spiritual ecstasy: Touristic consumption of shamanic mushroom rituals in Huautla de Jimenez, Mexico. In Millennials, Spirituality and Tourism (pp. 179-200). Routledge.

Perkins, D., Ruffell, S. G., Day, K., Pinzon Rubiano, D., & Sarris, J. (2023). Psychotherapeutic and neurobiological processes associated with ayahuasca: A proposed model and implications for therapeutic use. *Frontiers in Neuroscience*, *16*, 2255. doi:10.3389/fnins.2022.879221 PMID:36798604

Petersen, M. A., Feldes, S., & Cova, V. S. (2022). Ayahuasca Calling: Sacredness and the Emergence of Shamanic Vocations in Denmark and Peru. *Anthropology of Consciousness*, *33*(2), 255–278. doi:10.1111/anoc.12165

Pilecki, B., Luoma, J. B., Bathje, G. J., Rhea, J., & Narloch, V. F. (2021). Ethical and legal issues in psychedelic harm reduction and integration therapy. *Harm Reduction Journal*, *18*(1), 1–14. doi:10.118612954-021-00489-1 PMID:33827588

Rucker, J. J., Iliff, J., & Nutt, D. J. (2018). Psychiatry & the psychedelic drugs. Past, present & future. *Neuropharmacology*, *142*, 200–218. doi:10.1016/j.neuropharm.2017.12.040 PMID:29284138

Runnels, V., & Carrera, P. M. (2012). Why do patients engage in medical tourism? *Maturitas*, *73*(4), 300–304. doi:10.1016/j.maturitas.2012.08.011 PMID:23007007

Segev, L., Paz, A., & Potasman, I. (2005). Drug abuse in travelers to southeast Asia: An on-site study. *Journal of Travel Medicine*, *12*(4), 205–209. doi:10.2310/7060.2005.12406 PMID:16086895

Smith, W. R., & Appelbaum, P. S. (2022). Novel ethical and policy issues in psychiatric uses of psychedelic substances. *Neuropharmacology*, *216*, 109165. doi:10.1016/j.neuropharm.2022.109165 PMID:35718006

Subritzky, T. (2018). Beyond deficit and harm reduction: Incorporating the spectrum of wellness as an interpretive framework for cannabis consumption. *The International Journal on Drug Policy*, *60*, 18–23. doi:10.1016/j.drugpo.2018.07.013 PMID:30086481

Valdez, A., & Sifaneck, S. J. (1997). Drug tourists and drug policy on the US-Mexican border: An ethnographic investigation of the acquisition of prescription drugs. *Journal of Drug Issues*, *27*(4), 879–898. doi:10.1177/002204269702700413

Van Der Meer, P. B., Fuentes, J. J., Kaptein, A. A., Schoones, J. W., de Waal, M. M., Goudriaan, A. E., Kramers, K., Schellekens, A., Somers, M., Bossong, M. G., & Batalla, A. (2023). Therapeutic effect of psilocybin in addiction: A systematic review. *Frontiers in Psychiatry*, *14*, 1134454. doi:10.3389/fpsyt.2023.1134454 PMID:36846225

Vidriales, A. L., & Ovies, D. H. (2018). Psychedelic tourism in Mexico, a thriving trend. *Pasos (El Sauzal)*, *16*(4), 1037–1050. doi:10.25145/j.pasos.2018.16.072

Zafar, R., Siegel, M., Harding, R., Barba, T., Agnorelli, C., Suseelan, S., Roseman, L., Wall, M., Nutt, D. J., & Erritzoe, D. (2023). Psychedelic therapy in the treatment of addiction: The past, present and future. *Frontiers in Psychiatry*, *14*, 1183740. doi:10.3389/fpsyt.2023.1183740 PMID:37377473

Zeifman, R. J., Singhal, N., Breslow, L., & Weissman, C. R. (2021). On the relationship between classic psychedelics and suicidality: A systematic review. *ACS Pharmacology & Translational Science*, *4*(2), 436–451. doi:10.1021/acsptsci.1c00024 PMID:33860173

KEY TERMS AND DEFINITIONS

Drug Tourism: Drug tourism refers to the phenomenon where individuals travel to a particular destination with the intention of using drugs that are either legal or more easily accessible in that location. It involves the combination of recreational drug use with travel and tourism activities.

Indigenous Cultures: Indigenous cultures refer to the ethnic or native communities that have inhabited a specific geographic region for generations. These cultures often have a distinct set of customs, traditions, languages, and beliefs that are passed down through generations. Indigenous cultures are deeply connected to their lands, have unique knowledge systems, and may hold spiritual or healing practices that incorporate natural substances, including psychedelics.

Legalization: Legalization refers to the process by which a substance, activity, or practice that was previously illegal or restricted becomes lawful and regulated by the governing authorities. In the context of psychedelic substances, legalization may involve the removal of legal barriers and the establishment of regulations and frameworks for their controlled and responsible use.

Mental Health: Mental health refers to a person's emotional, psychological, and social well-being. It encompasses factors such as one's thoughts, feelings, behaviors, and the ability to cope with life's challenges. Good mental health is characterized by a state of well-being, effective functioning, and the ability to engage in productive activities and maintain healthy relationships.

Psychedelic Retreat: A psychedelic retreat is a structured program or organized event that provides individuals with a guided and supportive environment to engage in psychedelic experiences. These retreats typically involve the use of psychedelics in combination with therapeutic or spiritual practices, such as meditation, group therapy, or shamanic ceremonies. The aim is to facilitate personal growth, self-exploration, and healing under the guidance of experienced facilitators or practitioners.

Psychedelic Substance: A psychedelic substance refers to a type of psychoactive drug that induces altered states of consciousness, often characterized by profound changes in perception, thoughts, and emotions. Examples of psychedelic substances include LSD (lysergic acid diethylamide), psilocybin (found in certain species of mushrooms), and DMT (dimethyltryptamine).

Psychedelic Tourism: Psychedelic tourism refers to the practice of traveling to specific destinations or participating in organized retreats or experiences with the purpose of engaging in psychedelic substances and related activities. It involves seeking out opportunities to explore the potential therapeutic, spiritual, or recreational aspects of psychedelics in a supportive and controlled setting.

Therapeutic Effect: The therapeutic effect refers to the beneficial outcomes or positive changes that result from a particular treatment, intervention, or substance. In the context of psychedelics, therapeutic effects may include improvements in mental health symptoms, increased self-awareness, enhanced emotional processing, and transformative or healing experiences.

APPENDIX

Table 1. Most common drug tourism destinations

Destination	Drug tourism patterns
Australia	Interstate drug tourism to Australian Capital Territory and South Australia, owing to more liberal marijuana use
Bolivia	'Cocaine bar' in capital, La Paz
Czech Republic	Marijuana and ecstasy sold in 'drug-friendly' night clubs
Denmark	Tourist attraction of Freetown Christiania, a self-proclaimed autonomous neighborhood in Copenhagen
India	Organized crime groups supply drugs to tourists at rave dance parties in states such as Goa
Indonesia	Use of psilocybin mushrooms on islands such as Bali, despite strict national drug laws
Mexico	Consumption of psilocybin mushrooms in towns in the southern state of Oaxaca, despite federal health law prohibitions Ingestion of hallucinogenic peyote cactus in the Wirikuta desert
Morocco	Rif mountains recognized centre for hashish ('kif') production
Netherlands	'Smart shops' in Amsterdam sell hallucinogenic agents, stimulants and relaxants; coffee shops sell cannabis
Peru	Shaman-guided psychedelic Ayahuasca experiences
Portugal	Liberal drug policy reform laws; needle exchange and opioid substitution programs attract problem drug use tourists
Southeast Asia	Opiates, ecstasy, mushrooms and marijuana attract drug tourists to the Golden Triangle, connecting Thailand to Myanmar and Laos Full moon parties on Ko Pha-Ngan Island, Thailand
Spain	Use of recreational substances at electronic dance music parties on island of Ibiza
Uruguay	Possession of drugs for personal use not criminalised; cocaine paste also popular
USA	Legal, regulated marijuana dispensaries in California, Colorado and Washington

Source: Flaherty et al., 2017

Chapter 10
The Pathway of Critical Pedagogy in Tourism Education

Barış Çıvak
https://orcid.org/0000-0001-9334-3683
Anadolu University, Turkey

ABSTRACT

This study explains the roots of critical pedagogy and the ways and benefits of critical tourism pedagogy. By criticizing tourism education designed for the sector, it tries to open a space for students' enlightenment and liberation through critical pedagogy. The boundaries of critical pedagogy, which is a radical and willing pedagogy for transformation, are drawn to develop critical consciousness that will lead to the conversion of the individual, the learning environment, and society in general. Instead of the education system in which the didactic teaching method was adopted, a roadmap was presented for improving a culture of understanding, wisdom, and criticism with critical pedagogy. These are: (1) increasing critical tourism studies in terms of quality and quantity, (2) increasing critical courses in the education curriculum and enriching their content, (3) creating freedom and developing dialogue, and (4) combining different teaching methods.

INTRODUCTION

A high level of technical knowledge is required in the applied sciences, especially in engineering and medicine. Part of the curriculum can be devoted to technical knowledge and skills. Undergraduate education in tourism cannot be designed like

DOI: 10.4018/978-1-6684-9255-0.ch010

engineering (Belhassen and Caton, 2011). Of course, there may be technical and applied courses to gain professional skills. However, designing a purely pragmatist tourism education only strengthens the capitalist ideology and understanding of management. To balance this, critical pedagogy in tourism education seems essential. It is substantial to develop a culture of understanding and criticism in tourism undergraduate education.

It is seen that the tourism curriculum includes courses focused on gaining knowledge, skills, and competence in the field of tourism (Sheldon, Fesenmaier & Tribe, 2011; Airey & Johnson, 1999). It is seen that there are no courses on critical thinking in many tourism programs. Tourism education is aimed at the interests of the industry. This study challenges neoliberal policies and their impact on tourism pedagogy. This study presents the basic framework for initiating a transformation by revealing the purpose, nature, and ways of critical pedagogy in tourism. Tourism academics have a mission to give direction for moral and political stances. They should rethink their responsibility as public intellectuals to address social issues and true democracy (Giroux, 2021).

This study constitutes a resource for students' enlightenment and liberation through critical pedagogy in tourism education. Tourism education, created with a pragmatist perspective, focuses on the efficiency and productivity of businesses rather than focusing on employee problems. It is seen that the exploitation and domination practices in the sector are ignored. It is possible to see this from the subject distributions that academic studies focus heavily on (Çıvak & Sezerel, 2018). This study aims to fill this gap in the field. In doing so, a direction is drawn to realize the transformation in tourism education, following in the footsteps of the pioneers who steered critical pedagogy (see Freire, 1993; Bourdieu & Passeron, 1990). Thus, it will help to develop a critical consciousness in society. There is a need for a more radical and willing pedagogy for transformation (Abraham, 2014).

The following sections are given the roots of critical pedagogy, critical pedagogy in tourism, and a conclusion.

Origin of Critical Pedagogy

The historical roots of critical pedagogy go back to Marx and Engels. The critical tradition was later reconsidered and formulated by the Frankfurt school. Althusser, one of the Marxist thinkers, refers to the schools as an ideological apparatus of the state. At this point, the state places the ideology in the minds of individuals with the education process (Althusser, 1989). In addition, the researchers who made significant contributions to critical pedagogy are Pierre Bourdieu and Jean-Claude Passeron. Bourdieu and Passeron (1990) claim that the school makes and reproduces inequalities in education and deepens class segregation. Paulo Freire (1993) revealed how the education system reproduces inequalities and similar human typologies in Pedagogy of the Oppressed. It should be stated that the researchers, who have left their mark on critical thinking, are guiding the new generation of researchers and educators in teaching critical paradigms. Academics who followed Freire, such as Michael Apple, Henry Giroux, Donald Macedo, Peter McLaren, and Ira Shor, contributed to enriching critical pedagogy (Abraham, 2014).

One of the most helpful sources of critical pedagogy is the critical theory created by the Frankfurt School (Abraham, 2014). The Frankfurt School was founded in 1923. The school includes theorists such as Max Horkheimer, Theodor W. Adorno, Herbert Marcuse, Walter Benjamin, Erich Fromm, Friedrich Pollock, and Leo Löwenthal (Wiggershaus, 1994). The researchers of the school moved to the USA due to the Second World War. One of the prominent figures among the second-generation theorists, when they returned after the war, was Jürgen Habermas (Wiggershaus, 1994). These thinkers are very precious in the formation of the critical paradigm.

The critical paradigm questions the "usual" and "normal" states of all stereotyped values in the historical process, criticize ideology, and examines alienation, commodification, and oppression (Çıvak & Sezerel, 2018). In critical studies, related structures, mechanisms, and processes that ensure the continuity of distorted structures are tried to be revealed (Prasad, 2005). Critical researchers criticize the positivist perspective by targeting it. The critical paradigm is primarily linked to the Marxist ideology and the critical theory developed by the Frankfurt School (Neuman, 2014). Studies of critical theory criticize the historical and current conditions of oppression practices and seek ways to change them (Glesne, 2013, p. 12). Critical social science studies generally focus on oppressed and excluded individuals in society. It deals with the perspectives of the oppressed (Freire, 1993), the practices of domination (Scott, 1992), masculine domination (Bourdieu, 2018), the effect of the ideological apparatus of the state on individuals (Althusser, 2006), the characteristics of working under capitalist conditions (Braverman, 1974; Friedman, 1977; Burawoy, 1979). While these studies reveal the current despotic regime, a step is taken for the liberation of the oppressed and excluded (Neuman, 2014).

Although there is no standard set of methods in studies conducted with a critical paradigm, it has distinctive features specific to the qualitative research design (Glesne, 2013, p. 14); (1) Critical theorists focus on the perspectives of the oppressed and exploited sections of society, power-centered issues, and dominant elements in the use of power. (2) The content of the language is checked. The purpose of the developed discourses and which power relations they try to legitimize are revealed. (3) In critical theory, importance is given to the relations between thought and action, theory and practice. It explores ways to change existing structures by identifying what actions are accepted by society and questioning learned limitations, impulses, and lives through the eyes of the oppressed.

Brazilian philosopher and educator Paulo Reglus Neves Freire (1921-1997) focuses on educational opportunities to transform unjust social relations. It follows the traditions of the Frankfurt School. According to Kincheloe: "the critical pedagogy that emerged from Paulo Freire's work in the poverty of northeastern Brazil in the 1960s combined the ethics of freedom, and critical theory of Frankfurt School with progressive impulses in education..." (Kincheloe, 2007, p. 12). Freire analyzes the roles of teachers and students in his work "Pedagogy of the Oppressed." In the education system where the didactic teaching method is adopted, "the teacher teaches, and the students learn; the teacher knows everything, and the students know nothing; the teacher speaks, and the students listen meekly" (Freire, 1993, p. 54).

Freire made his analysis from his experiences in Latin American society. He especially questioned power relations, exploitation, and inequalities. Freire is a prominent academic who has influenced generations of teachers, organizers, and activists worldwide. One of the issues he focused on the most was "inequality in education." It revealed how inequalities in education reflect and reinforce other types of inequality. He has made education of adult and literacy education the most important part of his life (Freire, 2018). By examining the functioning of the social system and human typologies based on his life experiences, Freire revealed how the education system reproduces inequalities and similar human typologies. Describing the social order as operating in the oppressor-oppressed relationship, Freire objected to the system's functioning in a way that would serve the purposes of the oppressors and that the oppressed should internalize it. In his analysis, he revealed that the oppressors perceive the oppressed not as a human/subject but as an 'object' and "dehumanizes" them. The subordinate classes (the oppressed) who internalize this as a truth adopt the attitudes and actions of the oppressors when they find the opportunity and become free. One of the significant reasons for this is the current education system (Freire, 1993).

Pierre Bourdieu is a doyen guiding critical pedagogy and new-generation researchers. In addition to the field theory and reflexive sociology, he added to the scientific community, Bourdieu drew attention to the inequalities in education and how the dominant ideology was reproduced. Bourdieu and Passeron examined how the children of the working class and the middle class pass through the education system (exams, teacher judgment, symbolic violence). Besides, they questioned the literacy levels of children, their ability to use language, and the relationship of class - the academic success. Accordingly, schools are a critical point of cultural reproduction, in other words, the transition to social reproduction. Pedagogical action is symbolic violence in itself because it is imposed by power. The education system reproduces cultural structures and power relations in social classes (Bourdieu & Passeson, 1990). In this way, the imposed elements are engraved in the minds and become legitimate. Individuals do not even question these elements because they are consolidated and legitimized.

Bourdieu and Passeron 1990, state that social origin is as important as the gender variable and that these variables are efficient in-branch selection and success. According to the results of the field research they conducted, in 1965-66, the chance of the child of an agricultural worker to enter university was around 3%, while it was 36% for the children of middle-level managers and 72% for the children of industrialists (Bourdieu and Passeron, 1990). This is an indication that the unequal position of individuals creates inequalities again. When we evaluate the current state of the education system and the opportunities for different student categories to access university, we can understand the relations between school and social classes. For example, to meet the conditions for entering the conservatory of universities, the student must be prepared before the exam (going through a special education). This preparation is often done through private courses or trainers. A person with low economic capital can't access these courses, and the probability of entering a conservatory is also weak. Therefore, the possibility of people from the lower class becoming an actor is reduced.

Certificates or diplomas issued by educational institutions are the documents that express to their graduates that they are competent in the relevant field and that they are protected in a way. These documents perform insurance for them to get better jobs and higher wages in the educated/uneducated distinction. While well-educated people are employed in jobs that require mental labor, untrained people are employed in jobs that require physical strength in hotels. For instance, managers, receptionists, and office workers (sales and marketing, human resources, accounting, etc.) are educated people (with diplomas or certificates). On the other hand, the education level of the employees in departments such as housekeeping, laundry, and landscaping (primary, secondary, or high school, according to the schools they graduated from) is low.

A diploma or certificate provides an opportunity for individuals (according to someone who has not received a certificate or diploma) and changes their way of life. Because the school institution contributes to the distribution of cultural capital and thus to the re-creation of the social space (Bourdieu, 2015, p. 35). However, cultural capital is not only related to school but also to the cultural capital of families. Families tend to maintain and develop all their power and privileges in social space. This trend involves fertility, marriage, succession, economic, and educational strategies. If cultural capital is more important than economic ones, families give importance to education.

Bourdieu and Passeron draw attention to the fact that the school works like Maxwell's demon. The school continues its sorting function and maintains the difference between well-equipped children and low-equipped ones. This allows the distinction between children. The school establishes or protects social boundaries between children (Bourdieu, 2015, pp. 36-37). Graduates are members of a union and gain the power to dominate by separating from others. For example, an individual who is a business graduate who has a diploma from a good school will take the title of manager in the future and will settle in the oppressor position in the order of domination. Here is the arrangement developed by the school for inclusion in the nobility class. Gentrification through educational institutions is blessed with a diploma, certificate, or certificate of competence awarded to the individual. Diploma ceremonies also clearly reveal this blessing. From now on, the management function is handed over to the next generation (management candidates) with this ceremony. In short, school is both a constructive and structured structure (Bourdieu, 2015, p. 43).

The education system has previously stated that students are an authority in the teaching or imposition of legitimate practices. Apart from this, the school conveys the different cultural accumulation of which it is a part, according to disciplines (philosophy, mathematics, sociology, etc.), departments, and educational institutions. In other words, the predispositions of the individual who graduated from the Department of business administration at Harvard University and the person who graduated from the Department of Anatomy and Physiology at Cambridge University differ. Of course, social origin and demographic characteristics (gender, age, etc.) are effective in this. However, the school not only changes the living standards of individuals (the university, the department they attend) but also shapes their predispositions (thanks to the accumulation of economic, cultural, and social capital). Education capital makes a formal classification (engineer, doctor, teacher, etc.) and shapes individuals' capitals (economic, cultural, social) and habitus. It also has the function of assigning a status to its graduates thanks to the diplomas/certificates given by the education system (Bourdieu, 2017).

What is Critical Pedagogy?

Before critical theory was developed, critical thinking as a way of seeing reality was practiced among intellectuals of different generations (Abraham, 2014). It was a subject that Greek philosophers at first and then Renaissance intellectuals emphasized. Socrates emphasizes "a life that is not criticized, questioned, and thought about is not worth living" (Demircioğlu, 2018, p. 13). In other words, individuals can examine reality and correctly understand phenomena and events in social life by using critical thinking.

Critical pedagogy is "a way of thinking, negotiating, and transforming the relationship between classroom teaching, knowledge production, the institutional structure of the school, and the social and material relations of the wider community, society, and nation-state" (McLaren, 1998, p. 45). Critical pedagogy is a movement that includes teaching and learning relationships so that students gain critical self-awareness and social awareness and take appropriate measures against oppressive forces (McKernan, 2013). Giroux defines the democratic character of critical pedagogy as follows:

"The democratic character of critical pedagogy is largely defined through a series of fundamental assumptions that include knowledge, power, and values. All issues must be subject to critical scrutiny because critical pedagogy is more than a close reading of texts, radical classroom practices, or promoting critical literacy. It is also about combining social change with learning, education with democracy, knowledge and forms of intervention in public life" (Giroux, 2008, pp. 18-19).

Critical pedagogy argues that the struggles for education policies and practices are not just for the school but for society (Apple, 2011). Because education is deeply connected with the social context in which it exists. The oppressive role of social rules should be well-considered. Every society contains its cultural elements and is transmitted from family to children. For critical pedagogy, education and schools are fields where power relations are reproduced, and resistance is exhibited through various identities (İnal, 2009, p. 4). For a democratic society, resistance to the order of domination is a way out. Critical pedagogy is a significant tool to instill this.

The political nature of education positions educators as either agents of the state or agents of transformative change. Educators either maintain the status quo or create the context to be questioned (Ledwith, 2001). One of the biggest problems with critical pedagogy is that in many nations some academic pioneers of the 'critical pedagogy' movement, and critical and democratic education in general, are not adequately connected to the realities of schools (Apple, 2011, p. 24).

Critical pedagogy will be one of the main steps to allow students to meet the critical paradigm. In tourism education, courses created with a critical view should be opened in undergraduate and graduate programs. Besides, academicians working on critical issues have substantial responsibilities like disseminating critical scientific congresses and increasing their publications in terms of quality and quantity to spread the critical view. Apple (2010) states that critical educators should participate in movements to create more democratic institutions both in education and in all aspects of society. We can characterize this as a stepping stone in the dissemination of the critical tradition.

Capitalist ideology limits critical thinking and common sense in the daily lives of individuals. In addition, it is seen that individualism is emphasized at the expense of injustice toward large segments of society, and there is not enough focus on society (Abraham, 2014). One of the ways to overcome this is solidarity and transformation-based critical pedagogy. For this, first of all, it is necessary to understand the function and internal processes of the school institution. Although school is a field where the dominant ideology is taught, it is a field where interpersonal agreement and conflicts occur, and subjectivities and experiences are produced (Giroux, 1998).

Freire argues that there are two basic understandings of education. The first is Banker Education Model, and the other is Problem Defining Education Model. In the banking education model, the teacher is the subject, and the student is the object. In this education model, the teacher is the all-knowing authority. Freire explains this education model with a metaphor. According to him, educational activity is like an investment, with a teacher as an investor and a student as an investment object. The teacher's role is to make investments that students patiently listen to, memorize and repeat. The student's role is accepting, sorting, and repeating the information. A student who grows up in such an education model assumes the role of a 'representative' who cannot 'think' or 'seek their rights' and repeats the information marketed by the system. This understanding creates a multitude of societies (Freire, 1993). In Marcuse's words, he transforms into a "One-Dimensional Man" (Marcuse, 1976).

We are in a period where the content of education is created in favor of the market, and the labor force is trained for the market (Apple, 2012). The majority of academics have made their contributions to the field by adopting the dominant ideology. It is seen that the dominant paradigm in academic studies in the field of tourism is positivism, and the critical paradigm is adopted in very few studies (Çıvak & Sezerel, 2018; Çıvak & Emir, 2021). It is certain that the adoption of positivist/pragmatist ideology in schools. Therefore, it is possible to say critical pedagogy is limited or not given in most tourism schools. It is seen when the curricula of the schools are examined.

The curriculum in business schools includes a pragmatist view. There are courses such as management and organization, marketing, accounting, and finance. These courses focus on increasing efficiency and maximizing profits for business. This process ensures that the pragmatist pedagogy, which the trainer also goes through, is reproduced by transferring it from generation to generation. The notions of efficiency, productivity, and profitability become a doxa by forming the basic building block of education in business schools. Giroux, together with neoliberalism, expresses how the structural change in language in education, other institutions, and society was determined by the government with these words: "Market forces have radically changed human behavior and language. The language of citizenship gave way to the language of commercialization, privatization, and liberalization. In addition, individual and social representation is largely defined by market-oriented individualism, competition, and consumption" (Giroux, 2007, p. 68). According to this statement, we can say that the competitive element created by the capitalist ideology comes to the fore, especially in business schools. The way to gain a place and rise in the competitive labor market is to compete with others. Especially in courses such as human resources management, a clear presentation is made on how to differentiate from other candidates during the job application process.

It is possible to say that the distinction between social classes has deepened with the privatization of education. Families with high economic capital can take private lessons for their children. So, the children of the elite class can study at a good university. There is a significant difference between a student from the lower-income group and a student from the upper-income group. This will cause social status and economic differences among individuals in the future, and inequality will be reproduced (Bourdieu & Passeron, 1990). In other words, the children of the working class and the children of industrial bosses can't live under equal conditions. They are born in unequal environments. This system of inequalities also ensures the reproduction of inequality. One of the important ways to overcome this is a classless society. In building a classless society, it seems possible that individuals pass the social world through the filter of criticism and become conscious.

The Necessity of Critical Pedagogy in Tourism Education

The occurrence and spread of tourism education have emerged depending on the development of tourism activities. Tourism emerged as a field of study after the 1960s. In other words, it didn't have undergraduate discipline or research activity (Fidgeon, 2010, p. 700). The need and demand for tourism professionals in the sector played an important role (Airey, 2005). Airey (1979) states that tourism education in Britain began in the mid-1960s, and the first diplomas were awarded in the late 1960s. McKercher (2001) emphasizes that after a slow incubation period, a significant expansion period has been entered in the numbers of tourism education programs and student demand. The spread of travel movements has led to the need for more tourism professionals. This has been provided to increase tourism education programs (Daniel et al., 2017). At this point, since tourism education has been shaped according to the needs of the sector, the majority of the education curriculum has been designed according to the management view. Cooper and Shepherd (1997) emphasize that it is beneficial for both sides if education and industry act together in the tourism academy. Such studies show that the tourism curriculum has been shaped to meet the needs of the sector.

Students in the tourism sector need different skills, abilities, and knowledge to face these competitive and rough conditions (Sheldon, Fesenmaier & Tribe, 2011). Entrepreneurial skills are very important in this case (Daniel et al., 2017). Airey and Johnson (1999) stated that professional objectives are highly prevalent in tourism education. Table 1 clearly shows the rate of importance of professional goals in tourism education.

Table 1. Top twenty 'aims and objectives' of tourism degree courses

Aims and Objective	Mentions n = 99
1. Tourism offers exciting career opportunities	76
2. Tourism offers employment/employer link	56
3. Tourism is a large, global, growth industry	50
4. Tourism allows the application of skills/theory	48
5. Tourism offers international (career) opportunities	45
6. The course develops business/management skills	36
7. Opportunities exist in the private/public sectors	31
8. The course develops transferable skills	28
9. The course provides a sound academic education	25
10. Tourism provides a broad foundation for study	22
11. The course aims to meet the needs of the tourism industry	21
12. It develops analysis/judgment/understanding of decision-making	20
13. The course takes a European perspective	20
14. It broaches social/sustainable concerns	19
15. It develops professionalism and professional attitudes	17
16. The course emphasizes quality and excellence	17
17. It aims to develop flexibility	14
18. It broaches an understanding of service and service delivery	14
19. It aims to develop successful graduates	10
20. It seeks to develop an understanding of competitive forces	7

Source: Airey & Johnson (1999)

The School of Tourism Management program of The Hong Kong Polytechnic University (shtm, n.d.), which stands out in the field of tourism education, Ecole Hôtelière de Lausanne's Hospitality Management program (ehl.edu, n.d.), Swiss Hotel Management School's Bachelor of Arts (shms, n.d.) programs do not have any critical thinking or critical pedagogy when examined. Since the content of the programs is designed for the sector, it is seen that there are courses in management and organization, marketing, finance, accounting, and language. Most of these are lessons created with a pragmatist perspective built on effectiveness, efficiency, and profitability. When looking at Anadolu University Tourism Management program (turizm.anadolu.edu.tr, n.d.), it is seen that there is a 'Critical Thinking' course. In addition, it is noteworthy that there are courses on critical topics such as "Critical Tourism Studies" and "Labor Issues in Tourism" in the Tourism Management graduate program. Academicians who conduct critical studies can be the starting point for transformation by opening critical courses in the universities where they work.

There are significant problems for workers in the tourism industry. The inelasticity of touristic demand affects employment. Economic recession, epidemics, natural disasters, or terrorist events restrict touristic activities (Ünlüönen & Çeti, 2019). In this case, tourism employment also shrinks. Although destinations try to find solutions to the seasonality problem (Baum & Luntrop, 2001), it is difficult to change (Butler, 1998, p. 19). Seasonal employment precariates tourism workers (Çıvak, 2021; Yıldırım, 2021, Çelik & Erkuş Öztürk, 2016).

Working conditions in the tourism sector remain below the determined standards. Low wages, labor-intensive work with long periods, non-payment of overtime wages, seasonal or flexible employment, informality, inadequate accommodation conditions, and high staff turnover (Baum, 1999; Tired, 2013; ILO, 2010) are among the most significant problems. In addition, there are domination, harassment, discrimination and work-life imbalance, stress and health problems, high labor turnover, and seniority problems (Yorgun, 2013; Kaya, 2012; Chalkiti & Sigala, 2010; Poulston, 2008; Silva, 2006).

All these adverse conditions evoke a new tourism education paradigm (Sheldon et al., 2011). At this point, it is argued that critical pedagogy should get involved. One of the major responsibilities of tourism educators is to show the ways of resistance by explaining the adverse conditions, exploitation, and domination (Çıvak, 2021). It is now imperative for critical theorists to meticulously fulfill this great responsibility. Ledwith (2001) argues that people are located at the very core of their lives, at the interface of liberation and domination. Knowing how individual and organized resistance mechanisms work will increase a tourism student's chances of survival.

The Pathway of Critical Pedagogy in Tourism

Is a critical tourism pedagogy possible? It has been stated that critical pedagogy is possible, and its benefits are listed in the literature. Rouzrokh et al. (2017) identified five pedagogical themes in their study on critical pedagogy in tourism. These; (1) build safe spaces and develop trust, (2) create empathy, (3) engage tourism literature with 'real life', (4) open the door to constant thinking, and (5) decentralize power and knowledge. Fullagar and Wilson (2012) question how one can reflexively work in and across normalized thinking to involve those who are not yet more critically thinking or to explore new ways of understanding. They state that critical academics cannot take the risk of disappearing by waiting for the revolutionary overthrow of the "managerial regime." They should remain open to the future, dialogue, and discussion by using 'soft power' in a post-disciplinary manner. Belhassen and Caton (2011) explore the theoretical and philosophical roots of critical pedagogy and discuss the benefits of critical pedagogy in tourism education. They claim that critical tourism pedagogy has benefits in terms of individual freedom, social justice, and work productivity. Giroux (2021) emphasizes that pedagogy is always a moral and political practice and does not only address agency and power struggle but also includes issues of economic and social justice and discourses of criticism and possibility as part of a broader democratic project. Jamal, Kircher & Donaldson (2021) state that critical pedagogy should be included to help students break free from deeply entrenched assumptions and deliberately develop pluralistic, relational solutions to address injustices. However, it should not be forgotten that transformative learning is not a semester but a lifelong process (Werry, 2008).

In applied sciences, especially engineering, and medicine, a high level of technical knowledge is required for the practice of the profession, and this is reasonably acceptable. Therefore, a part of the curriculum can be devoted to skills development in such areas. However, undergraduate education in tourism cannot be designed like engineering (Belhassen & Caton, 2011). Of course, foreign language knowledge and practical knowledge must be present in tourism education. However, a pure pragmatist vision in education ensures to gain of strength in the understanding of business administration. At this point, it seems essential to develop a culture of understanding, wisdom, and criticism through a critical pedagogy in tourism. The basic principles necessary to provide an infrastructure for critical tourism pedagogy can be listed as follows.

Increasing Critical Tourism Studies in Terms of Quality and Quantity

It is seen that it is handled with a critical paradigm that is little learned in tourism literature. The mainstream debate in tourism magazines in Turkey is positivism as it is (Çıvak & Sezerel, 2018; Çıvak & Emir, 2021; Yayla & Ergün, 2020). Similarly, it is possible to say roughly the contents of positivism in the most prestigious tourism magazines. One significant way to overcome this is to encourage the critical work of journal editors and boards. Since the academic field is also the career field of a researcher, it is expected to increase the beauty and limitlessness of publication. The demands of critical observers can be curbed by controlling the chances of critical work being published. Each critical study will make a difference in tourism students and tourism education. It is possible to say that it will give an understanding that a different way of thinking is possible.

Increasing Critical Courses in the Curriculum and Enriching Their Content

When the tourism undergraduate programs are examined, it is seen that the courses for technical skills, efficiency, and productivity are dominant. The main reason is that the area is designed directly for the sector. Tourism faculties make several concrete promises to attract prospective students and prepare their curriculum accordingly. There are exciting career opportunities, employment opportunities in the public and private sectors, skill development, freedom to work in different fields, and professionalism in the promises (Airey & Johnson, 1999). To overcome this, it is necessary to create courses such as critical thinking, critical management studies, critical tourism studies, labor issues, social problems, state-individual relationships, and ideologies in undergraduate and graduate curricula. However, while the content of the course is being created, the educator should reflectively review his practices. The collision of non-critical information (information produced with a pragmatist perspective) and critical information in tourism education may result in erroneous, interrupted, and confusing information pollution.

Creation of a Free Atmosphere and Development of Dialogue

Didactic education, as in Freire's (1993) Banking education model example, transforms tourism students into "One-Dimensional Humans" in Marcuse's words (Marcuse, 1976). Instead, a culture of discussion needs to be created. It should be ensured that the student can establish a comfortable dialogue with other students and educators. Even the design of the classroom can significantly affect this. Therefore, seeds of wisdom can sprout in an environment where the student can feel comfortable and not afraid of being judged. It is essential to create a free atmosphere for improving critical thinking. The free sharing of ideas in all public areas will contribute significantly to the level of inquiry, understanding, and wisdom.

Combining Different Teaching Methods

Several teaching methods, in addition to classical teaching methods, increase interest and awareness of critical thinking. For example, besides the classical student assignments and in-class discussions, film analysis, book criticism, and discussion of the historical context of social events and phenomena can add a different perspective. Ayıkoru and Park (2019) state that films play an important role in facilitating the critical analysis of the phenomenon under study, showing and problematizing abstract concepts and ideas. They also emphasize that it facilitates students' ability to distinguish multiple and alternative discourses about management. In addition, education with drama techniques can provide a considerable advantage.

Critical pedagogy in tourism will lead to the deterioration of the status quo in the field by criticizing the practices that reproduce inequality and oppression. Thus, the tradition of criticism transforms the field and society in line with social justice (McLean, 2006; Hytten, 2006).

CONCLUSION

Critical pedagogy doesn't seem possible to be included in tourism programs at present. When the literature and tourism curricula are examined, it is seen that critical pedagogy is at a limited level in the field of tourism; It has been observed that most studies and programs focus on issues such as efficiency, productivity, and career development. For critical pedagogy, primarily an academic background is required. Therefore, critical tourism academics are needed. Critical thinking will likely become widespread as the number of academics engaged in critical studies increases. Because critical tourism studies need to be extensive to instill critical consciousness. One of the ways to do this is a free environment. Building more democratic regimes everywhere instead of oppressive regimes opens a door for critical thinking. The structured curriculum for critical pedagogy should include courses in philosophy and sociology. A theoretical background should be built for students to think critically.

Critical pedagogy is an opportunity to create social transformation outside of individual freedom, social justice, and work productivity (Belhassen & Caton, 2011). Critical pedagogy enables students to gain self-awareness and social awareness and to take appropriate measures against oppressive forces (McKernan, 2013). Studies show that working conditions in the tourism sector are quite abrasive (Çıvak, 2021; Çıvak & Besler, 2022; Yıldırım, 2021; Yorgun, 2013; Kaya, 2012; Chalkiti & Sigala, 2010). These problems and the methods of coping with them should be transferred to the students in tourism education. Showing the ways of organized resistance against adverse working conditions, labor exploitation, and domination (Çıvak, 2021) in tourism will enable tourism students to survive, transform the field, and gain their rights.

This study conveys the roots and scope of critical pedagogy and the tourism education curriculum. Four major steps for critical tourism pedagogy are presented. These steps are essential for a critical program in tourism education and constitute the starting point for critical pedagogy. The new generation who thinks, questions, and resists instead of a one-dimensional human being formed by a didactic education

(Marcuse, 1979) will ensure social development. In this respect, critical pedagogy is a way out.

REFERENCES

Anatolian University. (n.d.). *Course Catalog and Orientation Tables.* AU. turizm. anadolu.edu.tr. (n.d.). https://turizm.anadolu.edu.tr/ogrenci/ders-katalog-ve-intibak-tablolari

Abraham, G. Y. (2014). Critical pedagogy: Origin, vision, action & consequences. *KAPET, 10*(1), 90–98.

Airey, D. (1979). Tourism education in the United Kingdom. *Revue de Tourisme, 2*(79), 13–15.

Airey, D. (2005). Growth and development. In D. Airey & J. Tribe (Eds.), *An international handbook of tourism education.* Elsevier. doi:10.1016/B978-0-08-044667-7.50005-7

Airey, D., & Johnson, S. (1999). The content of tourism degree courses in the UK. *Tourism Management*, 20.

Althusser, L. (1989). *İdeoloji ve Devletin İdeolojik Aygıtları* (Y. Alp & M. Özışık, Trans.). İletişim Yayınları.

Althusser, L. (2006). Ideology and ideological state apparatuses (notes towards an investigation). *The anthropology of the state. REAd (Porto Alegre), 9*(1), 86–98.

Apple, M. W. (2010). *Global crises, social justice and education.* Routledge.

Apple, M. W. (2011). Democratic education in neoliberal and neoconservative times. *International Studies in Sociology of Education, 21*(1), 21–31. doi:10.1080/09620214.2011.543850

Apple, M. W. (2012). *Education and power.* Routledge. doi:10.4324/9780203143124

Ayikoru, M., & Park, H. Y. (2019). Films and critical pedagogy in management education: A tourism studies context. *Academy of Management Learning & Education, 18*(3), 414–432. doi:10.5465/amle.2015.0134

Baum, T. (1999). Seasonality in tourism: Understanding the challenges: Introduction. *Tourism Economics, 5*(1), 5–8. doi:10.1177/135481669900500101

Baum, T., & Lundtorp, S. (2001). *Seasonality in tourism: An introduction. Seasonality in tourism.* Elsevier. doi:10.4324/9780080516806

Belhassen, Y., & Caton, K. (2011). On the need for critical pedagogy in tourism education. *Tourism Management, 32*(6), 1389–1396. doi:10.1016/j.tourman.2011.01.014

Bourdieu, P. (2015). *Pratik nedenler* (H. U. Tanrıöver, Trans.). Hil Yayın.

Bourdieu, P. (2017). *Ayrım: Beğeni yargısının toplumsal eleştirisi* (D. Fırat & F. Berkkurt, Trans.). Heretik.

Bourdieu, P. (2018). *Eril tahakküm* (B. Yılmaz, Trans.). Bağlam.

Bourdieu, P., & Passeron, J. C. (1990). Reproduction in education, society and culture. *Sage (Atlanta, Ga.).*

Braverman, H. (1998). *Labor and monopoly capital: The degradation of work in the twentieth century.* Monthly Review Press.

Burawoy, M. (1979). *Manufacturing consent.* The University of California Press.

Butler, R. (1998). Seasonality in tourism: Issues and implications. *Tourism Review, 53*(3), 18–24.

Çelik, C., & Erkuş-Öztürk, H. (2016). Role of precariousness and space on class formation process: The case of Antalya's tourism workers. *Capital and Class, 40*(3), 419–445. doi:10.1177/0309816816653879

Chalkiti, K., & Sigala, M. (2010). Staff turnover in the Greek tourism industry. *International Journal of Contemporary Hospitality Management, 22*(3), 335–359. doi:10.1108/09596111011035945

Çıvak, B. ve Sezerel, H. (2018). Araştırma paradigmaları ve turizm yazını. *Turizm Akademik Dergisi, 5*(1), 1–14.

Çıvak, B. (2021). *Otel işletmelerinde emek sömürüsü, tahakküm ve direniş ilişkilerine dair niteliksel bir araştırma* [Unpublished doctoral dissertation, Anadolu University, Eskişehir].

Çıvak, B., & Emir, O. (2021). Boş zaman ve rekreasyon konulu lisansüstü tezlerde araştırma paradigması eğilimi. *Journal of Tourism & Gastronomy Studies, 9*(3), 2186–2202.

Cooper, C., & Shepherd, R. (1997). The relationship between tourism education and the tourism industry: Implications for tourism education. *Tourism Recreation Research, 22*(1), 34–47. doi:10.1080/02508281.1997.11014784

Daniel, A. D., Costa, R. A., Pita, M., & Costa, C. (2017). Tourism education: What about entrepreneurial skills? *Journal of Hospitality and Tourism Management, 30*, 65–72. doi:10.1016/j.jhtm.2017.01.002

Demircioğlu, A. (2018). *Eleştirel düşünme eğitimi.* Gece Kitaplığı. ehl.edu. https://www.ehl.edu/en/study/bachelor-degree-in-hospitality

Fidgeon, P. R. (2010). Tourism education and curriculum design: A time for consolidation and review? *Tourism Management, 31*(6), 699–723. doi:10.1016/j.tourman.2010.05.019

Freire, P. (1993). *Pedagogy of the oppressed.* Penguin Books.

Freire, P. (2018). Education as a practice of freedom. In *The Brazil Reader* (pp. 396–398). Duke University Press.

Friedman, A. (1977). *Industry and labour: Class struggles at work under monopoly capitalism.* MacMillan. doi:10.1007/978-1-349-15845-4

Fullagar, S., & Wilson, E. (2012). Critical pedagogies: A reflexive approach to knowledge creation in tourism and hospitality studies. *Journal of Hospitality and Tourism Management, 19*(1), 1–6. doi:10.1017/jht.2012.3

Giroux, H. (1998). Giriş. okuma-yazma ve siyasal güçlenme eğitbilimi. In *Okuryazarlık, sözcükleri ve dünyayı okuma* (S. Ayhan, Trans.). İmge Yayınları.

Giroux, H. (2021). Critical pedagogy. In U. Bauer, U. H. Bittlingmayer, & A. Scherr (Eds.), *Handbuch bildungs- und erziehungssoziologie*. Springer. doi:10.1007/978-3-658-31395-1_19-1

Giroux, H. A. (2007). *Elestirel pedagoji ve neoliberalizm* (B. Baysal, Trans.). Kalkedon.

Giroux, H. A. (2008). *Elestirel pedagojinin vaadi* (U. D. Tuna, Trans.). Kalkedon Yayinlari.

Glesne, C. (2013). *Nitel araştırmaya giriş* (A. Ersoy & P. Yalçınoğlu, Trans.). Anı Yayıncılık.

Hytten, K. (2006). Education for social justice: Provocations and challenges. *Educational Theory, 56*(2), 221–236. doi:10.1111/j.1741-5446.2006.00013.x

İnal, K. (2009). Eleştirel pedagoji: Ezilenler için eğitim. *Journal of Critical Pedagogy, 1*(1), 2–10.

International Labour Organization. (2010). *Developments and challenges in the hospitality and tourism sector.* ILO. https://www.ilo.org/wcmsp5/groups/public/---ed_norm/---relconf/documents/meetingdocument/wcms_166938.pdf

Jamal, T., Kircher, J., & Donaldson, J. P. (2021). Re-visiting design thinking for learning and practice: Critical pedagogy, conative empathy. *Sustainability (Basel), 13*(2), 964. doi:10.3390u13020964

Kaya, İ. (2012). Türk turizm sektöründe istihdamın temel özelliklerinin ve sorunlarının ücretli çalışanlar açısından coğrafi bölgelere göre analizi. *Balıkesir Üniversitesi Sosyal Bilimler Enstitüsü Dergisi, 15*(28), 241–257.

Kincheloe, J. L. (2007). Critical pedagogy in the Twenty-first century. *Evolution; International Journal of Organic Evolution.*

Ledwith, M. (2001). Community work as critical pedagogy: Re-envisioning Freire and Gramsci. *Community Development Journal: An International Forum, 36*(3), 171–182. doi:10.1093/cdj/36.3.171

Marcuse, H. (1976). One-dimensional man. *Persona y Derecho, 3*, 690.

McKercher, B. (2002). The future of tourism education: An Australian scenario? *Tourism and Hospitality Research, 3*(3), 199–210. doi:10.1177/146735840200300302

McKernan, J. A. (2013). The origins of critical theory in education: Fabian socialism as social reconstructionism in nineteenth-century Britain. *British Journal of Educational Studies, 61*(4), 417–433. doi:10.1080/00071005.2013.824947

McLaren, P. (1998). *Life in schools: An introduction to critical pedagogy in the foundations of education.* Longman.

McLean, M. (2006). *Pedagogy and the university: Critical theory and practice.* Bloomsbury Publishing.

Neuman, W. L. (2014). *Social research methods: Qualitative and quantitative approaches.* Pearson.

Poulston, J. (2008). Hospitality workplace problems and poor training: A close relationship. *International Journal of Contemporary Hospitality Management, 20*(4), 412–427. doi:10.1108/09596110810873525

Prasad, P. (2005). *Crafting qualitative research: Working in the postpositivist traditions.* ME Shape.

Rouzrokh, M., Muldoon, M., Torabian, P., & Mair, H. (2017). The memory-work sessions: Exploring critical pedagogy in tourism. *Journal of Hospitality, Leisure, Sport and Tourism Education, 21*, 163–173. doi:10.1016/j.jhlste.2017.08.006

Scott, J. (1992). *Domination and the arts of resistance: Hidden transcripts.* Yale University Press.

Sheldon, P. J., Fesenmaier, D. R., & Tribe, J. (2011). The tourism education futures initiative (TEFI): Activating change in tourism education. *Journal of Teaching in Travel & Tourism, 11*(1), 2–23. doi:10.1080/15313220.2011.548728

SHTM (school of Hotel & Tourism Management) (n.d.). *Undergraduate Programmes.* SHTM. https://www.polyu.edu.hk/shtm/study/undergraduate-programmes/undergraduate-programme/bachelor-of-science-scheme-in-hotel-and-tourism-management/specialism-or-secondary-major/

Silva, P. (2006). Effects of disposition on hospitality employee job satisfaction and commitment. *International Journal of Contemporary Hospitality Management, 18*(4), 317–328. doi:10.1108/09596110610665320

Ünlüönen, K. ve Çeti, B. (2019). Salgın hastalıklar sebebiyle oluşan krizlerin turizm sektörü üzerindeki etkisinin değerlendirilmesi. *Ankara Hacı Bayram Veli Üniversitesi Turizm Fakültesi Dergisi, 22*(2), 109–128.

Werry, M. (2008). Pedagogy of/as/and tourism: Or, shameful lessons. *Review of Education, Pedagogy & Cultural Studies, 30*(1), 14–42. doi:10.1080/10714410701566207

Wiggershaus, R. (1994). *The Frankfurt School.* Polity Press.

Yayla, Ö., & Ergün, G. S. (2020). Sosyal bilimlerde paradigma ikilemi: Turizm alanında yazılmış lisansüstü tezler üzerine bir inceleme. *Türk Turizm Araştırmaları Dergisi, 4*(2), 1624–1639. doi:10.26677/TR1010.2020.416

Yıldırım, M. (2021). Tracing the consent, adaptation and resistance practices of an 'unsustainable' workforce: The governmentality of workplaces in tourism industry. *Tourism Management, 84*, 104260. doi:10.1016/j.tourman.2020.104260

Yorgun, S. (2013). Türkiye'de turizm sektöründe esnek istihdam uygulamaları ve toplu iş sözleşmelerinde yer alan düzenlemeler. *Turar Turizm & Araştırma Dergisi, 2*(2), 41–56.

ADDITIONAL READING

Bourdieu, P., & Passeron, J. C. (1990). Reproduction in education, society and culture. *Sage (Atlanta, Ga.).*

Freire, P. (1993). *Pedagogy of the oppressed.* Penguin Books.

Giroux, H. (2021). Critical Pedagogy. In U. Bauer, U. H. Bittlingmayer, & A. Scherr (Eds.), *Handbuch Bildungs- und Erziehungssoziologie.* Springer. doi:10.1007/978-3-658-31395-1_19-1

KEY TERMS AND DEFINITIONS

Critical Management Studies: Critical management studies advocate an approach to management theory and practice that encompasses much more than a narrow focus on improving efficiency, effectiveness, and profitability by combining broader issues such as justice, human development, and community well-being.

Critical Paradigm: The critical paradigm questions the status of stereotyped historical, economic, social, and cultural values that are human products; It examines alienation, commodification, and oppression by making a general criticism of ideology.

Critical Pedagogy: Critical pedagogy is a movement that involves teaching and learning relationships so that students acquire self-awareness and social awareness and take appropriate measures against oppressive forces.

Critical Theory: Critical theory claims that scientific knowledge can only be produced through criticism.

Compilation of References

Aarseth, H., & Olsen, B. M. (2008). Food and masculinity in dual-career couples. *Journal of Gender Studies, 17*(4), 277–287. doi:10.1080/09589230802419922

Abou-Shouk, M. A., Mannaa, M. T., & Elbaz, A. M. (2021). Women's empowerment and tourism development: A cross-country study. *Tourism Management Perspectives, 37.* https://doi.org/doi:10.1016/j.tmp.2020.100782

Abraham, G. Y. (2014). Critical pedagogy: Origin, vision, action & consequences. *KAPET, 10*(1), 90–98.

Abril, E. P., Tyson, K., & Morefield, K. (2022). SnapChat this, Instagram that the interplay of motives and privacy affordances in college students' sharing of food porn. *Telematics and Informatics, 74,* 1–12. doi:10.1016/j.tele.2022.101889

Achieng, O. L. (2021). *Effect of training on employee commitment at Prideinn Azure Hotel.* School of Hospitality and Tourism Management.

Aday, J. S., Mitzkovitz, C. M., Bloesch, E. K., Davoli, C. C., & Davis, A. K. (2020). Long-term effects of psychedelic drugs: A systematic review. *Neuroscience and Biobehavioral Reviews, 113,* 179–189. doi:10.1016/j.neubiorev.2020.03.017 PMID:32194129

Addo, E. (2011). European heritage and cultural diversity: The bricks and mortar of Ghana's tourism industry. *Journal of Contemporary African Studies, 29*(4), 405–425. doi:10.1080/02589001.2011.607017

Adebayo, P. F., & Ojo, E. O. (2012). Food security in Nigeria: An overview. *European Journal of Sustainable Development, 1*(2), 199–199. doi:10.14207/ejsd.2012.v1n2p199

Adiyia, B., Vanneste, D., & Van Rompaey, A. (2017). The poverty alleviation potential of tourism employment as an off-farm activity on the local livelihoods surrounding Kibale National Park, western Uganda. *Tourism and Hospitality Research, 17*(1), 34–51. doi:10.1177/1467358416634156

Adoni, H. (1979). The functions of mass media in the political socialization of adolescents. *Communication Research, 6*(1), 84–106. doi:10.1177/009365027900600106

Adu-Ampong, E. A. (2019). Historical Trajectories of Tourism Development Policies and Planning in Ghana, 1957–2017. *Tourism Planning & Development*, *16*(2), 124–141. doi:10.1080/21568316.2018.1537002

Adu-Ampong, E. A., & Mensah, C. (2021). Tourism and hospitality research in the peripheries: Thematic focus and a research agenda from Ghana. *African Journal of Hospitality, Tourism and Leisure*, *10*(3), 1092–1109. doi:10.46222/ajhtl.19770720-151

AFP. (2015). *Boxing orangutans a throwback as Thailand warms to animal rights*. AFP. https://tribune.com.pk/story/948705/boxing-orangutans-a-throwback-as-thailand-warms-to-animal-rights

Afshari-Jouybari, H., & Farahnaky, A. (2011). Evaluation of Photoshop software potential for food colorimetry. *Journal of Food Engineering*, *106*(2), 170–175. doi:10.1016/j.jfoodeng.2011.02.034

Agoston, G. (1999). *European warfare 1453–1815*. Springer.

Agustina, R., Yusuf, M., Sutiyan, O. S. J., Ardianto, R., & Norvadewi, N. (2022). Employee performance mediated quality of work-life relationship satisfaction on the job and organizational commitment. *Jurnal Darma Agung*, *30*(2), 589–605.

Ahmed, M. A., & Khalid, S. (2019). Empowering Leadership and Proactive Behavior: Mediating Role of Psychological Empowerment and Moderating Role of Leader-Follower Distance. *Abasyn University Journal of Social Sciences, 12*(1).

Airey, D., & Johnson, S. (1999). The content of tourism degree courses in the UK. *Tourism Management*, 20.

Airey, D. (1979). Tourism education in the United Kingdom. *Revue de Tourisme*, *2*(79), 13–15.

Airey, D. (2005). Growth and development. In D. Airey & J. Tribe (Eds.), *An international handbook of tourism education*. Elsevier. doi:10.1016/B978-0-08-044667-7.50005-7

Akanle, O., & Olutayo, O. (2011). Kinship construction variability among Nigerian international migrants: The context of contemporary Diaspora. *Human Affairs*, *21*(4), 470–480. doi:10.247813374-011-0044-2

Akena, F. A. (2022). The "Miss Curvy Uganda" pageant: Representation, commodification, and exploitation of women's bodies. *Social Semiotics*, *32*(2), 224–239. doi:10.1080/10350330.2020.1743524

Akoğlan-Kozak, M. (1996). Konaklama endüstrisinde kadının konumu. *Anatolia Turizm Araştırmaları Dergisi*, *7*(3-4), 16–23.

Akyeampong, O. (2009). Tourism development in Ghana, 1957-2007. *Legon Journal of Sociology*, *3*(2), 1–23.

Akyeampong, O. A. (2011). Pro-poor tourism: Residents' expectations, experiences and perceptions in the Kakum National Park Area of Ghana. *Journal of Sustainable Tourism, 19*(2), 197–213. doi:10.1080/09669582.2010.509508

Aladwan, K., Bhanugopan, R., & D'Netto, B. (2015). The effects of human resource management practices on employees' organizational commitment. *The International Journal of Organizational Analysis, 23*(3), 472–492. doi:10.1108/IJOA-11-2014-0822

Alber, E., Häberlein, T., & Martin, J. (2010). Changing Webs of Kinship: Spotlights on West Africa. *Africa Spectrum, 45*(3), 43–67. doi:10.1177/000203971004500303

Allen, N. J., & Meyer, J. P. (1990). The measurement and antecedents of affective, continuance, and normative commitment to the organization. *Journal of Occupational Psychology, 63*(1), 1–18. doi:10.1111/j.2044-8325.1990.tb00506.x

Allen, P., & Sachs, C. (2007). Women and food chains: The gendered politics of food. *International Journal of Sociology of Agriculture and Food, 15*(1), 1–23. doi:10.48416/ijsaf.v15i1.424

Alrawadieh, Z., Demirdelen Alrawadieh, D., Olya, H. G. T., Erkol Bayram, G., & Kahraman, O. C. (2022). Sexual harassment, psychological well-being, and job satisfaction of female tour guides: The effects of social and organizational support. *Journal of Sustainable Tourism, 30*(7), 1639–1657. doi:10.1080/09669582.2021.1879819

Alruwaily, A., Mangold, C., Greene, T., Arshonsky, J., Cassidy, O., Pomeranz, J. L., & Bragg, M. (2020). Child social media influencers and unhealthy food product placement. *Pediatrics, 146*(5), 1–10. doi:10.1542/peds.2019-4057 PMID:33106342

Althusser, L. (1989). *İdeoloji ve Devletin İdeolojik Aygıtları* (Y. Alp & M. Özışık, Trans.). İletişim Yayınları.

Althusser, L. (2006). Ideology and ideological state apparatuses (notes towards an investigation). *The anthropology of the state. REAd (Porto Alegre), 9*(1), 86–98.

Amable, B. (2003). *The diversity of modern capitalism.* Oxford University Press. doi:10.1093/019926113X.001.0001

Amankwaa, A., Susomrith, P., & Seet, P.-S. (2022). Innovative behavior among service workers and the importance of leadership: Evidence form an emerging economy. *The Journal of Technology Transfer, 47*(2), 506–530. doi:10.100710961-021-09853-6

Amehule, S. (2022). Delegation of duty and employee commitment in hotels in Port Harcourt. *BW Academic Journal, 1*(1), 10.

Anagnostakis, I. (2013). Byzantine diet and cuisine. In between ancient and modern gastronomy. In I. Anagnostakis (Ed.), *Flavors and delights tastes and pleasures of ancient and Byzantine cuisine* (pp. 43–64).

Anatolian University. (n.d.). *Course Catalog and Orientation Tables.* AU. turizm.anadolu.edu.tr. (n.d.). https://turizm.anadolu.edu.tr/ogrenci/ders-katalog-ve-intibak-tablolari

Anderson, B. (2010). Migration, immigration controls, and the fashioning of precarious workers. *Work, Employment and Society*, *24*(2), 300–317. doi:10.1177/0950017010362141

Ankomah, P. K. (1991). Tourism skilled labor: The case of sub-Saharan Africa. *Annals of Tourism Research*, *18*(3), 433–442. doi:10.1016/0160-7383(91)90050-L

Annor, F., & Burchell, B. (2017). A cross-national comparative study of work demands/support, work-to-family conflict and job outcomes. *International Journal of Cross Cultural Management*, *18*(1), 53–72. doi:10.1177/1470595817746195

Ap, J., & Wong, K. K. (2001). Case study on tour guiding: Professionalism, issues and problems. *Tourism Management*, *22*(5), 551–563. doi:10.1016/S0261-5177(01)00013-9

Apple, M. W. (2010). *Global crises, social justice and education*. Routledge.

Apple, M. W. (2011). Democratic education in neoliberal and neoconservative times. *International Studies in Sociology of Education*, *21*(1), 21–31. doi:10.1080/09620214.2011.543850

Apple, M. W. (2012). *Education and power*. Routledge. doi:10.4324/9780203143124

Aria, M., & Cuccurullo, C. (2017). Bibliometrix: An R-tool for comprehensive science mapping analysis. *Journal of Informetrics*, *11*(4), 959–975. doi:10.1016/j.joi.2017.08.007

Arjona, R. (2017). John Calvin on the Lord's Supper: Food, rest, and healing for shivering souls. *Pastoral Psychology*, *66*(2), 177–190. doi:10.100711089-016-0740-2

Aronsson, K., & Gottzén, L. (2011). Generational positions at a family dinner: Food morality and social order. *Language in Society*, *40*(4), 405–426. doi:10.1017/S0047404511000455

Arslan, A., & Şimşek, G. (2018). Kadın turist rehberlerinin yaşadıkları mesleki sorunlar: Aydın turist rehberleri örneği. *Journal of Travel and Tourism Research*, (13), 23–49.

Aryeetey, E., Baffour, P. T., & Turkson, F. E. (2021). *Employment creation potential, labor skills requirements, and skill gaps for young people: Ghana case study*. African Growth Initiative at Brookings.

Ashley, C., Boyd, C., & Goodwin, H. (2000). Pro-poor tourism: Putting poverty at the heart of the tourism agenda. *Natural Resource Perspectives*, *51*. http://www.odi.org.uk/nrp/51.pdf

Ateş, M., Kazan, İ., & Soybalı, H. H. (2022). Turist rehberliğinde ayrımcılık: Kadın turist rehberlerine yönelik bir uygulama. [TURED]. *Turist Rehberliği Dergisi*, *5*(1), 1–16. doi:10.34090/tured.1064528

Ayaz, N., & Demir, C. (2019). Perceived work stress factors: A study on tour guidesç. *Journal of Tourism and Gastronomy Studies*, *7*(1), 415–427. doi:10.21325/jotags.2019.370

Ayikoru, M., & Park, H. Y. (2019). Films and critical pedagogy in management education: A tourism studies context. *Academy of Management Learning & Education*, *18*(3), 414–432. doi:10.5465/amle.2015.0134

Aykaç, A. (2006). Linking scale of operation and labor transformation on tourism: Comparative analysis on three Turkish cases. *Anatolia*, *17*(2), 257–277. doi:10.1080/13032917.2006.9687189

Aykaç, A. (2007). Labor transformation through the prism of the household: Evidence from the Turkish tourism industry. *New Perspectives on Turkey*, *37*, 87–113. doi:10.1017/S089663460000474X

Aykaç, A. (2009). *Yeni işler, yeni işçiler: Turizm sektöründe emek*. İletişim Yayınları.

Aynalem, S., Birhanu, K., & Tesefay, S. (2016). Employment opportunities and challenges in tourism and hospitality sectors. *Journal of Tourism & Hospitality (Los Angeles, Calif.)*, *5*(6), 1–5. doi:10.4172/2167-0269.1000257

Bae, S. (2021). Evaluating hospitality employees' various relationships and the effects on organizational commitment. *International Hospitality Review*. doi:10.1108/IHR-01-2021-0005

Bagwell, L. S., & Bernheim, B. D. (1996). Veblen effects in a theory of conspicuous consumption. *The American Economic Review*, *86*(3), 349–373. https://www.jstor.org/stable/2118201

Bakker, M. (2019). A Conceptual Framework for Identifying the Binding Constraints to Tourism-Driven Inclusive Growth. *Tourism Planning & Development*, *16*(5), 575–590. doi:10.1080/21568316.2018.1541817

Bakker, M., & Messerli, H. R. (2017). Inclusive growth versus pro-poor growth: Implications for tourism development. *Tourism and Hospitality Research*, *17*(4), 384–391. doi:10.1177/1467358416638919

Bastian, M., Heymann, S., & Jacomy, M. (2009). Gephi: An open source software for exploring and manipulating networks. *International AAAI Conference on Weblogs and Social Media*. AAAI. 10.1609/icwsm.v3i1.13937

Baum, T. (1999). Seasonality in tourism: Understanding the challenges: Introduction. *Tourism Economics*, *5*(1), 5–8. doi:10.1177/135481669900500101

Baum, T. (2015). Human resources in tourism: Still waiting for change? *Tourism Management*, *50*, 204–212. doi:10.1016/j.tourman.2015.02.001

Baum, T., & Lundtorp, S. (2001). *Seasonality in tourism: An introduction. Seasonality in tourism*. Elsevier. doi:10.4324/9780080516806

Bays, J. C. (2009). *Mindful Eating: A guide to rediscovering a healthy and joyful relationship with food--includes CD*. Shambhala Publications.

Bays, J. C. (2009). *Mindful Eating: A guide to rediscovering a healthy and joyful relationship with food--includes CD*. Shambhala Publications. Mindful Eating - Google Books

Belhassen, Y., & Caton, K. (2011). On the need for critical pedagogy in tourism education. *Tourism Management*, *32*(6), 1389–1396. doi:10.1016/j.tourman.2011.01.014

Benson, E. P. (2001). Why sacrifice? In E. P. Benson & A. G. Cook (Eds.), *Ritual sacrifice in ancient Peru* (pp. 1–20). University of Texas Press., doi:10.7560/708938-002

Bergen, R. K., & Bogle, K. A. (2000). Exploring the connection between pornography and sexual violence. *Violence and Victims*, *15*(3), 227–234. doi:10.1891/0886-6708.15.3.227 PMID:11200099

Berke-Berga, A., Dovladbekova, I., Sumilo, E., & Baumane-Vitolina, I. (2015). Development of employee integration: Share ownership perspective. *Proceedings of Economic Science for Rural Development*, *39*, 79–88.

Berthoud, H. R., Zheng, H., & Shin, A. C. (2012). Food reward in the obese and after weight loss induced by calorie restriction and bariatric surgery. *Annals of the New York Academy of Sciences*, *1264*(1), 36–48. doi:10.1111/j.1749-6632.2012.06573.x PMID:22616827

Beugré, C. D., & Offodile, O. F. (2001). Managing for organizational effectiveness in sub-Saharan Africa: A culture-fit model. *International Journal of Human Resource Management*, *12*(4), 535–550. doi:10.1080/713769650

Bhaskar, R. (2013). *A realist theory of science*. Routledge. doi:10.4324/9780203090732

Bies, A. R. (2019). *Organizational Socialization Tactics and Learning in the United States Hotel Industry* [Doctoral dissertation, Purdue University Graduate School].

Bigo, D. (2019). Shared secrecy in a digital age and a transnational world. *Intelligence and National Security*, *34*(3), 379–394. doi:10.1080/02684527.2019.1553703

Blackwood, E. (1997). Women, Land, and Labor: Negotiating Clientage and Kinship in a Minangkabau Peasant Community. *Ethnology*, *36*(4), 277–293. doi:10.2307/3774038

Bohman, J. (2005). Critical theory. Edward N. Zalta (ed.). Stanford Encyclopedia of Philosophy. Stanford Press.

Bonny-Noach, H., & Sagiv-Alayoff, M. (2020). Cannabis tourist destinations: Risk for vulnerable travellers with pre-existing mental disorders. *Journal of Travel Medicine*, *27*(4), taz098. doi:10.1093/jtm/taz098 PMID:31821507

Booyens, I. (2022). Social innovation for sustainable tourism development. In I. Booyens & P. Brouder (Eds.), *Handbook of innovation for sustainable tourism* (pp. 193–209). Edward Elgar. doi:10.4337/9781800372740.00020

Bound Alberti, F. (2021). Fat shaming, feminism and Facebook: What ‘women who eat on tubes’ reveal about social media and the boundaries of women’s bodies. *European Journal of Cultural Studies*, *24*(6), 1304–1318. doi:10.1177/13675494211055499

Bourdieu, P. (1994). *Distinction: A social critique of the judgment of taste* (R. Niece, Trans.). Harvard University Press.

Bourdieu, P. (2015). *Pratik nedenler* (H. U. Tanrıöver, Trans.). Hil Yayın.

Bourdieu, P. (2017). *Ayrım: Beğeni yargısının toplumsal eleştirisi* (D. Fırat & F. Berkkurt, Trans.). Heretik.

Bourdieu, P. (2018). *Eril tahakküm* (B. Yılmaz, Trans.). Bağlam.

Bourdieu, P., & Passeron, J. C. (1990). Reproduction in education, society and culture. *Sage (Atlanta, Ga.)*.

Boxill, I. (2000). Overcoming social problems in the Jamaican tourism industry. *Tourism in the Caribbean*, 17-40.

Bratley, K. J., & Aloysius, S. M. (2019). Transparency in managerial practices and affective commitment. *The Journal of Business*, *6*(2), 2019–2061.

Braverman, H. (1998). *Labor and monopoly capital: The degradation of work in the twentieth century*. Monthly Review Press.

Brooks, G. (2012). Misbehavior, its dimensions, and relationship to commitment in organizations. *Rethinking misbehavior and resistance in organizations, 19*, 237-257.

Bryce, T. (2005). *The kingdom of the Hittites*. OUP Oxford. doi:10.1093/acprof:oso/9780199281329.001.0001

Bryon, J. (2012). Tour guides as storytellers–from selling to sharing. *Scandinavian Journal of Hospitality and Tourism*, *12*(1), 27–43. doi:10.1080/15022250.2012.656922

Bugdol, M., Jedynak, P., Bugdol, M., & Jedynak, P. (2015). Integration in Different Organizational Situations. *Integrated Management Systems*, 161-192.

Burawoy, M. (1979). *Manufacturing consent*. The University of California Press.

Burgoyne, C., & Mearns, K. (2017). Managing stakeholder relations, natural resources and tourism: A case study from Ololosokwan, Tanzania. *Tourism and Hospitality Research*, *17*(1), 68–78. doi:10.1177/1467358416639088

Butler, R. (1998). Seasonality in tourism: Issues and implications. *Tourism Review*, *53*(3), 18–24.

Büyüktepe, H. S., Gökdemir, S., & Korkmaz, H. (2019). Turist rehberlerinin sorunları üzerine nitel bir araştırma: Çanakkale örneği. *Journal of Travel and Tourism Research*, (14), 94–117.

Byrne, E., Kearney, J., & MacEvilly, C. (2017). The role of influencer marketing and social influencers in public health. *Proceedings of the Nutrition Society, 76*(OCE3), E103, 21-23. 10.1017/S0029665117001768

Byrtek, G. J. (2000). Creating shared values: An exploratory examination of methodology. [Doctoral dissertation, Walden University].

Çakmak, T. F. (2020). Turist rehberlerini yasa dışı çalışmaya yönlendiren etmenler üzerine bir araştırma. *Journal of Hospitality and Tourism Issues*, *2*(2), 91–107.

Caldwell, D. F., Chatman, J. A., & O'Reilly, C. A. (1990). Building organizational commitment: A multi-firm study. *Journal of Occupational Psychology*, *63*(3), 245–261. doi:10.1111/j.2044-8325.1990.tb00525.x

Caldwell, R. (1996). Creating commitment to change: From a closed to open communication style. In D. Hussey (Ed.), *The Implementation Challenge* (pp. 229–239).

Çalık, A. Ö., & Çiftçi, G. (2013). Animal ethics in tourism. *International Journal of Business and Management Studies*, *1*(5), 43–53.

Callon, M., Courtial, J. P., Turner, W. A., & Bauin, S. (1983). From translations to problematic networks: An introduction to co-word analysis. *Social Sciences Information. Information Sur les Sciences Sociales*, *22*(2), 191–235. doi:10.1177/053901883022002003

Calvet, N.A., Conde, C.I., Ballart, A.L., & Almela, M.S. (2021). *Gender inequalities in tourism labor market*. Alba sud editorial.

Cang, V. (2018). Japan's Washoku as intangible heritage: The role of national food traditions in UNESCO's cultural heritage scheme. *International Journal of Cultural Property*, *25*(4), 491–513. doi:10.1017/S0940739118000267

Cankül, D., & Güzeloğlu, F. T. (2021). Profesyonel turist rehberlerinin yurtdışı turlarında yaşadıkları sorunlar. *Journal of Humanities and Tourism Research*, *11*(4), 844–862. doi:10.14230/johut1110

Cannas, R. (2012). An overview of tourism seasonality: Key concepts and policy. *AlmaTourism*, *3*(5), 40–58.

Carhart-Harris, R. L., Bolstridge, M., Day, C. M., Rucker, J., Watts, R., Erritzoe, D. E., Kaelen, M., Giribaldi, B., Bloomfield, M., Pilling, S., Rickard, J. A., Forbes, B., Feilding, A., Taylor, D., Curran, H. V., & Nutt, D. J. (2018). Psilocybin with psychological support for treatment-resistant depression: Six-month follow-up. *Psychopharmacology*, *235*(2), 399–408. doi:10.100700213-017-4771-x PMID:29119217

Carhart-Harris, R. L., Bolstridge, M., Rucker, J., Day, C. M., Erritzoe, D., Kaelen, M., Bloomfield, M., Rickard, J. A., Forbes, B., Feilding, A., Taylor, D., Pilling, S., Curran, V. H., & Nutt, D. J. (2016). Psilocybin with psychological support for treatment-resistant depression: An open-label feasibility study. *The Lancet. Psychiatry*, *3*(7), 619–627. doi:10.1016/S2215-0366(16)30065-7 PMID:27210031

Carod-Artal, F. J., & Vazquez-Cabrera, C. B. (2006). Mescaline and the San Pedro cactus ritual: Archaeological and ethnographic evidence in northern Peru. *Revista de Neurología*, *42*(8), 489–498. doi:10.33588/rn.4208.2006040 PMID:16625512

Carr, N. (2021). Redefining the work of working animals in the tourism industry: An animal-centric reflection. In C. Kline & J. Rickly (Eds.), *Exploring non-human work in tourism: From beasts of burden to animal ambassadors* (pp. 37–50). De Gruyter Oldenbourg. doi:10.1515/9783110664058-003

Carr, N., & Broom, D. M. (2018). *Tourism and animal welfare*. CABI International. doi:10.1079/9781786391858.0000

Carvalho, I., Costa, C., Lykke, N., & Torres, A. (2019). Beyond the glass ceiling: Gendering tourism management. *Annals of Tourism Research*, *75*, 79–91. doi:10.1016/j.annals.2018.12.022

Casanueva, C., Gallego, Á., & García-Sánchez, M. R. (2016). Social network analysis in tourism. *Current Issues in Tourism*, *19*(12), 1190–1209. doi:10.1080/13683500.2014.990422

Castañeda, A., Doan, D., Newhouse, D., Nguyen, M. C., Uematsu, H., & Azevedo, J. P. (2018). A new profile of the global poor. *World Development*, *101*, 250–267. doi:10.1016/j.worlddev.2017.08.002

Castro, C. B., Periñan, M. M. V., & Bueno, J. C. (2008). Transformational leadership and followers' attitudes: The mediating role of psychological empowerment. *International Journal of Human Resource Management*, *19*(10), 1842–1863. doi:10.1080/09585190802324601

Cave, P., & Kılıç, S. (2010). The role of women in tourism employment with special reference to Antalya, Turkey. *Journal of Hospitality Marketing & Management*, *19*(3), 280–292. doi:10.1080/19368621003591400

Celidwen, Y., Redvers, N., Githaiga, C., Calambás, J., Añaños, K., Chindoy, M. E., Vitale, R., Rojas, J. N., Mondragón, D., Rosalío, Y. V., & Sacbajá, A. (2023). Ethical principles of traditional Indigenous medicine to guide western psychedelic research and practice. *Lancet Regional Health. Americas*, *18*, 100410. doi:10.1016/j.lana.2022.100410 PMID:36844020

Çelik Uğuz, S., & Kaya, İ. (2016). Informal employment in tourism. In U. Yedipınar & E. Yıldız (Eds.), *Global Issues and Trends in Tourism*. St. Kliment Ohridski University Press.

Çelik, C., & Erkuş-Öztürk, H. (2016). Role of precariousness and space on class formation process: The case of Antalya's tourism workers. *Capital and Class*, *40*(3), 419–445. doi:10.1177/0309816816653879

Çetin, A. (2022). The extraordinary show of food: A sociological evaluation on some Instagram content. *Turkish Studies-Social Sciences*, *17*(4), 517–532.

Ceylan, U. (2021). Turist rehberlerinin mesleki motivasyonlarında tur üyelerinin yanlış davranışlarının etkisi. *Anemon Muş Alparslan Üniversitesi Sosyal Bilimler Dergisi*, *9*(1), 39–49. doi:10.18506/anemon.683064

Chalkiti, K., & Sigala, M. (2010). Staff turnover in the Greek tourism industry. *International Journal of Contemporary Hospitality Management*, *22*(3), 335–359. doi:10.1108/09596111011035945

Chan, A., Hsu, C. H., & Baum, T. (2015). The impact of tour service performance on tourist satisfaction and behavioral intentions: A study of Chinese tourists in Hong Kong. *Journal of Travel & Tourism Marketing*, *32*(1-2), 18–33. doi:10.1080/10548408.2014.986010

Chang, K. C. (2014). Examining the effect of tour guide performance, tourist trust, tourist satisfaction, and flow experience on tourists' shopping behavior. *Asia Pacific Journal of Tourism Research*, *19*(2), 219–247. doi:10.1080/10941665.2012.739189

Chang, T. Y., Kung, S. F., & Luh, D. B. (2012). The innovative service model: A study on tour guide service design in Taiwan by ICT application. *Academic Research International*, *2*(1), 192–198. doi:10.1300/J162v01n04_06

Cha, S. E., & Edmondson, A. C. (2006). When values backfire: Leadership, attribution, and disenchantment in a values-driven organization. *The Leadership Quarterly*, *17*(1), 57–78. doi:10.1016/j.leaqua.2005.10.006

Chathoth, P. K., Chan, E. S., Harrington, R. J., Okumus, F., & Song, Z. (2022). The Role of Organizational Factors on Employee Engagement and Hospitality Service Co-creation. *Journal of China Tourism Research*, *18*(4), 756–777. doi:10.1080/19388160.2021.1972068

Chathoth, P. K., Harrington, R. J., Chan, E. S., Okumus, F., & Song, Z. (2020). Situational and personal factors influencing hospitality employee engagement in value co-creation. *International Journal of Hospitality Management*, *91*, 102687. doi:10.1016/j.ijhm.2020.102687

Chauhan, J., Hedaoo, R., & Patil, M. (2022). Gastroporn on Social Media and Its Association with Food Choices and Body Mass Index Among Youth. *Advanced Studies in Biology*, *14*(1), 137–151. doi:10.12988/asb.2022.91588

Chen, Z, Huang, S, Liu, C, Min, M, & Zhou, L. (2018). Fit between Organizational Culture and Innovation Strategy: Implications for Innovation Performance. *Sustainability*, *10*(10), 3378. MDPI. doi:10.3390/su10103378

Chen, C. (2004). Searching for intellectual turning points: Progressive knowledge domain visualization. *Proceedings of the National Academy of Sciences of the United States of America*, *101*(suppl_1), 5303–5310. doi:10.1073/pnas.0307513100 PMID:14724295

Cheong, S., & Miller, M. (2000). Power and Tourism. A Foucauldian Observation. *Annals of Tourism Research*, *27*(2), 371–390. doi:10.1016/S0160-7383(99)00065-1

Chiang, C. Y., & Chen, W. C. (2014). The impression management techniques of tour leaders in group package tour service encounters. *Journal of Travel & Tourism Marketing*, *31*(6), 747–762. doi:10.1080/10548408.2014.889641

Chichaibelu, B. B., Bekchanov, M., Von Braun, J., & Torero, M. (2021). The global cost of reaching a world without hunger: Investment costs and policy action opportunities. *Food Policy*, *104*(12), 102151. doi:10.1016/j.foodpol.2021.102151

Chin, K. L. (2017). *The golden triangle: Inside southeast Asia's drug trade*. Cornell University Press.

Chiwawa, N., & Wissink, H. (2021). Determinants of employee engagement in the South African hospitality industry during COVID-19 lockdown epoch: Employee perception. *African Journal of Hospitality, Tourism and Leisure*, *10*(2), 487–499. doi:10.46222/ajhtl.19770720.113

Chowdhary, N., & Prakash, M. (2008). Challenges of tourist guiding-An assessment of situation in India. *Tourism Infrastructure. Technology & Operations*, (Part VII), 287–302.

Choy, D. J. (1995). The quality of tourism employment. *Tourism Management, 16*(2), 129–137. doi:10.1016/0261-5177(94)00023-4

Christie, M. F., & Mason, P. A. (2003). Transformative tour guiding: Training tour guides to be critically reflective practitioners. *Journal of Ecotourism, 2*(1), 1–16. doi:10.1080/14724040308668130

Ciobanu, E., & Florescu, C. M. (2019). Food Porn in, and (). *East-West Cultural Passage, 19*(2), 96-120. . https://doi.org/ doi:10.2478/ewcp-2019-0014

Çıvak, B. (2021). *Otel işletmelerinde emek sömürüsü, tahakküm ve direniş ilişkilerine dair niteliksel bir araştırma* [Unpublished doctoral dissertation, Anadolu University, Eskişehir].

Çıvak, B. (2021). *Otel işletmelerinde emek sömürüsü, tahakküm ve direniş ilişkilerine dair niteliksel bir araştırma.* [Unpublished doctoral dissertation, Anadolu Üniversitesi, Sosyal Bilimler Enstitüsü, Eskişehir, Türkiye].

Çıvak, B. ve Sezerel, H. (2018). Araştırma paradigmaları ve turizm yazını. *Turizm Akademik Dergisi, 5*(1), 1–14.

Çıvak, B., & Besler, S. (2022). A Critical Analysis of Managerial Control Mechanisms in Hotel Businesses. *SAGE Open, 12*(1). doi:10.1177/21582440221088850

Çıvak, B., & Emir, O. (2021). Boş zaman ve rekreasyon konulu lisansüstü tezlerde araştırma paradigması eğilimi. *Journal of Tourism & Gastronomy Studies, 9*(3), 2186–2202.

Civitello, L. (2004). *Cuisine and Culture: A History of food and people.* John Wiley&Sons.

Claudio-Quiroga, G., Gil-Alana, L. A., Gil-López, A., & Babinger, F. (2022). A gender approach to the impact of COVID-19 on tourism employment. *Journal of Sustainable Tourism.* doi:10.1080/09669582.2022.2068021

Cloke, P., & Perkins, H. C. (2005). Cetacean performance and tourism in Kaikoura, New Zealand. *Environment and Planning. D, Society & Space, 23*(6), 903–924. doi:10.1068/d57j

Cobbinah, P. B., & Darkwah, R. M. (2016). Reflections on tourism policies in Ghana. *International Journal of Tourism Sciences, 16*(4), 170–190. doi:10.1080/15980634.2016.1212595

Cohen, E. (1985). The tourist guide: The origins, structure and dynamics of a role. *Annals of Tourism Research, 12*(1), 5–29. doi:10.1016/0160-7383(85)90037-4

Cohen, S. A., Prayag, G., & Moital, M. (2014). Consumer behavior in tourism: Concepts, influences and opportunities. *Current Issues in Tourism, 17*(10), 872–909. doi:10.1080/13683500.2013.850064

Collins, R. (2019). *Developing a lean internationalisation strategy through open innovation, entrepreneurship and cross-cultural intelligence* [Doctoral dissertation, University of Newcastle].

Collins, C., & McCorkle, W. (2023). Decriminalization of Drugs and the Support for Free Markets in the United States. *Journal of Advanced Research in Social Sciences*, *6*(1), 34–53. doi:10.33422/jarss.v6i1.870

Conlon, C., Timonen, V., Elliott-O'Dare, C., O'Keeffe, S., & Foley, G. (2020). Confused about theoretical sampling? Engaging theoretical sampling in diverse grounded theory studies. *Qualitative Health Research*, *30*(6), 947–959. doi:10.1177/1049732319899139 PMID:31959073

Cooper, C., & Shepherd, R. (1997). The relationship between tourism education and the tourism industry: Implications for tourism education. *Tourism Recreation Research*, *22*(1), 34–47. doi:10.1080/02508281.1997.11014784

Costa, C., & Murphy, M. (2015). *Bourdieu, habitus and social research: The art of application*. Springer.

Costa, H. A., Costa, C., Breda, Z., Çacador, S., Carvalho, I., & Vıeira, R. (2013). Women in Brazil with higher education in tourism: Preliminary results of the Gentour Project. *Revista Turismo & Desenvolvimento (Aveiro)*, *19*, 53–66.

Costen, W. M., & Salazar, J. (2011). The impact of training and development on employee job satisfaction, loyalty, and intent to stay in the lodging industry. *Journal of Human Resources in Hospitality & Tourism*, *10*(3), 273–284. doi:10.1080/15332845.2011.555734

Coulter, K. (2016). *Animals, work, and the promise of interspecies solidarity* (Vol. 26). Palgrave Macmillan. doi:10.1057/9781137558800

Curtin, M., Hossain, T., & Verghese-Choudbury, A. (1997). Children living, learning and working in the city. In J. Beall (Ed.), *A city for all. Valuing difference and working with diversity* (pp. 66–78). Zed Books Ltd.

Dahles, H. (2002). The politics of tour guiding: Image management in Indonesia. *Annals of Tourism Research*, *29*(3), 783–800. doi:10.1016/S0160-7383(01)00083-4

Dalby, A. (2013). *Siren feasts: a history of food and gastronomy in Greece*. Routledge.

Danforth, A. L., Struble, C. M., Yazar-Klosinski, B., & Grob, C. S. (2016). MDMA-assisted therapy: A new treatment model for social anxiety in autistic adults. *Progress in Neuro-Psychopharmacology & Biological Psychiatry*, *64*, 237–249. doi:10.1016/j.pnpbp.2015.03.011 PMID:25818246

Daniel, A. D., Costa, R. A., Pita, M., & Costa, C. (2017). Tourism education: What about entrepreneurial skills? *Journal of Hospitality and Tourism Management*, *30*, 65–72. doi:10.1016/j.jhtm.2017.01.002

Dar, N., & Rahman, W. (2022). Interactional Justice and Deviant Workplace Behavior: The Mediating Role of Perceived Supervisor Support in Hotel Industry of KP-Pakistan. *Journal of Entrepreneurship, Management, and Innovation*, *4*(1), 149–166. doi:10.52633/jemi.v4i1.121

Darvish, H., & Tonta, Y. (2016). Diffusion of nanotechnology knowledge in Turkey and its network structure. *Scientometrics*, *107*(2), 569–592. doi:10.100711192-016-1854-0

Dasgupta, A. (2019). Abuse of magic mushroom, peyote cactus, LSD, khat, and volatiles. In *Critical issues in alcohol and drugs of abuse testing* (pp. 477–494). Academic Press. doi:10.1016/B978-0-12-815607-0.00033-2

Dashper, K. (2021). Conceptualizing non-human animals as "workers" within the tourism industry. In C. Kline & J. Rickly (Eds.), *Exploring non-human work in tourism: From beasts of burden to animal ambassadors* (pp. 21–36). De Gruyter Oldenbourg. doi:10.1515/9783110664058-002

Davids, J. (2012). Have a Heart: Challenges for Lead Vocational Teachers in the Changing VET Landscape. National Centre for Vocational Education Research Ltd.

Dawood Hristova, J. J., & Pérez-Jover, V. (2023). Psychotherapy with Psilocybin for Depression: Systematic Review. *Behavioral Sciences (Basel, Switzerland)*, *13*(4), 297. doi:10.3390/bs13040297 PMID:37102811

Dawson, M., & Abbott, J. (2011). Hospitality culture and climate: A proposed model for retaining employees and creating competitive advantage. *International Journal of Hospitality & Tourism Administration*, *12*(4), 289–304. doi:10.1080/15256480.2011.614533

De Graaf, C., & Kok, F. J. (2010). Slow food, fast food and the control of food intake. *Nature Reviews. Endocrinology*, *6*(5), 290–293. https://www.nature.com/doifinder/10.1038/nrendo.2010.41. doi:10.1038/nrendo.2010.41 PMID:20351697

De Morais Sato, P., Gittelsohn, J., Unsain, R. F., Roble, O. J., & Scagliusi, F. B. (2016). The use of Pierre Bourdieu's distinction concepts in scientific articles studying food and eating: A Narrative review. *Appetite*, *96*, 174–186. doi:10.1016/j.appet.2015.09.010 PMID:26368578

De Zulueta, T. (2003). Migrants in irregular employment in the agricultural sector of Southern European countries. Report for the debate in the standing committee. Council of Europe.

Deale, C. S., & Lee, S. H. (2023). Employee benefits in the hospitality industry: What do employees want and need? *Journal of Human Resources in Hospitality & Tourism*, 1–27. doi:10.1080/15332845.2023.2207998

Dedeoğlu, S. (2004). Dünya ekonomisi ve hayatta kalmanın feminizasyonu: Toplumsal cinsiyet, kalkınma ve küreselleşme. In S. Dedeoğlu & T. Subaşat (Eds.), *Kalkınma ve Küreselleşme*. Bağlam Yayıncılık.

Dejmanee, T. (2016). "Food porn" as postfeminist play: Digital femininity and the female body on food blogs. *Television & New Media*, *17*(5), 429–448. doi:10.1177/1527476415615944

Demircioğlu, A. (2018). *Eleştirel düşünme eğitimi*. Gece Kitaplığı. ehl.edu. https://www.ehl.edu/en/study/bachelor-degree-in-hospitality

Dempster, H., & Zimmer, C. (2020). Migrant workers in the tourism industry: How has COVID-19 affected them, and what does the future hold. *CGD Policy Paper, 173*.

Derviş, H. (2019). Bibliometric Analysis using Bibliometrix an R Package. *Journal of Scientometric Res.*, *8*(3), 156–160. doi:10.5530/jscires.8.3.32

Dewald, S. (2022). Detectives and technological frames: Integrating technology and social media into everyday work. *Policing and Society, 33*(1), 111,118. doi:10.1080/10439463.2022.2064858

Dglawi, M., & Hassel, M. (2021). *Integration of new employees: A study of integration challenges for international companies with homework environment.*

Dietler, M. (1996). Feasts and commensal politics in the political economy: food, power, and status in prehistoric Europe. In P. Wiessner & W. Schiefenhövel (Eds.), *Food and the status quest: An interdisciplinary perspective* (pp. 87–125). Berghahn Books.

Dilek, S. E. (2017). Turizmin ahlaki sorumluluğu ve hayvanlar sorunsalı. In E. Günlü-Küçükaltan & S. E. Dilek (Eds.), *Metalaşan turizm-metalaştırılan hayvanlar* (pp. 9–16). Nobel Akademik Yayıncılık.

Dolatabadi, H. R., & Safa, M. (2010). The effect of directive and participative leadership style on employees' commitment to service quality. *International Bulletin of Business Administration, 9*(1), 31–42.

Donohue, G. A., Tichenor, P. J., & Olien, C. N. (1973). Mass media functions, knowledge and social control. *The Journalism Quarterly, 50*(4), 652–659. doi:10.1177/107769907305000404

Downward, P., & Mearman, A. (2007). Retroduction as mixed-methods triangulation in economic research: Reorienting economics into social science. *Cambridge Journal of Economics, 31*(1), 77–99. doi:10.1093/cje/bel009

Dudley, K. D., Duffy, L. N., Terry, W. C., & Norman, W. C. (2021). The historical structuring of the US tourism workforce: A critical review. *Journal of Sustainable Tourism, 30*(12), 2823–2838. doi:10.1080/09669582.2021.1952417

Düz, B. (2017). Turist rehberliğinde meslek etiği ilkelerinin faydacı etik yaklaşımı ile belirlenmesi. *Hitit Üniversitesi Sosyal Bilimler Enstitüsü Dergisi, 10*(2), 1639–1662. doi:10.17218/hititsosbil.286973

Economists at Large. (2013). The $200 million question: How much does trophy hunting really contribute to African communities? *A report for the African Lion Coalition.* Economists at Large, Melbourne, Australia. https://iwbond.org/2015/09/03/the-200-million-question/

ECPAT. (2019a). *The link between voluntourism, orphanage tourism and child sexual exploitation.* ECPAT. https://ecpat.org/voluntourism-orphanage-tourism-sexual-abuse/

ECPAT. (2019b). *ECPAT country overview: Uganda.* ECPAT. https://ecpat.org/wp-content/uploads/2021/08/ECPAT-Country-Overview-Report-Uganda-April-2019-2.pdf

ECPAT. (2020). *Nepal – Country Overview.* ECPAT. https://ecpat.org/resource/nepal-country-overview/

ECPAT. (2021). *Sexual Exploitation of Children In Travel And Tourism.* ECPAT. https://ecpat.org/wp-content/uploads/2021/05/ECPAT-Summary-paper-on-Sexual-Exploitation-of-Children-in-Travel-and-Tourism-2020.pdf

Edelstein, S. (2010). *Food, cuisine, and cultural competency for culinary, hospitality, and nutrition professionals*. Jones & Bartlett Publishers.

Efendi, M., & Karakaş Tandoğan, G. (2019). Yerli turistlerin paket turlarda yaptıkları etik dışı davranışlar üzerine turist rehberleri gözüyle bir değerlendirme. *Seyahat ve Otel İşletmeciliği Dergisi, 16*(1), 21–33. doi:10.24010oid.451222

Ehrmann, T., Meiseberg, B., & Ritz, C. (2009). Superstar effects in deluxe gastronomy-An empirical analysis of value creation in German quality restaurants. *Kyklos, 62*(4), 526–541. doi:10.1111/j.1467-6435.2009.00449.x

EIGE. (2023). *Tourism & Issues of gender inequalities in the policy area*. EIGE. https://eige.europa.eu/gender-mainstreaming/policy-areas/tourism

Elshaer, A. M., & Marzouk, A. M. (2019). *Labor in the tourism and hospitality industry: Skills, ethics, issues, and rights*. Apple Academic Press. doi:10.1201/9780429465093

Emperger, W. (2004). Customer Relationship Management in the Hospitality Industry. *Erişim Tarihi, 10*. https://www.hsyndicate.org/file/152001440.pdf

Erkol Bayram, G., & Zengin, B. (2017). Turist Rehberlerinin Verimliliğini Etkileyen Psikolojik Faktörler. In Güzel, F. Ö., Altıntaş, V., Şahin. İ. (Eds.), Turist Rehberliği Araştırmaları Öngörüler ve Uygulamalar (pp. 197-218). Detay Yayıncılık.

Erkol Bayram, G. (2016). Turist rehberlerinin mobbinge maruz kalma durumunun demografik değişkenler kapsamında incelenmesine yönelik bir araştırma. *Journal of International Social Research, 9*(47), 1185–1194. doi:10.17719/jisr.2016.1465

Ersoy, A. & Ehtiyar, R. (2021). Determination of managerial resourcefulness in the Turkish hospitality industry: a grounded theory study. *Journal of Hospitality and Tourism Technology*. doi:10.1108/JHTT-03-2021-0083

Ertorer, S. E., Long, J., Fellin, M., & Esses, V. M. (2022). Immigrant perceptions of integration in the Canadian workplace. *Equality, Diversity and Inclusion, 41*(7), 1091–1111. doi:10.1108/EDI-02-2019-0086

Eser, S., & Şahin, S. (2020). Turist rehberliği mesleğinde yaşanan güncel sorunlar üzerine bir inceleme. *Türk Turizm Araştırmaları Dergisi, 4*(2), 1344–1355. doi:10.26677/TR1010.2020.399

Eshun, G., & Tichaawa, T. M. (2019). Reconsidering participation for local community well-being in ecotourism in Ghana. *Geo Journal of Tourism and Geosites, 27*(4), 1184–1200. doi:10.30892/gtg.27406-425

Eshun, G., & Tichaawa, T. M. (2020). Community participation, risk management and ecotourism sustainability issues in Ghana. *Geo Journal of Tourism and Geosites, 28*(1), 313–331. doi:10.30892/gtg.28125-472

Estigoy, E., Sulasula, J., & Guodu, X. A. (2020). Factors affecting employee commitment in the workplace: An analysis. *Journal of Education and Practice, 11*(27), 160–171.

EUROSTAT. (2018). *Tourism industries – employment.* Eurostat. https://ec.europa.eu/eurostat/statistics-explained/index.php?title=Tourism_industries_-_employment

EUROSTAT. (2023). *Tourism industries – employment.* Eurostat. https://ec.europa.eu/eurostat/statistics-explained/index.php?title=Tourism_industries_-_employment&oldid=445425

Fafchamps, M. (2004). *Market institutions in Sub-Saharan Africa: Theory and evidence.* MIT Press.

Farrington, J., Carney, D., Ashley, C., & Turton, C. (1999). Sustainable livelihoods in practice: Early applications of concepts in rural areas. *Natural Resource Perspectives*, *42*, 1–15.

Fatehi, K., Priestley, J. L., & Taasoobshirazi, G. (2020). The expanded view of individualism and collectivism: One, two, or four dimensions? *International Journal of Cross Cultural Management*, *20*(1), 7–24. doi:10.1177/1470595820913077

Feinberg, B. (2018). Undiscovering the pueblo magico: Lessons from Huautla for the psychedelic renaissance. *Plant Medicines, Healing and Psychedelic Science: Cultural Perspectives*, 37-54.

Fennell, D. A. (2012). *Tourism and animal ethics*. Routledge.

Fennell, D. A. (2022). An animal welfare literacy framework for tourism. *Annals of Tourism Research*, *96*, 103461. doi:10.1016/j.annals.2022.103461

Ferguson, P. P. (2010). Culinary nationalism. *Gastronomica*, *10*(1), 102–109. doi:10.1525/gfc.2010.10.1.102 PMID:21539054

Fidgeon, P. R. (2010). Tourism education and curriculum design: A time for consolidation and review? *Tourism Management*, *31*(6), 699–723. doi:10.1016/j.tourman.2010.05.019

Figueroa-Domecq, C., Palomo, J., Flecha-Barrio, M. D., & Segovia-Pérez, M. (2020). Technology double gender gap in tourism business leadership. *Information Technology & Tourism*, *22*(1), 75–106. doi:10.100740558-020-00168-0

Filieri, R. (2013). Consumer co-creation and new product development: A case study in the food industry. *Marketing Intelligence & Planning*, *31*(1), 40–53. doi:10.1108/02634501311292911

Fine, E. C., & Speer, J. H. (1985). Tour guide performances as sight sacralization. *Annals of Tourism Research*, *12*(1), 73–95. doi:10.1016/0160-7383(85)90040-4

Fink, S. L. (1992). High Commitment Workplaces. Greenwood publishing group, NY.

Flaherty, G. T., Maxemous, K. K., Nossier, R. E., & Bui, Y. G. (2017). The highs and lows of drug tourism: A travel medicine perspective. *Journal of Travel Medicine*, *24*(6). Advance online publication. doi:10.1093/jtm/tax068 PMID:29088481

Fleming, S. S. (2015). Déjà Vu? An updated analysis of the gender wage gap in the U.S. hospitality sector. *Cornell Hospitality Quarterly*, *56*(2), 180–190. doi:10.1177/1938965514567680

Fletcher, A. J. (2017). Applying critical realism in qualitative research: Methodology meets method. *International Journal of Social Research Methodology*, *20*(2), 181–194. doi:10.1080/13645579.2016.1144401

Florea, N. V. (2014). Good practices of integration for new employees in roman organizations. *Buletin Stiintific*, *37*(1), 13–23.

Flyvbjerg, B. (2006). Five Misunderstandings About Case-Study Research. *Qualitative Inquiry*, *12*(2), 219–245. doi:10.1177/1077800405284363

Folakemi, O., Anthonia, A., & Dayo, A. I. (2017). Human resource management practices and organisational commitment: Employees' perspective in the Nigerian hospitality sector. *New Trends and Issues Proceedings on Humanities and Social Sciences*, *4*(10), 246–255. doi:10.18844/prosoc.v4i10.3090

Fortainier, F., & van Wijk, J. (2010). Sustainable tourism industry development in sub-Saharan Africa: Consequences of foreign hotels for local employment. *International Business Review*, *19*(2), 191–205. doi:10.1016/j.ibusrev.2009.11.007

Foucault, M. (1977). *Discipline and punish*. Pantheon.

Francione, G. L. (2008). *Hayvan haklarına giriş: Çocuğunuz mu köpeğiniz mi?* (R. Akman & E. Gen, Trans.). İletişim Yayınları.

Frankfurt, H. G. (1958). Peirce's notion of abduction. *The Journal of Philosophy*, *55*(14), 593–597. doi:10.2307/2021966

Freeman, D., Cromwell, C., Aarenau, D., Hazelton, M., & Lapointe, M. (2005). Factors leading to successful workplace integration of employees who have experienced mental illness. *Employee Assistance Quarterly*, *19*(4), 51–58. doi:10.1300/J022v19n04_04

Freire, P. (1993). *Pedagogy of the oppressed*. Penguin Books.

Freire, P. (2018). Education as a practice of freedom. In *The Brazil Reader* (pp. 396–398). Duke University Press.

Friedman, A. (1977). *Industry and labour: Class struggles at work under monopoly capitalism*. MacMillan. doi:10.1007/978-1-349-15845-4

Frost, W., Laing, J., Best, G., Williams, K., Strickland, P., & Lade, C. (2016). *Gastronomy, tourism and the media* (*Vol. 74*). Channel View Publications.

Fullagar, S., & Wilson, E. (2012). Critical pedagogies: A reflexive approach to knowledge creation in tourism and hospitality studies. *Journal of Hospitality and Tourism Management*, *19*(1), 1–6. doi:10.1017/jht.2012.3

Furunes, T. (2005). Training paradox in the hotel industry. *Scandinavian Journal of Hospitality and Tourism*, *5*(3), 231–248. doi:10.1080/15022250510014372

Gaines, J., & Lee, C. V. (2019). Medical tourism. In *Travel medicine* (pp. 371–375). Elsevier. doi:10.1016/B978-0-323-54696-6.00039-2

Gambetti, A., & Han, Q. (2022). Camera eats first: Exploring food aesthetics portrayed on social media using deep learning. *International Journal of Contemporary Hospitality Management*, *34*(9), 3300–3331. doi:10.1108/IJCHM-09-2021-1206

Garber, A. K., & Lustig, H. R. (2011). Is fast food addictive? *Current Drug Abuse Reviews*, *4*(3), 146–162. doi:10.2174/1874473711104030146 PMID:21999689

García-Rosell, J., & Tallberg, L. (2021). Animals as tourism stakeholders: Huskies, reindeer, and horses working in Lapland. In C. Kline & J. Rickly (Eds.), *Exploring non-human work in tourism: From beasts of burden to animal ambassadors* (pp. 103–122). De Gruyter Oldenbourg. doi:10.1515/9783110664058-007

Geisen, T. (2014). Workplace integration through disability management. In *Handbook of Vocational Rehabilitation and Disability Evaluation: Application and Implementation of the ICF* (pp. 55–71). Springer International Publishing.

Geisen, T., & Harder, H. G. (Eds.). (2011). *Disability management and workplace integration: International research findings*. Gower Publishing, Ltd.

George, B. P., & Panko, T. R. (2011). Child sex tourism: Facilitating conditions, legal remedies, and other interventions. *Vulnerable Children and Youth Studies*, *6*(2), 134–143. doi:10.1080/17450128.2010.521599

Geva, A., & Goldman, A. (1991). Satisfaction measurement in guided tours. *Annals of Tourism Research*, *18*(2), 177–185. doi:10.1016/0160-7383(91)90002-S

Gezici, F. (2006). Components of sustainability: Two Cases from Turkey. *Annals of Tourism Research*, *33*(2), 442–455. doi:10.1016/j.annals.2005.11.002

Ghaderi, Z., Tavakoli, R., Bagheri, F., & Pavee, S. (2023). The role of gender equality in Iranian female tourism entrepreneurs' success. *Tourism Recreation Research*, 1–15. doi:10.1080/02508281.2023.2168857

Giroux, H. (1998). Giriş. okuma-yazma ve siyasal güçlenme eğitbilimi. In *Okuryazarlık, sözcükleri ve dünyayı okuma* (S. Ayhan, Trans.). İmge Yayınları.

Giroux, H. (2021). Critical pedagogy. In U. Bauer, U. H. Bittlingmayer, & A. Scherr (Eds.), *Handbuch bildungs- und erziehungssoziologie*. Springer. doi:10.1007/978-3-658-31395-1_19-1

Giroux, H. A. (2007). *Elestirel pedagoji ve neoliberalizm* (B. Baysal, Trans.). Kalkedon.

Giroux, H. A. (2008). *Elestirel pedagojinin vaadi* (U. D. Tuna, Trans.). Kalkedon Yayinlari.

Glesne, C. (2013). *Nitel araştırmaya giriş* (A. Ersoy & P. Yalçınoğlu, Trans.). Anı Yayıncılık.

Gonzalez, J. A., & Simpson, J. (2021). The workplace integration of veterans: Applying diversity and fit perspectives. *Human Resource Management Review*, *31*(2), 100775. doi:10.1016/j.hrmr.2020.100775

Görgenyi, E. H., Mate, B., Vafaei, S. A., & Fekete, M. F. (2017). The role of social media in gastronomy industry. *Apstract, Applied Studies in Agribusiness and Commerce*, *11*(3-4), 95–103. doi:10.19041/APSTRACT/2017/3-4/14

Gorman-Murray, A. (2008). Masculinity and the home: A critical review and conceptual framework. *The Australian Geographer*, *39*(3), 367–379. doi:10.1080/00049180802270556

Graciela, C. J. Rocío del C. S. Barquín, Lilia Z. V. & Elva, E. V. M. (2022). Child labor and child work in the touristic sector of Cozumel and Valle de Bravo, Mexico. International Journal of Hospitality & Tourism Administration, 23(3), 599-622.

Granovetter, M. (1983). The Strength of Weak Ties: A Network Theory Revisited. *Sociological Theory*, *1*, 201–233. doi:10.2307/202051

Greenwood, D. J. (1989). Culture by the pound: An anthropological perspective on tourism as cultural commoditization. In V. L. Smith (Ed.), *Hosts and guests: the anthropology of tourism* (pp. 129–138). University of Pennsylvania Press. doi:10.9783/9780812208016.169

Gremillion, K. J. (2011). *Ancestral appetites: Food in prehistory*. Cambridge University Press. doi:10.1017/CBO9780511976353

Griffith, C. J., Livesey, K. M., & Clayton, D. (2010). The assessment of food safety culture. *British Food Journal*, *112*(4), 439–456. http://www.emeraldinsight.com/0007-070X.htm. doi:10.1108/00070701011034448

Griffiths, R. R., Johnson, M. W., Carducci, M. A., Umbricht, A., Richards, W. A., Richards, B. D., Cosimano, M. P., & Klinedinst, M. A. (2016). Psilocybin produces substantial and sustained decreases in depression and anxiety in patients with life-threatening cancer: A randomized double-blind trial. *Journal of Psychopharmacology (Oxford, England)*, *30*(12), 1181–1197. doi:10.1177/0269881116675513 PMID:27909165

Grimm, M., Gubert, F., Koriko, F. O., Lay, J., & Nordman, C. J. (2013). Kinship-ties and entrepreneurship in Western Africa. *Journal of Small Business and Entrepreneurship*, *26*(2), 125–150. Advance online publication. doi:10.1080/08276331.2013.771854

Griswold, A. (2013). Digital detectives and virtual volunteers: Integrating emergent online communities into disaster response operations. [Digital detectives and virtual volunteers: Integrating emergent o...: Ingenta Connect]. *Journal of Business Continuity & Emergency Planning*, *7*(1), 13–25. PMID:24113633

Grönroos, C. (1978). A Service Oriented Approach to Marketing of Services. *European Journal of Marketing*, *12*(3), 588–601. doi:10.1108/EUM0000000004985

Guimaraes, F. F. C. R., & Silva, J. R. (2016). Pay gap by gender in the tourism industry of Brazil. *Tourism Management*, *52*, 440–450. doi:10.1016/j.tourman.2015.07.003

Gupta, P. (2014). Frozen vodka and white skin in tourist Goa. *Tourism and the Power of Otherness: Seductions of Difference*, *34*, 95–109. doi:10.21832/9781845414177-007

Güzel, F. Ö., Atilla Gök, G., & Büyüker İşler, D. (2013). Duygusal emek ve işten ayrılma niyeti ilişkisi: Turist rehberleri üzerinde bir araştırma. *Seyahat ve Otel İşletmeciliği Dergisi*, *10*(3), 107–123.

Gvion, L. (2011). Cooking, food, and masculinity: Palestinian men in Israeli society. *Men and Masculinities*, *14*(4), 408–429. doi:10.1177/1097184X11411269

Haden, M., Emerson, B., & Tupper, K. W. (2016). A public-health-based vision for the management and regulation of psychedelics. *Journal of Psychoactive Drugs*, *48*(4), 243–252. doi:10.1080/02791072.2016.1202459 PMID:27430375

Haig, B. D. (2005). An Abductive Theory of Scientific Method. *Psychological Methods*, *10*(4), 371–388. doi:10.1037/1082-989X.10.4.371 PMID:16392993

Hall, D., & Brown, F. (1996). Towards a welfare focus for tourism research. *Progress in Tourism and Hospitality Research*, *1*(2), 41–57. doi:10.1002/(SICI)1099-1603(199603)2:1<41::AID-PTH23>3.0.CO;2-6

Hall, S. (1997). The work of representation. In S. Hall (Ed.), *Representation: Cultural representations and signifying practices* (pp. 13–74). Sage.

Harianto, A. (2021, December). Wages Problems of Local Wisdom-based Tourism Development During Covid-19 Pandemic. In *2nd International Conference on Business Law and Local Wisdom in Tourism* (ICBLT 2021) (pp. 31-35). Atlantis Press. https://www.atlantis-press.com/proceedings/icblt-21/125965331

Harun, A. F., Ruslan, N., Adnan, W. A. W., Suliman, S. I., Ismail, J., & Baharin, H. (2020). Digitizing food experience: Food taste perception on digital image and true form using hashtags. *Bulletin of Electrical Engineering and Informatics*, *9*(5), 2141–2153. doi:10.11591/eei.v9i5.2252

Haruna, P. F. (2009). Revising the Leadership Paradigm in Sub-Saharan Africa: A Study of Community-Based Leadership. *Public Administration Review*, *69*(5), 941–950. doi:10.1111/j.1540-6210.2009.02043.x

Hasan, R., & Yu, Z. (2020). Capitalism leads to unhealthy food consumption. *Strategic Change*, *29*(6), 633–643. doi:10.1002/jsc.2379

Hatjidis, D., & Parker, A. (2018). The role of relationship quality in raising organizational change acceptance: The case of the Greek hotel industry. *Journal of Human Resources in Hospitality & Tourism*, *17*(3), 340–359. doi:10.1080/15332845.2017.1406284

Hechler, D. (1995). *Child sex tourism*. Geocities. http://www.geocities.com/CapitolHilllSenate/893 11 dbt-rep.html.

Helms, M. M., & Stern, R. (2001). Exploring the factors that influence employees' perceptions of their organization's culture. *Journal of Management in Medicine*, *15*(6), 415–429. doi:10.1108/EUM0000000006502 PMID:11811793

Henderson, R. M. (1979). *Employee integration and automation: a study at Caterpillar Tractor Company, Peoria, Illinois* [Doctoral dissertation, Massachusetts Institute of Technology].

Hirst, G., Curtis, S., Nielsen, I., Smyth, R., Newman, A., & Xiao, N. (2021). Refugee recruitment and workplace integration: An opportunity for human resource management scholarship and impact. *Human Resource Management Journal*, 1748-8583.12349. doi:10.1111/1748-8583.12349

Hollander, D. B. (2018). *Farmers and agriculture in the Roman economy*. Routledge Publish. doi:10.4324/9781315103884

Holloway, J. C. (1981). The guided tour: A sociological approach. *Annals of Tourism Research*, *8*(3), 377–402. doi:10.1016/0160-7383(81)90005-0

Holmberg, C., Chaplin, J. E., Hillman, T., & Berg, C. (2016). Adolescents' presentation of food in social media: An explorative study. *Appetite*, *99*, 121–129. doi:10.1016/j.appet.2016.01.009 PMID:26792765

Hoque, B. L., Lovelock, B., & Carr, A. (2022). Alleviating Indigenous poverty through tourism: The role of NGOs. *Journal of Sustainable Tourism*, *30*(10), 2333–2351. doi:10.1080/09669582.2020.1860070

Horowitz, M. D., Rosensweig, J. A., & Jones, C. A. (2007). Medical tourism: Globalization of the healthcare marketplace. *Medscape General Medicine*, *9*(4), 33. PMID:18311383

Huang, L., & Kao, P. H. (2011). How to tell a good tour guide under different strategic orientations. *African Journal of Business Management*, *5*(27), 11146. doi:10.5897/AJBM11.1291

Huang, S. (2010). A revised importance–performance analysis of tour guide performance in China. *Tourism Analysis*, *15*(2), 227–241. doi:10.3727/108354210X12724863327803

Huang, S., Hsu, C. H., & Chan, A. (2009). Tour guide performance and tourist satisfaction: A study of the package tours in Shanghai. *Journal of Hospitality & Tourism Research (Washington, D.C.)*, *34*(1), 3–33. doi:10.1177/1096348009349815

Hummel, J., & van der Duim, R. (2012). Tourism and development at work: 15 years of tourism and poverty reduction within the SNV Netherlands Development Organisation. *Journal of Sustainable Tourism*, *20*(3), 319–338. doi:10.1080/09669582.2012.663381

Hutchings, K., Moyle, C., Chai, A., Garofano, N., & Moore, S. (2020). Segregation of women in tourism employment in the APEC region. *Tourism Management Perspectives*, *34*, 100655. doi:10.1016/j.tmp.2020.100655

Hu, W., & Wall, G. (2012). Interpretative guiding and sustainable development: A framework. *Tourism Management Perspectives*, *4*, 80–85. doi:10.1016/j.tmp.2012.05.003

Hu, W., & Wall, G. (2013). Tour guide management in Hainan, China: Problems, implications and solutions. *Asia Pacific Journal of Tourism Research, 18*(3), 205–219. doi:10.1080/10941665.2011.640699

Hyman, J., & Mason, B. (1995). Managing employee involvement and participation. *Sage (Atlanta, Ga.).*

Hytten, K. (2006). Education for social justice: Provocations and challenges. *Educational Theory, 56*(2), 221–236. doi:10.1111/j.1741-5446.2006.00013.x

Ibanez, J., Aylett, R., & Ruiz-Rodarte, R. (2003). Storytelling in virtual environments from a virtual guide perspective. *Virtual Reality (Waltham Cross), 7*(1), 30–42. doi:10.100710055-003-0112-y

Ibrahim, A. M. (2020). Psychological empowerment and organizational commitment among employees in the lodging industry. *Journal of Human Resources in Hospitality & Tourism, 19*(3), 277–295. doi:10.1080/15332845.2020.1737766

Ibrahim, Y. (2015). Food porn and the invitation to gaze: Ephemeral consumption and the digital spectacle. [IJEP]. *International Journal of E-Politics, 6*(3), 1–12. doi:10.4018/IJEP.2015070101

İlhan, Y., & Soybali, H. H. (2018). Turist rehberlerinin mesleki sorunları üzerine bir araştırma. [TURED]. *Turist Rehberliği Dergisi, 1*(1), 13–23. doi:10.34090/tured.443615

ILO. (2011). *Stimulating rural growth, jobs and income through tourism, rural policy brief.* ILO. https://www.ilo.org/wcmsp5/groups/public/---ed_emp/documents/publication/wcms_161153.pdf

ILO. (2012). Migrant Workers in the International Hotel Industry. *International Migration Paper No. 112.* ILO. https://www.ilo.org/sector/Resources/publications/WCMS_180596/lang--en/index.htm

ILO. (2015). *The Contribution of Labour Mobility to Economic Growth.* ILO. https://www.ilo.org/wcmsp5/groups/public/---europe/---ro-geneva/---ilo-ankara/documents/legaldocument/wcms_398770.pdf

ILO. (2017). *ILO guidelines on decent work and socially responsible tourism.* ILO. https://www.ilo.org/wcmsp5/groups/public/---ed_dialogue/---sector/documents/normativeinstrument/wcms_546337.pdf

ILO. (2019). *Küresel yabancı göçmen işçi tahminleri sonuç raporu.* ILO. https://www.ilo.org/wcmsp5/groups/public/---europe/---ro-geneva/---ilo-ankara/documents/genericdocument/wcms_811865.pdf

ILO. (2020a). *ILO Monitor: COVID-19 and the world of work. Third edition, Updated estimates and analysis.* ILO. https://www.ilo.org/wcmsp5/groups/public/@dgreports/@dcomm/documents/briefingnote/wcms_743146.pdf

ILO. (2020b). *COVID-19 crisis and the informal economy, immediate responses and policy challenges.* ILO. https://www.ilo.org/wcmsp5/groups/public/---ed_protect/---protrav/---travail/documents/briefingnote/wcms_743623.pdf

ILO. (2022a). *The future of work in the tourism sector: Sustainable and safe recovery and decent work in the context of the COVID Report for the Technical Meeting on COVID 19 pandemic*. ILO. https://www.ilo.org/wcmsp5/groups/public/---ed_dialogue/---sector/documents/meetingdocument/wcms_840403.pdf

ILO. (2022a). *World employment and social outlook | Trends 2022*. ILO. https://www.ilo.org/wcmsp5/groups/public/---dgreports/---dcomm/---publ/documents/publication/wcms_834081.pdf

ILO. (2022b). *Application of international labour standards 2022: Report of the committee of experts on the application of conventions and recommendations*. ILO. ilo.org/wcmsp5/groups/public/---ed_norm/---relconf/documents/meetingdocument/wcms_836653.pdf

ILO. (2022b). *The future of work in the tourism sector: Sustainable and safe recovery and decent work in the context of the COVID-19 pandemic*. ILO. https://www.ilo.org/wcmsp5/groups/public/---ed_dialogue/---sector/documents/meetingdocument/wcms_840403.pdf

ILO. (2023). *What is child labor*. ILO. https://www.ilo.org/ipec/facts/lang--en/index.htm

ILO. (2023). *Women's entrepreneurship development*. ILO. https://www.ilo.org/empent/areas/womens-entrepreneurship-development-wed/lang--en/index.htm

ILOSTAT. (2018). *Total employed by sex*. ILO. http://ilostat.ilo.org

İnal, K. (2009). Eleştirel pedagoji: Ezilenler için eğitim. *Journal of Critical Pedagogy*, *1*(1), 2–10.

Inglis, D., & Gimlin, D. (2015). The globalization of food. Nova York.

Inouye, A., & Wolfgang, A. (2022). 3, 4-Methylenedioxymethamphetamine (MDMA)-Assisted Therapy in Hawaii: A Brief Review. *Cureus*, *14*(6). doi:10.7759/cureus.26402 PMID:35915689

International Labour Organization. (2010). *Developments and challenges in the hospitality and tourism sector*. ILO. https://www.ilo.org/wcmsp5/groups/public/---ed_norm/---relconf/documents/meetingdocument/wcms_166938.pdf

IPEA. (2008). *Sistema de informações sobre o mercado de trabalhono setor turismo: Caracterização da ocupaçãoformal noturismo, com base nos dados da RAIS de 2002 e 2006, e daocupação informal no turismo, com base nos dados da PNADde 2002 e 2006*. IPEA. http://agencia.ipea.gov.br/images/stories/15_caracterizacao_da_ocupacao_formal_e_informal_2002_2006_setembro_2008.pdf

Irshad, H. (2010). *Rural tourism- An overview*. Government of Alberta, Rural Development Division.

Işın, P. M. (2018). *Bountiful empire: A History of ottoman cuisine*. Reaktion Books.

Jaafar, M., Rasoolimanesh, S. M., & Ismail, S. (2017). Perceived sociocultural impacts of tourism and community participation: A case study of Langkawi Island. *Tourism and Hospitality Research*, *17*(2), 123–134. doi:10.1177/1467358415610373

Jamal, T., Kircher, J., & Donaldson, J. P. (2021). Re-visiting design thinking for learning and practice: Critical pedagogy, conative empathy. *Sustainability (Basel)*, *13*(2), 964. doi:10.3390u13020964

Jeffreys, S. (1999). Globalizing sexual exploitation: Sex tourism and the traffic in women. *Leisure Studies*, *18*(3), 179–186. doi:10.1080/026143699374916

Johnson, M. W., & Griffiths, R. R. (2017). Potential therapeutic effects of psilocybin. *Neurotherapeutics; the Journal of the American Society for Experimental NeuroTherapeutics*, *14*(3), 734–740. doi:10.100713311-017-0542-y PMID:28585222

Jucan, M. S., & Jucan, C. N. (2013). Gender trends in tourism destinations. *Social and Behavioral Sciences*, *92*, 437–444.

Kalyoncu, M., & Yuksek, G. (2017). The Effect of impression management techniques which tourist guides use in package tours on personal interaction quality. *Journal of Tourism and Hospitality Management*, *5*(1), 15–25. doi:10.15640/jthm.v5n1a2

Kapera, I. (2018). Sustainable tourism development efforts by local governments in Poland. Sustainable Cities and Society, 40, 581–588. doi:10.1016/j.scs.2018.05.001

Karamustafa, K., & Çeşmeci, N. (2006). Paket tur operasyonunda turist rehberlerinin karşılaştıkları yönetsel sorunlar üzerine bir araştırma. *Anatolia: Turizm Araştırmaları Dergisi*, *17*(1), 70–86.

Karamustafa, K., & Kılıçhan, B. (2019). Delphi yöntemi ile profesyonel turist rehberliği mesleğine ilişkin sorunların tespit edilmesi. *Afyon Kocatepe Üniversitesi Sosyal Bilimler Dergisi*, *21*(4), 1348–1365. doi:10.32709/akusosbil.597628

Karatepe, O. M. (2011). Job resourcefulness as a moderator of the work-family conflict – job satisfaction relationship: A study of hotel employees in Nigeria. *Journal of Hospitality and Tourism Management*, *18*(1), 10–17. doi:10.1375/jhtm.18.1.10

Karatepe, O. M., & Olugbade, O. A. (2009). The effects of job and personal resources on hotel employees' work engagement. *International Journal of Hospitality Management*, *28*(4), 504–512. doi:10.1016/j.ijhm.2009.02.003

Kavak, M. (2020). Frig kentleri. In T. F. Çakmak (Ed.), *Turist rehberleri için Anadolu'da antik kent ve merkezler* (pp. 123–151). Detay Yayıncılık.

Kaya, İ. (2012). Türk turizm sektöründe istihdamın temel özelliklerinin ve sorunlarının ücretli çalışanlar açısından coğrafi bölgelere göre analizi. *Balıkesir Üniversitesi Sosyal Bilimler Enstitüsü Dergisi*, *15*(28), 241–257.

Kaya, Ş. (2017). Turizmde kadın emeği. *Karatahta İş Yazıları Dergisi*, *9*, 1–22.

Kazlauskaite, R., Buciuniene, I., & Turauskas, L. (2006). Building employee commitment in the hospitality industry. *Baltic Journal of Management*, *1*(3), 300–314. doi:10.1108/17465260610690944

Kc, B., Morais, D. B., Peterson, M. N., Seekamp, E., & Smith, J. W.KC. (2019). Social network analysis of wildlife tourism microentrepreneurial network. *Tourism and Hospitality Research*, *19*(2), 158–169. doi:10.1177/1467358417715679

Kesten, D. (2011). *The healing secrets of food: A practical guide for nourishing body, mind, and soul*. New World Library.

Khoa, B., & Hoa, L. (2021). The effect of knowledge management process on the employee commitment: Evidence from digital marketing industry. *Management Science Letters*, *11*(5), 1557–1564. doi:10.5267/j.msl.2020.12.017

Kidido, J. K., & Baah, K. (2020). Sharecropping arrangement in the contemporary agricultural economy of Ghana: A study of Techiman North District and Sefwi Municipality, Ghana. *Journal of Planning and Land Management*, *1*(2), 50–62. doi:10.36005/jplm.v1i2.22

Kimbu, A. N., Afenyo Agbe, E. A., & Adam, I. (2019). *These African women are driving the continent's tourism industry – but they need support*. Research Gate. https://www.researchgate.net/publication/337480264_These_African_women_are_driving_the_continents_tourism_industry_-_but_they_need_support

Kim, Y. H. (2019). Organic shoppers' involvement in organic foods: Self and identity. *British Food Journal*, *121*(1), 139–156. doi:10.1108/BFJ-03-2018-0202

Kincheloe, J. L. (2007). Critical pedagogy in the Twenty-first century. *Evolution; International Journal of Organic Evolution*.

Kindan, S. A. (2022). *Türkiye'de turizm sektöründe göçmen emeği: Antalya ili örneği* [Unpublished master's thesis, Ordu Üniversitesi, Sosyal Bilimler Enstitüsü].

King, C. (2010). "One size doesn't fit all": Tourism and hospitality employees' response to internal brand management. *International Journal of Contemporary Hospitality Management*, *22*(4), 517–534. doi:10.1108/09596111011042721

Kirca-Demirbaga, K. (2021). *A study on the character development and education of gifted pupils using a critical realist approach* [Doctoral dissertation, Durham University]. http://etheses.dur.ac.uk/14348/

Kontogeorgopoulos, N. (2009). The role of tourism in elephant welfare in Northern Thailand. *Journal of Tourism*, *X*(2), 1–19.

Koops, T., Turner, D., Neutze, J., & Briken, P. (2017). Child sex tourism–prevalence of and risk factors for its use in a German community sample. *BMC Public Health*, *17*(1), 1–8. doi:10.118612889-017-4270-3 PMID:28427370

Köroğlu, Ö. (2012). İçsel ve dışsal iş doyum düzeyleri ile genel iş doyum düzeyi arasındaki ilişkinin belirlenmesi: Turist rehberleri üzerinde bir araştırma. *Doğuş Üniversitesi Dergisi*, *13*(2), 275–289. doi:10.31671/dogus.2018.131

Köroğlu, Ö., Ulusoy Yıldırım, H., & Balıkoğlu, A. (2020). Kadın turist rehberlerinin mesleki sorunlarının belirlenmesi. *Mehmet Akif Ersoy Üniversitesi Sosyal Bilimler Enstitüsü Dergisi*, (31), 26–40. doi:10.20875/makusobed.586054

Kortt, M. A., Sinnewe, E., & Pervan, S. J. (2018). The gender wage gap in the tourism industry: Evidence from Australia. *Tourism Analysis*, *23*(1), 137–149. doi:10.3727/108354217X15143857878697

Kosoy, N., & Corbera, E. (2010). Payments for ecosystem services as commodity fetishism. *Ecological Economics*, *69*(6), 1228–1236. doi:10.1016/j.ecolecon.2009.11.002

Kotsev, E. (2008). Integration of the Individual and the Group in the Organization. A Grup, Bulgaria.

Kotsev, E. (2022). Knowing When and How to Trust Superiors' Decisions: Toward a Conceptual Model of Subordinate Managers' Behavior. *Organizacija*, *55*(1), 50–63. doi:10.2478/orga-2022-0004

Kremers, A. (2018). The relationship between knowledge hiding and turnover intentions: The mediating role of affective commitment and the moderating effect of commitment HR attributions. Human Resource Studies, 1-51. Tilburg University.

Kuhn, T. S. (1970). *The Structure of Scientific Revolutions*. University of Chicago Press.

Kuo, N. T., Chang, K. C., Cheng, Y. S., & Lin, J. C. (2015). Effects of tour guide interpretation and tourist satisfaction on destination loyalty in Taiwan's Kinmen Battlefield Tourism: Perceived playfulness and perceived flow as moderators. *Journal of Travel & Tourism Marketing, 33*(sup1), 103-122. doi:10.1080/10548408.2015.1008670

Kurt Dündar, A., & Arslantürk, Y. (2022). Paket turlarda etik dışı davranışlar: Yabancı turistler üzerine nitel bir değerlendirme. *Güncel Turizm Araştırmaları Dergisi*, *6*(2), 517–535. doi:10.32572/guntad.1101091

Kusluvan, S. (2003). *Managing employee attitudes and behaviors in the tourism and tourism industry*. Nova Science Publishers.

Kutlu, D. (2012). Türkiye işgücü piyasasında güncel gelişmeler ve güvencesizleşme örüntüleri. In Ö. Göztepe (Ed.), *Güvencesizleştirme: Süreç, Yanılgı, Olanak* (pp. 61–116). Nota Bene Yayınları.

Ladkin, A. (2011). Exploring tourism labor. *Annals of Tourism Research*, *38*(3), 1135–1155. doi:10.1016/j.annals.2011.03.010

Lashley, C. (1996). Research issues for employee empowerment in hospitality organisations. *International Journal of Hospitality Management*, *15*(4), 333–346. doi:10.1016/S0278-4319(96)00025-4

Lashley, C. (1999). Employee empowerment in services: A framework for analysis. *Personnel Review*, *28*(3), 169–191. doi:10.1108/00483489910264570

Lasrado, F., & Lasrado, F. (2018). Awards and Assessments. *Achieving Organizational Excellence: A Quality Management Program for Culturally Diverse Organizations*, 121-141. Springer.

Lawson, T. (1997). *Economics and reality*. Routledge. doi:10.4324/9780203195390

Lazaridis, G. (2007). Les infirmières exclusives and migrant Quasi-Nurses in Greece. *European Journal of Women's Studies*, *14*(3), 227–245. doi:10.1177/1350506807079012

Leclerc, D., & Martin, J. N. (2004). Tour guide communication competence: French, German and American tourists' perceptions. *International Journal of Intercultural Relations*, *28*(3-4), 181–200. doi:10.1016/j.ijintrel.2004.06.006

Ledwith, M. (2001). Community work as critical pedagogy: Re-envisioning Freire and Gramsci. *Community Development Journal: An International Forum*, *36*(3), 171–182. doi:10.1093/cdj/36.3.171

Lee, C., & Chien, S. (2008). Structural breaks, tourism development, and economic growth: Evidence from Taiwan. *Mathematics and Computers in Simulation*, *77*(4), 358–368. doi:10.1016/j.matcom.2007.03.004 PMID:32288113

Lee, H. J., & Peccei, R. (2007). Organizational-Level Gender Dissimilarity and Employee Commitment. *British Journal of Industrial Relations*, *45*(4), 687–712. doi:10.1111/j.1467-8543.2007.00644.x

Lee-Ross, D. (2005). Perceived job characteristics and internal work motivation: An exploratory cross-cultural analysis of the motivational antecedents of hotel workers in Mauritius and Australia. *Journal of Management Development*, *24*(3), 253–266. doi:10.1108/02621710510584062

Lercari, D. (2023). Sandy beaches: Publication features, thematic areas and collaborative networks between 2009 and 2019. *Estuarine, Coastal and Shelf Science*, 281. doi:10.1016/j.ecss.2023.108211

Lerma-Cabrera, J. M., Carvajal, F., & Lopez-Legarrea, P. (2015). Food addiction as a new piece of the obesity framework. *Nutrition Journal*, *15*(1), 1–5. doi:10.118612937-016-0124-6 PMID:26758504

Leslie, D., & Catungal, J. P. (2012). Social justice and the creative city: Class, gender and racial inequalities. *Geography Compass*, *6*(3), 111–122. doi:10.1111/j.1749-8198.2011.00472.x

Levy, D. E., & Lerch, P. B. (1991). Tourism as a factor in development implications for gender and work in Barbados. *Gender & Society*, *5*(1), 67–85.

Lewchuk, W., de Wolff, A., King, A., & Polanyi, M. (2003). From job strain to employment strain: Health effects of precarious employment. *Just Labour*, *3*. doi:10.25071/1705-1436.165

Lewis, T. (2018). Digital food: From paddock to platform. *Communication Research and Practice*, *4*(3), 212–228. doi:10.1080/22041451.2018.1476795

Liang, J., Yin, Z., Yang, J., Li, Y., Xu, M., Li, J., Yang, M., & Niu, L. (2022). Bibliometrics and Visualization Analysis of Research in the Field of Sustainable Development of the Blue Economy (2006–2021). *Frontiers in Marine Science*, *9*, 936612. Https://Doi.Org/10.3389/Fmars.2022.936612. doi:10.3389/fmars.2022.936612

Lin, H., Tian, J., Kong, Y., & Gao, J. (2022). Impact of tour guide humor on tourist pro-environmental behavior: Utilizing the conservation of resources theory. *Journal of Destination Marketing & Management*, *25*, 100728. doi:10.1016/j.jdmm.2022.100728

Little, D. C., Newton, R. W., & Beveridge, M. C. M. (2016). Aquaculture: A rapidly growing and significant source of sustainable food? Status, transitions, and potential. *The Proceedings of the Nutrition Society*, *75*(3), 274–286. doi:10.1017/S0029665116000665 PMID:27476856

Li, Y., Liu, B., Zhang, R., & Huan, T. C. (2020). News information and tour guide occupational stigma: Insights from the stereotype content model. *Tourism Management Perspectives*, *35*, 100711. doi:10.1016/j.tmp.2020.100711

López-López, Á., & Quintero-Venegas, G. (2021). Animal dark tourism in Mexico: Bulls performing their own slaughter. In C. Kline & J. Rickly (Eds.), *Exploring non-human work in tourism: From beasts of burden to animal ambassadors* (pp. 69–82). De Gruyter Oldenbourg. doi:10.1515/9783110664058-005

Lowe, H., Toyang, N., Steele, B., Grant, J., Ali, A., Gordon, L., & Ngwa, W. (2022). Psychedelics: Alternative and potential therapeutic options for treating mood and anxiety disorders. *Molecules (Basel, Switzerland)*, *27*(8), 2520. doi:10.3390/molecules27082520 PMID:35458717

Lowe, M. R., & Butryn, M. L. (2007). Hedonic hunger: A new dimension of appetite? *Physiology & Behavior*, *91*(4), 432–439. doi:10.1016/j.physbeh.2007.04.006 PMID:17531274

Lundberg, C., Gudmundson, A., & Andersson, T. D. (2009). Herzberg's two-factor theory of work motivation tested empirically on seasonal workers in hospitality and tourism. *Tourism Management*, *30*(6), 890–899. doi:10.1016/j.tourman.2008.12.003

Luo, C. C., Wang, Y. C., & Tai, Y. F. (2019). Effective training methods for fostering exceptional service employees. *Journal of Hospitality and Tourism Insights*, *2*(4), 469–488. doi:10.1108/JHTI-01-2019-0013

Lutkajtis, A. (2020). Lost saints: Desacralization, spiritual abuse and magic mushrooms. *Fieldwork in Religion*, *14*(2), 118–139. doi:10.1558/firn.40554

Lyneham, S., & Facchini, L. (2019). Benevolent harm: Orphanages, voluntourism and child sexual exploitation in South-East Asia. In A. P. M. Michael Phelan (Ed.), *Trends & issues in crime and criminal justice* (pp. 76–88). Australian Institute of Criminology.

Macdonald, D. (1953). A theory of mass culture. *Diogenes*, *1*(3), 1–17. doi:10.1177/039219215300100301

Mackenzie, S. H., Boudreau, P., & Raymond, E. (2020). Women's adventure tour guiding experiences: Implications for well-being. *Journal of Hospitality and Tourism Management*, *45*, 410–418. doi:10.1016/j.jhtm.2020.09.006

Mackenzie, S. H., & Kerr, J. H. (2013). Stress and emotions at work: An adventure tourism guide's experiences. *Tourism Management*, *36*, 3–14. doi:10.1016/j.tourman.2012.10.018

Mael, F., Wyatt, W., & Iyer, U. J. (2022). Veterans to workplace: Keys to successful transition. *Military Psychology*, *34*(5), 516–529. doi:10.1080/08995605.2021.2016307

Magablih, K., & Naamneh, M. (2010). Child labor in the tourism industry in Jordan. *Tourism Analysis*, *15*(1), 89–97. doi:10.3727/108354210X12724734223711

Magdoff, F., & Magdoff, H. (2004). Disposable workers: Today's reserve army of labor. *Monthly Review (New York, N.Y.)*, *55*(11), 18. Advance online publication. doi:10.14452/MR-055-11-2004-04_2

Magee, R. M. (2007). Food puritanism and food pornography: The gourmet semiotics of Martha and Nigella. *Americana: The Journal of American Popular Culture, 6*(2).

Mahmud, S., Alam, Q., & Härtel, C. (2014). Mismatches in skills and attributes of immigrants and problems with workplace integration: A study of IT and engineering professionals in Australia. *Human Resource Management Journal*, *24*(3), 339–354. doi:10.1111/1748-8583.12026

Mak, A. H. N., Wong, K. K. F., & Chang, R. C. Y. (2010). Factors affecting the service quality of the tour guiding profession in Macau. *International Journal of Tourism Research*, *12*(3), 205–218. doi:10.1002/jtr.746

Mak, A. H. N., Wong, K. K. F., & Chang, R. C. Y. (2011). Critical issues affecting the service quality and professionalism of the tour guides in Hong Kong and Macau. *Tourism Management*, *32*(6), 1442–1452. doi:10.1016/j.tourman.2011.01.003

Mäkinen, A. (2018). Development of internal communication policy in growth enterprise. International Business, 1-75. Oulu Business School.

Mak, J. (2003). *Tourism and the economy*. University of Hawaii Press.

Mancini, M. (1990). *Conducting tours: A practical guide*. Southwestern Publishing.

Mansour, A., Rowlands, H., Al-Gasawneh, J. A., Al-Qudah, S., Shrouf, H., & Al-smadi, L. A. (2021). The mediation effect of employees' affective organizational commitment: The impact of employees' perceptions of the social support for training programs on enhancing their organizational citizenship behavior in Jordanian hotels sector. *Academy of Strategic Management Journal*, *20*, 1–23.

Marcuse, H. (1976). One-dimensional man. *Persona y Derecho*, *3*, 690.

Marks, M. (2022). The varieties of psychedelic law. *Neuropharmacology*, 109399. PMID:36565855

Marshall, T., Mottier, E. M., & Lewis, R. A. (2015). Motivational factors and the hospitality industry: A case study examining the effects of changes in the working environment. [JBCS]. *Journal of Business Case Studies*, *11*(3), 123–132. doi:10.19030/jbcs.v11i3.9289

Marx, K. (2011). Kapital, I. Cilt, (Çev. Mehmet Selik ve Nail Satlıgan). İstanbul: Yordam Kitap.

Mash, D. C. (2023). Ibogaine: Ibogaine therapy as a method of care treatment for opioid dependence. *Psychedelics as Psychiatric Medications*, 75.

McBrien, B., O'Brien, F., Shetty, S. R., Eustace-Cook, J., & Byrne, G. (2022). Assessment of workplace integration of migrant nurses and midwives within international health care settings: A scoping review protocol. *JBI Evidence Synthesis*, *20*(7), 1799–1805. doi:10.11124/JBIES-21-00239 PMID:35131992

McDonnell, E. M. (2016). *Food porn: The conspicuous consumption of food in the age of digital reproduction. Food, media and contemporary culture*. Palgrave Macmillan., doi:10.1057/9781137463234_14

McDowell, L. (2002). Masculine discourses and dissonances: Strutting 'lads', protest masculinity, and domestic respectability. *Environment and Planning. D, Society & Space*, *20*(1), 97–119. doi:10.1068/d43j

McDowell, L., Batnitzky, A., & Dyer, S. (2009). Precarious work and economic migration: Emerging immigrant divisions of labour in Greater London's service sector. *International Journal of Urban and Regional Research*, *33*(1), 3–25. doi:10.1111/j.1468-2427.2009.00831.x

McKercher, B. (2002). The future of tourism education: An Australian scenario? *Tourism and Hospitality Research*, *3*(3), 199–210. doi:10.1177/146735840200300302

McKernan, J. A. (2013). The origins of critical theory in education: Fabian socialism as social reconstructionism in nineteenth-century Britain. *British Journal of Educational Studies*, *61*(4), 417–433. doi:10.1080/00071005.2013.824947

McLaren, P. (1998). *Life in schools: An introduction to critical pedagogy in the foundations of education*. Longman.

McLean, M. (2006). *Pedagogy and the university: Critical theory and practice*. Bloomsbury Publishing.

Meclisi, İ. S. İ. G. (2022). *Çocuk işçiliği*. ISIGMECLISI. https://isigmeclisi.org/20806-cocuk-isciligi-mahmut-boyunegmez

Medina-Muñoz, D. R., Medina-Muñoz, R. D., & Gutiérrez-Pérez, F. J. (2016). A Sustainable development approach to assessing the engagement of tourism enterprises in poverty alleviation. *Sustainable Development (Bradford)*, *24*(4), 220–236. doi:10.1002d.1624

Meintjes, H., & Giese, S. (2006). Spinning the epidemic: The making of mythologies of orphanhood in the context of AIDS. *Childhood*, *13*(3), 407–430. doi:10.1177/0907568206066359

Mejova, Y., Abbar, S., & Haddadi, H. (2016). Fetishizing food in digital age: Fodporn around the world. In *Proceedings of the International AAAI Conference on Web and Social Media* (Vol. 10, No. 1, pp. 250-258). 10.1609/icwsm.v10i1.14710

Mejova, Y., Haddadi, H., Noulas, A., & Weber, I. (2015). Foodporn: Obesity patterns in culinary interactions. In *Proceedings of the 5th International Conference on Digital Health 2015* (pp. 51-58). 10.1145/2750511.2750524

Mensah, I. (2017). Benefits and challenges of community-based ecotourism in park-fringe communities: The case of *Mesomagor* of Kakum National Park, Ghana. *Tourism Review International, 21*(1), 81–98. doi:10.3727/154427217X14866652018947

Meyer, S. B., & Lunnay, B. (2013). The application of abductive and retroductive inference for the design and analysis of theory-driven sociological research. *Sociological Research Online, 18*(1), 86–96. doi:10.5153ro.2819

Mhlanga, O., & Tichaawa, T. M. (2017). Influence of social media on customer experiences in restaurants: A South African study. *Tourism: An International Interdisciplinary Journal, 65*(1), 45-60. https://hrcak.srce.hr/178622

Michael, N., & Fuste-Forne, F. (2022). Marketing of luxurious gastronomic experiences on social media: The visual storytelling of luxury hotels. *International Journal of Tourism Research, 24*(6), 827–838. doi:10.1002/jtr.2548

Mi, L., Gan, X., Xu, T., Long, R., Qiao, L., & Zhu, H. (2019). A new perspective to promote organizational citizenship behaviour for the environment: The role of transformational leadership. *Journal of Cleaner Production, 239*, 118002. doi:10.1016/j.jclepro.2019.118002

Milgram, S. (1967). The small world problem. *Psychology Today, 1*(1), 61–67.

Min, J. C. (2012). A short-form measure for assessment of emotional intelligence for tour guides: Development and evaluation. *Tourism Management, 33*(1), 155–167. doi:10.1016/j.tourman.2011.02.014

Mintz, S. W., & Du Bois, C. M. (2002). The anthropology of food and eating. *Annual Review of Anthropology, 31*(1), 99–119. https://www.jstor.org/stable/4132873. doi:10.1146/annurev.anthro.32.032702.131011

Mitchell, T. R., Holtom, B. C., Lee, T. W., Sablynski, C. J., & Erez, M. (2001). Why People Stay: Using Job Embeddedness to Predict Voluntary Turnover. *Academy of Management Journal, 44*(6), 1102–1121. doi:10.2307/3069391

Mokomane, Z. (2014). Introduction. In Z. Mokomane (Ed.), *Work–Family Interface in Sub-Saharan Africa. International Perspectives on Social Policy, Administration, and Practice*. Springer. doi:10.1007/978-3-319-01237-7_1

Mokomane, Z., & Chilwane, D. (2014). A Review of Work–Family Research in Sub-Saharan Africa. In Z. Mokomane (Ed.), *Work–Family Interface in Sub-Saharan Africa. International Perspectives on Social Policy, Administration, and Practice*. Springer. doi:10.1007/978-3-319-01237-7_12

Monterrubio, C. (2022). The informal tourism economy, COVID-19 and socioeconomic vulnerability in Mexico. *Journal of Policy Research in Tourism, Leisure & Events, 14*(1), 20–34. doi:10.1080/19407963.2021.2017726

Mooney, S. K. (2014). *How the intersections of age, gender, ethnicity and class influence the longevity of a hospitality career in New Zealand* [Doctoral dissertation, Auckland University of Technology].

Moreno, J. L. (1934). *Who Shall Survive?* Beacon House.

Moreno, J. L., & Jennings, H. H. (1938). Statistics of social constellations. *Sociometry*, *1*(3), 342–374. doi:10.2307/2785588

Mulheir, G., & Cavanagh, M. (2016). *Orphanage entrepreneurs: The trafficking of Haiti's invisible children.* Lumos. https://lumos.contentfiles.net/media/documents/document/2017/12/Haiti_Trafficking_Report_ENG_WEB_NOV16.pdf

Muñoz-Bullón, F. (2009). The gap between male and female pay in the Spanish tourism industry. *Tourism Management*, *30*(5), 638–649. doi:10.1016/j.tourman.2008.11.007

Mura, P., & Wijesinghe, S. N. R. (2023). Critical theories in tourism – a systematic literature review. *Tourism Geographies*, *25*(2-3), 2–3, 487–507. doi:10.1080/14616688.2021.1925733

NAE. (2018). *New American economy, hospitality & tourism, immigrant workers*. NAE. https://www.newamericaneconomy.org/issues/hospitality-tourism/

Naef, P. (2018). "Narco-heritage" and the Touristification of the Drug Lord Pablo Escobar in Medellin, Colombia. *Journal of Anthropological Research*, *74*(4), 485–502. doi:10.1086/699941

Naqvi, F. (2017). The Relationship between Employee Empowerment and Organizational Commitment in the Hotel Industry and its Implications for the Management. *International Journal of Economic Research*, *14*(12), 17–35.

Narvanen, E., Mesiranta, N., Sutinen, U. M., & Mattila, M. (2018). Creativity, aesthetics and ethics of food waste in social media campaigns. *Journal of Cleaner Production*, *195*, 102–110. doi:10.1016/j.jclepro.2018.05.202

Nassani, A. A., Aldakhil, A. M., Qazi Abro, M. M., Islam, T., & Zaman, K. (2019). The impact of tourism and finance on women empowerment. *Journal of Policy Modeling*, *41*(2), 234–254. doi:10.1016/j.jpolmod.2018.12.001

Nath, J. (2011). Gendered fare? A qualitative investigation of alternative food and masculinities. *Journal of Sociology (Melbourne, Vic.)*, *47*(3), 261–278. doi:10.1177/1440783310386828

Naughton, P., McCarthy, M., & McCarthy, S. (2015). Acting to self-regulate unhealthy eating habits. An investigation into the effects of habit, hedonic hunger, and self-regulation on sugar consumption from confectionery foods. *Food Quality and Preference*, *46*, 173–183. doi:10.1016/j.foodqual.2015.08.001

Naumov, K. (2020). Tourist perceptions and satisfaction regarding tour guiding in The Republic of North Macedonia. *European Journal of Multidisciplinary Studies*, *5*(29), 12–26. doi:10.26417/797jhv49a

Ncube, F., & Jerie, S. (2012). Leveraging employee engagement for competitive advantage in the hospitality industry. A comparative study of hotels A and B in Zimbabwe. *Journal of Emerging Trends in Economics and Management Science*, *3*(4), 380–388.

Nettekoven, L. (1979). Mechanism of cultural interaction. In E. De Kadt (Ed.), *Tourism: Passport to development* (pp. 135–145). Oxford University Press.

Neuman, W. L. (2014). *Social research methods: Qualitative and quantitative approaches*. Pearson.

Ngqangashe, Y., Maldoy, K., De Backer, C. J., & Vandebosch, H. (2022). Exploring adolescents' motives for food media consumption using the theory of uses and gratifications. *Communications*, *47*(1), 73–92. doi:10.1515/commun-2019-0164

Nicholas, C. R., Wang, J. B., Coker, A., Mitchell, J. M., Klaire, S. S., Yazar-Klosinski, B., Emerson, A., Brown, R. T., & Doblin, R. (2022). The effects of MDMA-assisted therapy on alcohol and substance use in a phase 3 trial for treatment of severe PTSD. *Drug and Alcohol Dependence*, *233*, 109356. doi:10.1016/j.drugalcdep.2022.109356 PMID:35286849

Northup, T. (2014). Truth, lies, and packaging: How food marketing creates a false sense of health. *Food Studies: An Interdisciplinary Journal, 3*(1), 9-18. ww.food-studies.com,

Northup, T. (2014). Truth, lies, and packaging: How food marketing creates a false sense of health. *Food Studies: An Interdisciplinary Journal, 3*(1), 9-18. ww.food-studies.com.

Novak, M. (1982). *The Spirit of Capitalism*. Madison Books. The Spirit of Democratic Capitalism - Google Books.

Nyahunzvi, D. K., & Njerekai, C. (2013). Tour guiding in Zimbabwe: Key issues and challenges. *Tourism Management Perspectives*, *6*, 3–7. doi:10.1016/j.tmp.2012.11.003

O'Mahoney, J. (2016). Critical realism and qualitative research: An introductory overview. In C. Cassell, A. L. Grandy, & G. Cunliffe (Eds.), *Sage handbook of qualitative research methods*.

Obadic, A. (2016). Gender discrimination and pay gap on tourism labor market. *International Journal of Mechanical and Industrial Engineering*, *10*(3), 823–828.

Ocha, W. (2023). (Diplomacy and International Studies, Rangsit University, Pathumthani, Thailand) Buddhism, gender, and sexualities: Queer spiritualties in Thailand. *Equality, Diversity and Inclusion*, *42*(5). doi:10.1108/EDI-01-2020-0025

Oddy, D. J. (2003). *From plain fare to fusion food: British diet from the 1890s to the 1990s*. Boydell Press.

OECD. (2022). *OECD tourism trends and policies 2022*. OECD Publishing.

Okazaki, E. (2008). A Community-Based Tourism Model: Its Conception and Use. *Journal of Sustainable Tourism*, *16*(5), 511–529. doi:10.1080/09669580802159594

Okello, M. M., & Novelli, M. (2014). Tourism in the East African Community (EAC): Challenges, opportunities, and ways forward. *Tourism and Hospitality Research*, *14*(1–2), 53–66. doi:10.1177/1467358414529580

Oliver, X., & Sard, M. (2021). Gender wage gap in hospitality. *Journal of Hospitality & Tourism Research (Washington, D.C.)*, *45*(2), 345–372. doi:10.1177/1096348020909423

Omay, U. (2011). Yedek işgücü ordusu olarak kadınlar. *Çalışma ve Toplum, 30*, 137-166.

Ong, C. E., Ryan, C., & McIntosh, A. (2014). Power-knowledge and tour-guide training: Capitalistic domination, utopian visions and the creation and negotiation of UNESCO's Homo Turismos in Macao. *Annals of Tourism Research, 48*, 221–234. doi:10.1016/j.annals.2014.06.010

Onorati, M. G., & Giardullo, P. (2020). Social media as taste re-mediators: Emerging patterns of food taste on TripAdvisor. *Food, Culture, & Society, 23*(3), 347–365. doi:10.1080/15528014.2020.1715074

Orozco, M., & Harris, S. (2023). Psilocybin and the Meaning Response: Exploring the Healing Process in a Retreat Setting in Jamaica. *Anthropology of Consciousness, 34*(1), 130–160. doi:10.1111/anoc.12162

Ortega, A. A. C. (2022). Bodies of transnational island urbanism: Spatial narratives of inclusion/exclusion of Filipinas in Philippine islands. *Urban Studies (Edinburgh, Scotland), 59*(16), 3365–3381. doi:10.1177/00420980211067941

Osareh, F. (1996). Bibliometrics, Citation Analysis and Co-Citation Analysis: A Review of Literature I. *Libri, 46*(3), 149–158. doi:10.1515/libr.1996.46.3.149

Öter, Z., & Sonuç, N. (2014). Cultural heritage interpretation in gender context: Case of female tourist guides in Turkey. *In Proceedings of the Heritage, Tourism and Hospitality International Conference (HTHIC),* İstanbul, Türkiye, 66-77.

Overbeek, H. (2002). Neo liberalism and the regulation of global labor mobility. *The Annals of the American Academy of Political and Social Science, 581*(May), 74–90. doi:10.1177/000271620258100108

Ovies, D. H. (2021). Millennials in the search for spiritual ecstasy: Touristic consumption of shamanic mushroom rituals in Huautla de Jimenez, Mexico. In Millennials, Spirituality and Tourism (pp. 179-200). Routledge.

Oygard, L. (2000). Studying food tastes among young adults using Bourdieu's theory. *Journal of Consumer Studies & Home Economics, 24*(3), 160-169.

Özuğurlu, M. (2006). Türkiye'den güvencesiz çalışma ve sınıflar mücadelesinin yeni gündemi. In C. Gürkan, Ö. Taştan, & O. Türel (Eds.), *Küreselleşmeye güneyden tepkiler*. Dipnot Kitabevi.

Pagés, C. (2010). The age of productivity. In *The age of productivity: Transforming economies from the bottom up*. Palgrave Macmillan US.

Panko, T. R., & George, B. P. (2018). Animal sexual abuse and the darkness of touristic immorality. In M. Korstanje & B. George (Eds.), Virtual traumascapes and exploring the roots of dark tourism. (pp. 175-189). IGI Global. doi:10.4018/978-1-5225-2750-3.ch009

Park, C. K. (2019). Transformational leadership and employee engagement from the HRD perspective. *Korean Review of Corporation Management, 10*(3), 19–36. doi:10.20434/KRICM.2019.10.10.3.19

Patel, R. C. (2012). Food sovereignty: Power, gender, and the right to food. *PLoS Medicine, 9*(6), 1–5. doi:10.1371/journal.pmed.1001223 PMID:22745604

Pearce, P. L. (1984). Tourist-guide interaction. *Annals of Tourism Research, 11*(1), 129–146. doi:10.1016/0160-7383(84)90100-2

Pekerşen, Y., & Tugay, O. (2020). Professional satisfaction as a key factor in employee retention: A case of the service sector. *Journal of Tourism and Services, 11*(20), 1–27. doi:10.29036/jots.v11i20.123

Perkins, D., Ruffell, S. G., Day, K., Pinzon Rubiano, D., & Sarris, J. (2023). Psychotherapeutic and neurobiological processes associated with ayahuasca: A proposed model and implications for therapeutic use. *Frontiers in Neuroscience, 16*, 2255. doi:10.3389/fnins.2022.879221 PMID:36798604

PETA. (2022). *Santorini donkey rides*. PETA. https://investigations.peta.org/santorini-donkey-rides/

Peters, P. E. (2019). Revisiting the Social Bedrock of Kinship and Descent in the Anthropology of Afrika. R.R. Grinker, R.R, Lubkemann, S.C., Steiner, C.B. & Gonçalves, E. (eds). A Companion to the Anthropology of Africa. John Wiley & Sons. doi:10.1002/9781119251521

Peter, S., & Park, L. S. C. (2018). Theoretical research: Changing research methodology: Two case studies of critical realism informing social work doctoral research. *Aotearoa New Zealand Social Work, 30*(1), 65–70. doi:10.11157/anzswj-vol30iss1id426

Petersen, M. A., Feldes, S., & Cova, V. S. (2022). Ayahuasca Calling: Sacredness and the Emergence of Shamanic Vocations in Denmark and Peru. *Anthropology of Consciousness, 33*(2), 255–278. doi:10.1111/anoc.12165

Peters, J. C., Wyatt, H. R., Donahoo, W. T., & Hill, J. O. (2002). From instinct to intellect: The challenge of maintaining healthy weight in the modern world. *Obesity Reviews, 3*(2), 69–74. doi:10.1046/j.1467-789X.2002.00059.x PMID:12120422

Petit, O., Cheok, A. D., & Oullier, O. (2016). Can food porn make us slim? How brains of consumers react to food in digital environments. *Integrative Food, Nutrition and Metabolism, 3*(1), 251–255. doi:10.15761/IFNM.1000138

Peyton, T., Gip, H., Pasamehmetoglu, A., & Guchait, P. (2023). How authentic leadership cultivates trust and desirable workplace behaviors in hotels: Commitment and leader-follower value congruence matters. *Journal of Human Resources in Hospitality & Tourism*, 1–28. doi:10.1080/15332845.2023.2207987

Pham, P. T. (2020). *The Impact of Servitisation on Businesses*. Metropolia University of Applied Sciences.

Phillips, L. (2006). Food and globalization. *Annual Review of Anthropology, 35*(1), 37–57. doi:10.1146/annurev.anthro.35.081705.123214

Pilecki, B., Luoma, J. B., Bathje, G. J., Rhea, J., & Narloch, V. F. (2021). Ethical and legal issues in psychedelic harm reduction and integration therapy. *Harm Reduction Journal, 18*(1), 1–14. doi:10.118612954-021-00489-1 PMID:33827588

Piluso, F. (2021). # Foodporn: Fetishized sharing of food and its images. *Digital Age in Semiotics & Communication, 4*(1), 31–41. doi:10.33919/dasc.20-21.3-4.3

Pizam, A., Jansen-Verbeke, M., & Steel, L. (1997). Are all tourists alike, regardless of nationality? The perceptions of Dutch tour-guides. *Journal of International Hospitality, Leisure &. Tourism Management, 1*(1), 19–38. doi:10.1300/J268v01n01_03

Pizam, A., & Jeong, G. H. (1996). Cross-cultural tourist behavior: Perceptions of Korean tour-guides. *Tourism Management, 17*(4), 277–286. doi:10.1016/0261-5177(96)00019-2

Pizam, A., & Sussmann, S. (1995). Does nationality affect tourist behavior? *Annals of Tourism Research, 22*(4), 901–917. doi:10.1016/0160-7383(95)00023-5

Platteau, J.-P. (2014). Redistributive Pressures in Sub-Saharan Africa: Causes, Consequences, and Coping Strategies. Akyeampong, E., Bates, R.H., Nunn, N. & Robinson, J. (eds) Africa's Development in Historical Perspective, pp. 153 – 207. Cambridge University Press. doi:10.1017/CBO9781139644594.008

Ploesser, K., & Russell, N. (2009). The people integration challenge. In *Handbook of research on business process modeling* (pp. 274–298). IGI Global. doi:10.4018/978-1-60566-288-6.ch013

Pohjanheimo, T., Paasovaara, R., Luomala, H., & Sandell, M. (2010). Food choice motives and bread liking of consumers embracing hedonistic and traditional values. *Appetite, 54*(1), 170–180. doi:10.1016/j.appet.2009.10.004 PMID:19835923

Poulin, R. (2003). Globalization and the sex trade: Trafficking and the commodification of women and children. *Canadian Women's Studies, 22*(3/4), 38–47.

Poulston, J. (2008). Hospitality workplace problems and poor training: A close relationship. *International Journal of Contemporary Hospitality Management, 20*(4), 412–427. doi:10.1108/09596110810873525

Pourbarkhordari, A., Zhou, E. H., & Pourkarimi, J. (2016). Role of transformational leadership in creating a healthy work environment in business setting. *European Journal of Business and Management, 8*(3), 57–70.

Prakash, M., & Chowdhary, N. (2010). Tour guides: Roles, challenges and desired competences a review of literature. *International Journal of Hospitality and Tourism Systems, 3*(1), 1–12.

Prasad, P. (2005). *Crafting qualitative research: Working in the postpositivist traditions*. ME Shape.

Preteceille, E., & Terrail, J. P. (1986). Capitalism, consumption, and needs. *Capital and Class, 10*(2), 226–229. doi:10.1177/030981688602900115

Price, L. (2016). Using retroduction to address wicked problems. In *Crisis System* (pp. 109–129). Routledge. doi:10.4324/9781315563138-7

Pritchard, A. (1969). Statistical bibliography or bibliometrics? *The Journal of Documentation, 25*(4), 348–349.

PUBLICO. (2017*). La discriminación de género en un hotel de Tenerife: ellos 640 euros y ellas 139.* Publico. https://www.publico.es/sociedad/discriminacion-genero-hotel-tenerife-640-euros-139.html

Punaks, M., & Feit, K. (2014). *The paradox of orphanage volunteering: Combating child trafficking through ethical voluntourism.* Next Generation Nepal. http://www.nextgenerationnepal.org/File/The-Paradox-of-Orphanage-Volunteering.pdf

Purkis, S. (2008). Turizmle kalkınmada enformel emek kullanımı: Marmaris örneği. *Ekonomik Yaklaşım Dergisi, 19*(69), 107–126. doi:10.5455/ey.10678

Putnik, K., Houkes, I., Jansen, N., Nijhuis, F., & Kant, I. (2018). Work-home interface in a cross-cultural context: A framework for future research and practice. *International Journal of Human Resource Management, 31*(13), 1–18. doi:10.1080/09585192.2017.1423096

Qian, X., Zhang, M., & Jiang, Q. (2020). Leader humility, and subordinates' organizational citizenship behavior and withdrawal behavior: Exploring the mediating mechanisms of subordinates' psychological capital. *International Journal of Environmental Research and Public Health, 17*(7), 2544. doi:10.3390/ijerph17072544 PMID:32276358

Quinn, R., & Rohrbaugh, J. (1983). A Spatial Model of Effectiveness Criteria: Toward a Competing Values Approach to Organizational Analysis. *Management Science, 29*(3), 363–377. doi:10.1287/mnsc.29.3.363

Rajendran, D., Farquharson, K., & Hewege, C. (2017). Workplace integration: The lived experiences of highly skilled migrants in Australia. *Equality, Diversity and Inclusion, 36*(5), 437–456. doi:10.1108/EDI-11-2016-0094

Ramesh, A., & Gelfand, M. J. (2010). Will they stay or will they go? The role of job embeddedness in predicting turnover in individualistic and collectivistic cultures. *The Journal of Applied Psychology, 95*(5), 807–823. doi:10.1037/a0019464 PMID:20836585

Reas, P. J. (2013). 'Boy, have we got a vacation for you': Orphanage tourism in Cambodia and the commodification and objectification of the orphaned child. *Thammasat Review, 16*(1), 121–139.

Reas, P. J. (2020). "Children that are cute enough to eat": The commodification of children in volunteering vacations to orphanages and childcare establishments in Siem Reap, Cambodia. *Tourism, Culture & Communication, 20*(2-3), 83–93. doi:10.3727/109830420X15894802540151

Regan, T. (2007). Kafesler boşalsın: Hayvan haklarıyla yüzleşmek. Çev. Serpil Çağlayan, İstanbul: İletişim Yayınları.

Reisinger, Y., & Steiner, C. (2006). Reconceptualising interpretation: The role of tour guides in authentic tourism. *Current Issues in Tourism*, *9*(6), 481–498. doi:10.2167/cit280.0

Rickly, J., & Kline, C. (2021). Introduction: Working for the (hu)man in the tourism industry. In C. Kline & J. Rickly (Eds.), *Exploring non-human work in tourism: From beasts of burden to animal ambassadors* (pp. 1–18). De Gruyter Oldenbourg. doi:10.1515/9783110664058-001

Rieu, J. N., & Kamara, M. (2016). Determinants of Employee Commitment in Hotel Industry in Nairobi County: A Case of Rated Hotels in Nairobi Central Business District. *International Journal of Innovative Social Sciences & Humanities Research*, *4*(4), 1–21.

Ritzer, G., & Jurgenson, N. (2010). Production, consumption, presumption: The nature of capitalism in the age of the digital 'prosumer'. *Journal of Consumer Culture*, *10*(1), 13–36. doi:10.1177/1469540509354673

Robeva-Stoyanova, B. (2023). Integrating people in the organization: A neglected management function? *XIX. IBANESS Congress Series on Economics, Business and Management*, (pp. 50-54). Ibaness..

Robinson, R. N., Martins, A., Solnet, D., & Baum, T. (2019). Sustaining precarity: Critically examining tourism and employment. *Journal of Sustainable Tourism*, *27*(7), 1008–1025. doi:10.1080/09669582.2018.1538230

Rockower, P. S. (2012). Recipes for gastrodiplomacy. *Place Branding and Public Diplomacy*, *8*(3), 235–246. doi:10.1057/pb.2012.17

Rose, N., & Lourival, I. (2019). Hegemony, counter-hegemony, and food systems literacy: Transforming the global industrial food system. *Australian Journal of Environmental Education*, *35*(2), 110–122. doi:10.1017/aee.2019.9

Roth, C., Breckner, A., Wensing, M., Mahler, C., Krug, K., & Berger, S. (2023). Barriers and enabling factors for workplace integration of internationally qualified nurses: A qualitative study of perceptions of German nurses. *Nursing Open*, *00*(8), 1–11. doi:10.1002/nop2.1760 PMID:37060232

Rousseau, S. (2012). *Food and social media: You are what you tweet*. Rowman Altamira. doi:10.5040/9781350042193

Rouzrokh, M., Muldoon, M., Torabian, P., & Mair, H. (2017). The memory-work sessions: Exploring critical pedagogy in tourism. *Journal of Hospitality, Leisure, Sport and Tourism Education*, *21*, 163–173. doi:10.1016/j.jhlste.2017.08.006

Rozuel, C., & McGhee, P. (2012). Investigating the convergence of corporate social responsibility and spirituality at work. *Australian Journal of Professional and Applied Ethics*, *14*(1), 47–62.

Rucker, J. J., Iliff, J., & Nutt, D. J. (2018). Psychiatry & the psychedelic drugs. Past, present & future. *Neuropharmacology*, *142*, 200–218. doi:10.1016/j.neuropharm.2017.12.040 PMID:29284138

Runnels, V., & Carrera, P. M. (2012). Why do patients engage in medical tourism? *Maturitas*, *73*(4), 300–304. doi:10.1016/j.maturitas.2012.08.011 PMID:23007007

Rutsaert, P., Regan, A., Pieniak, Z., McConnon, Á., Moss, A., Wall, P., & Verbeke, W. (2013). The use of social media in food risk and benefit communication. *Trends in Food Science & Technology*, *30*(1), 84–91. doi:10.1016/j.tifs.2012.10.006

Salleh, S., Hashima, N. H., & Murphy, J. (2015). Instagram marketing: a content analysis of top Malaysian restaurant brands. *E-Review of Tourism Research, 6*, 1-5.

Saltik, Z. (2023). Behavioral Approaches to Leadership. In *Leadership Approaches in Global Hospitality and Tourism* (pp. 40–60). IGI Global. doi:10.4018/978-1-6684-6713-8.ch003

Sandjong, F. L. D. (2021). *Training Strategies for Increasing Employee Job Satisfaction and Retention in the Hospitality Industry* [Doctoral dissertation, Walden University].

Saper, C. B., Chou, T. C., & Elmquist, J. K. (2002). The need to feed: Homeostatic and hedonic control of eating. *Neuron*, *36*(2), 199–211. doi:10.1016/S0896-6273(02)00969-8 PMID:12383777

Sarmento, M., Pedro, M. I., & Pinto, E. (2016). Integration and socialization of new employees in the bank sector. *International Journal of Latest Trends in Finance & Economic Sciences*, *6*(2), 1098–1107.

Saunders, M., Lewis, P., & Thornhill, A. (2019). *Research methods for business students* (8th ed.). Pearson.

Schireup, C. U., Hansen, P., & Castles, S. (2006). *Migration, citizenship and the European welfare state: A European dilemma*. Oxford University Press. doi:10.1093/0198280521.001.0001

Schmidt, C. J. (1979). The guided tour: Insulated adventure. *Urban life, 7*(4), 441-467. https:// doi.org/ doi:10.1177/089124167900700402

Schmidt, W., & Müller, A. (2021). Workplace universalism and the integration of migrant workers and refugees in Germany. *Industrial Relations Journal*, *52*(2), 145–160. doi:10.1111/irj.12320

Schneider-Kamp, A. (2021). Inclusion of the excluded: Consumers' quest for hedonism in food consumption. *International Journal of Consumer Studies*, *45*(3), 320–334. doi:10.1111/ijcs.12612

Scott, J. (1992). *Domination and the arts of resistance: Hidden transcripts*. Yale University Press.

Seal, A., Gavaravarapu, S. M., & Konapur, A. (2022). Can foodporn prime healthy eating? Thinking beyond digital gazing and satiety. *European Journal of Clinical Nutrition*, *1493-1496*(11), 1493–1496. doi:10.103841430-022-01139-w PMID:35411027

Segal, L. (1990). Pornography and violence: What the 'experts' really say. *Feminist Review*, *36*(1), 29–41. doi:10.1057/fr.1990.43

Segev, L., Paz, A., & Potasman, I. (2005). Drug abuse in travelers to southeast Asia: An on-site study. *Journal of Travel Medicine*, *12*(4), 205–209. doi:10.2310/7060.2005.12406 PMID:16086895

Şenol, F., & Aliyev, İ. (2015). Türkiye'de Rus turist pazarına hâkim tur acentalarında istihdam edilen turist rehberleri üzerine bir inceleme. *Kastamonu Üniversitesi İktisadi ve İdari Bilimler Fakültesi Dergisi*, *8*, 167–181.

Sezgin, E., & Duz, B. (2018). Testing the proposed "GuidePerf" scale for tourism: Performances of tour guides in relation to various tour guiding diplomas. *Asia Pacific Journal of Tourism Research*, *23*(2), 170–182. doi:10.1080/10941665.2017.1410196

Sharpley, R. (2020). Tourism, sustainable development and the theoretical divide: 20 years on. *Journal of Sustainable Tourism*, *28*(11), 1932–1946. doi:10.1080/09669582.2020.1779732

Sheldon, P. J., Fesenmaier, D. R., & Tribe, J. (2011). The tourism education futures initiative (TEFI): Activating change in tourism education. *Journal of Teaching in Travel & Tourism*, *11*(1), 2–23. doi:10.1080/15313220.2011.548728

SHTM (school of Hotel & Tourism Management) (n.d.). *Undergraduate Programmes.* SHTM. https://www.polyu.edu.hk/shtm/study/undergraduate-programmes/undergraduate-programme/bachelor-of-science-scheme-in-hotel-and-tourism-management/specialism-or-secondary-major/

Sigala, M. (2008). A supply chain management approach for investigating the role of tour operators on sustainable tourism: The case of TUI. *Journal of Cleaner Production*, *16*(15), 1589–1599. doi:10.1016/j.jclepro.2008.04.021

Sijtsman, W. (2001). *Tourism training development for women's empowerment: An amendatory methodology* [Unpublished master's dissertation, Netherlands Institute of Tourism and Transport Studies, Netherlands].

Silva, M. (2014). El trabajo infantil rural en la agenda pública del desarrollo sustentable. *Espacio Abierto*, *23*(1), 23–44.

Silva, P. (2006). Effects of disposition on hospitality employee job satisfaction and commitment. *International Journal of Contemporary Hospitality Management*, *18*(4), 317–328. doi:10.1108/09596110610665320

Simeone, M., & Scarpato, D. (2020). Sustainable consumption: How does social media affect food choices? *Journal of Cleaner Production*, *277*, 1–6. doi:10.1016/j.jclepro.2020.124036

Simmons, D. G. (1994). Community participation in tourism planning. *Tourism Management*, *15*(2), 98–108. doi:10.1016/0261-5177(94)90003-5

Simpson, E., Conway, C., Erickson, E., Kramer, O., Oberbroeckling, M., O'Malley, C., & Ybarra, H. (2022). Examining the OT Student Perspective of Group Work in Graduate School. *The American Journal of Occupational Therapy*, *76*(Supplement_1), 76. doi:10.5014/ajot.2022.76S1-PO120

Singelmann, J. (1978). The sectoral transformation of the labor force in seven industrialized countries, 1920-1970. *American Journal of Sociology*, *83*(5), 1224–1234. doi:10.1086/226680

Singh, A., Sibi, P. S., & Bashir, A. (2023). The Journal of Convention and Event Tourism: A retrospective analysis using bibliometrics. *Journal of Convention & Event Tourism*, *24*(1), 87–108. doi:10.1080/15470148.2022.2150731

Smith, W. R., & Appelbaum, P. S. (2022). Novel ethical and policy issues in psychiatric uses of psychedelic substances. *Neuropharmacology*, *216*, 109165. doi:10.1016/j.neuropharm.2022.109165 PMID:35718006

Snyman, S. (2014). The impact of ecotourism employment on rural household incomes and social welfare in six southern African countries. *Tourism and Hospitality Research*, *14*(1–2), 37–52. doi:10.1177/1467358414529435

Sobal, J. (2005). Men, meat, and marriage: Models of masculinity. *Food & Foodways*, *13*(1-2), 135–158. doi:10.1080/07409710590915409

Sokolov, R. (2007). Piscinae and the myth of Roman decadence at table. In S. R. Friedland (Ed.), *Food & morality: Proceedings of the Oxford Symposium on Food And Cookery* (pp. 263-68). Prospect Books.

Sondaitė, J., & Keidonaitė, G. (2020). Experience of transformative leadership: Subordinate's perspective. *Business: Theory and Practice*, *21*(1), 373–378. doi:10.3846/btp.2020.11113

Soonsan, N., & Somkai, U. (2023). Dimensions of gastronomic experience affecting on sharing experience: Place attachment as a mediator and length of stay as a moderator. *Journal of Hospitality and Tourism Insights*, *6*(1), 344–361. doi:10.1108/JHTI-07-2021-0171

Spenceley, A., & Snyman, S. (2017). Can a wildlife tourism company influence conservation and the development of tourism in a specific destination? *Tourism and Hospitality Research*, *17*(1), 52–67. doi:10.1177/1467358416634158

Srivastava, S., & Singh, S. (2020). Linking personal growth initiative and organizational identification to employee engagement: Testing the mediating-moderating effects in Indian hotel industry. *Journal of Hospitality and Tourism Management*, *45*, 79–89. doi:10.1016/j.jhtm.2020.07.005

Stainback, K., Robinson, C. L., & Tomaskovic-Devey, D. (2005). Race and workplace integration: A politically mediated process? *The American Behavioral Scientist*, *48*(9), 1200–1228. doi:10.1177/0002764205274816

STATCAN. (2019). *Tourism Human Resource Module, 2019*. Statcan. https://www150.statcan.gc.ca/n1/daily-quotidien/201030/dq201030c-eng.htm

Stauth, G. (1984). Households, Modes of Living, and Production Systems. Smith, J., Wallerstein, I. & Evers, H.-D. (eds), Households and the World-Economy. Sage.

Sterckx, R. (2011). *Food, sacrifice, and sagehood in early China*. Cambridge University Press. doi:10.1017/CBO9780511736247

Stevens, T. M., Aarts, N., Termeer, C. J. A. M., & Dewulf, A. (2016). Social media as a new playing field for the governance of agro-food sustainability. *Current Opinion in Environmental Sustainability, 18*, 99–106. doi:10.1016/j.cosust.2015.11.010

Stutchbury, K. (2022). Critical realism: An explanatory framework for small-scale qualitative studies or an 'unhelpful edifice'? *International Journal of Research & Method in Education, 45*(2), 113–128. doi:10.1080/1743727X.2021.1966623

Suarez-Lledo, V., & Alvarez-Galvez, J. (2021). Prevalence of health misinformation on social media: Systematic review. *Journal of Medical Internet Research, 23*(1), e17187. doi:10.2196/17187 PMID:33470931

Subritzky, T. (2018). Beyond deficit and harm reduction: Incorporating the spectrum of wellness as an interpretive framework for cannabis consumption. *The International Journal on Drug Policy, 60*, 18–23. doi:10.1016/j.drugpo.2018.07.013 PMID:30086481

Su, M. M., Wall, G., Ma, J., Notarianni, M., & Wang, S. (2020). Empowerment of women through cultural tourism: Perspectives of Hui minority embroiderers in Ningxia, China. *Journal of Sustainable Tourism, 31*(2), 307–328. doi:10.1080/09669582.2020.1841217

Sutton, J., & Thompson, R. (1998). Multidisciplinary integration: A decision methodology and procedure for instruction. In *FIE'98. Proceedings of 28th Annual Frontiers in Education Conference,* (pp. 450-455). IEEE. 10.1109/FIE.1998.736891

Swancutt, D. M. (2007). Still before sexuality:" Greek" androgyny, the roman imperial politics of masculinity and the roman invention of the Tribas. In D. M. Swancutt (Ed.), *Mapping gender in ancient religious discourses* (pp. 11–61). Brill. doi:10.1163/ej.9789004154476.i-582.12

Takenaka, A. K., Villafuerte, J., Gaspar, R., & Narayanan, B. (2020). COVID-19 impact on international migration, remittances, and recipient households in developing Asia. *ADB Briefs*, (August). doi:10.22617/BRF200219-2

Tang, C. Y., & Chang, J. C. (2006). *An investigation of service quality between the licensed and un-licensed tour leaders. The 7th Biennial Conference on Tourism in Asia*, Jeonju, Korea.

Tariq, H., Abrar, M., & Ahmad, B. (2023). Humility breeds creativity: the moderated mediation model of leader humility and subordinates' creative service performance in hospitality. *International Journal of Contemporary Hospitality Management*. doi:10.1108/IJCHM-07-2022-0851

Taşpınar, O., & Temeloğlu, E. (2018). Influence of Tv cooking shows on the behavioral intention of participating in gastronomic tourism. *Journal of Tourism & Gastronomy Studies, 6*(2), 24–40. doi:10.21325/jotags.2018.200

Tavory, I., & Timmermans, S. (2014). *Abductive Analysis: Theorizing Qualitative Research. Cicago*. University of Chicago Press. doi:10.7208/chicago/9780226180458.001.0001

Taylor, C. Y. (2019). *Strategies to improve employee engagement in the hospitality industry* [Doctoral dissertation, Walden University].

Tejaswi, M. J. (2010). People Integration a Challenge: Experts. *The Times of India.*

Tetik Dinç, N., Balıklıoğlu, A., & Ulusoy Mutlu, H. (2022). Turist rehberlerine yönelik turist nezaketsizliği ve rehberlerin nezaketsiz davranışlara ilişkin tepkileri. *Selçuk Üniversitesi Sosyal Bilimler Enstitüsü Dergisi*, (49), 100–117. doi:10.52642usbed.1152188

The News Minute. (2020). The invisibility of women's labour in tourism and the impact of COVID-19. *The News Minute.* https://www.thenewsminute.com/article/invisibility-women-s-labour-tourism-and-impact-covid-19-123315

Thompson, R. J. (2016). *An investigation of work environment characteristics and work engagement of limited-and full-service hotel property front-line employees*, [Doctoral dissertation, The University of Southern Mississippi].

Thrift, N. (2006). Re-inventing invention: New tendencies in capitalist commodification. *Economy and Society*, *35*(2), 279–306. doi:10.1080/03085140600635755

Toksöz, G. (2011). *Kalkınmada Kadın Emeği.* Varlık Yayınları.

Tomassini, L., & Bertella, G. (2023). The human gaze at animals and the missing animal gaze in tourism studies. *Current Issues in Tourism*, 1–6. doi:10.1080/13683500.2023.2215920

Tosun, C. (2000). Limits to Community Participation in the Tourism Development Process in Developing Countries. *Tourism Management*, *21*(6), 613–633. doi:10.1016/S0261-5177(00)00009-1

Travelpulse. (2023). How women are playing a pivotal role in the African travel industry. *Travel Pulse.* https://www.travelpulse.com/News/Destinations/How-Women-are-Playing-a-Pivotal-Role-in-the-African-Travel-Industry

Tribe, J. (2008). Tourism: A Critical Business. *Journal of Travel Research*, *46*(3), 245–255. Advance online publication. doi:10.1177/0047287507304051

Tsaur, S. H., Cheng, T. M., & Hong, C. Y. (2019). Exploring tour member misbehavior in group package tours. *Tourism Management*, *71*, 34–43. doi:10.1016/j.tourman.2018.09.018

Tsaur, S. H., & Lin, W. R. (2014). Hassles of tour leaders. *Tourism Management*, *45*, 28–38. doi:10.1016/j.tourman.2014.03.017

Tucker, H. (2007). Undoing shame: Tourism and women's work in Turkey. *Journal of Tourism and Cultural Change*, *5*(2), 87–105. doi:10.2167/jtcc089.0

TÜİK. (2023). *Çocuk istatistikleri.* https://data.tuik.gov.tr/Bulten/Index?p=Istatistiklerle-Cocuk-2022-49674&dil=1

Turizm Databank. (2022). *Turizm sektörü istihdamı.* https://www.turizmdatabank.com/turizm-istatistikleri/otellerde-toplam-calisan-sayisi/

Türkmen, S., & Deniz, G. B. (2022). Seyahat acentalarının turist rehberlerine yönelik sergiledikleri etik dişi davranişlarin belirlenmesi: ÇARO örneği. *Pamukkale Üniversitesi Sosyal Bilimler Enstitüsü Dergisi*, (53), 367–377. doi:10.30794/pausbed.1145202

Twining-Ward, L. D., Li, W., Bhammar, H. M., & Wright, E. (2018). *Supporting Sustainable Livelihoods through Wildlife Tourism.* The World Bank Group. doi:10.1596/29417

Uçar, A. (2020). *Determination of hedonic hunger and mindful eating in high school students and its relationship with nutritional status* [Unpublished Masters' Thesis, Baskent University].

Uğuz, S. Ç., & Topbaş, F. (2016). Turizmde kadın istihdamı ve ücret ayrımcılığı: Karşılaştırmalı bir analiz. *Anatolia: Turizm Araştırmaları Dergisi*, *27*(1), 62–78.

Ugwu, K. E., Okoroji, L. I., & Chukwu, E. O. (2018). Participative decision making and employee performance in the hospitality industry: A study of selected hotels in Owerri Metropolis, Imo State. *Management Studies and Economic Systems*, *4*(1), 57–70. doi:10.12816/0053651

UNCTAD. (2023). *African tourists bring better jobs for women and youth.* UNCTAD. https://unctad.org/news/african-tourists-bring-better-jobs-women-and-youth

UNICEF. (1989). *Convention on the rights of the child.* UNICEF. https://www.unicef.org/turkiye/en/convention-rights-child

UNICEF. (2021). *Child labour: Global estimates 2020, trends and the road forward.* UNICEF. https://data.unicef.org/resources/child-labour-2020-global-estimates-trends-and-the-road-forward/

United Nations. (2013). *Combating child sex tourism.* UN. https://www.ohchr.org/en/stories/2013/04/combating-child-sex-tourism

United Nations. (2023). *Global Issues: Ending poverty.* UN. https://www.un.org/en/global-issues/ending-poverty, accessed on April 26, 2023.

Ünlüönen, K. ve Çeti, B. (2019). Salgın hastalıklar sebebiyle oluşan krizlerin turizm sektörü üzerindeki etkisinin değerlendirilmesi. *Ankara Hacı Bayram Veli Üniversitesi Turizm Fakültesi Dergisi*, *22*(2), 109–128.

UNWOMEN. (2022). *Advancing women's economic empowerment in the tourism sector in covıd-19 response and recovery.* UN Women. https://www.unwomen.org/sites/default/files/2022-05/Advancing-womens-economic-empowerment-in-the-tourism-sector-in-COVID-19-en.pdf

UNWTO. (2018a). *Sustainable development.* UNWTO. http://sdt.unwto.org/content/about-us-5,

UNWTO. (2018b) *Sustainable Development of Tourism, One Planet – Sustainable Tourism Programme*. UNWTO. http://sdt.unwto.org/about-10yfp-stp,

UNWTO. (2020). *2020: A year in review.* UNWTO. https://www.unwto.org/covid-19-and-tourism-2020

UNWTO. (2023). *Women take 'centre stage' in tourism development*. UN. https://www.unwto.org/news/women-take-centre-stage-in-tourism-development#:~:text=Fifty%2Dfour%20per%20cent%20of,higher%20than%20in%20other%20sectors

Urry, J. (1990). The tourist gaze: Leisure and travel in contemporary societies. *Sage (Atlanta, Ga.)*.

Urry, J. (2002). *The Tourist Gaze*. Sage Publications.

Uzochukwu, A. C., Nwankwo, E. C., & Okafor, S. O. (2023). Effect of reward system on employee's productivity in Nigeria hospitality industry. *World Journal of Advanced Research and Reviews*, *17*(01), 361–373. doi:10.30574/wjarr.2023.17.1.0035

Valdez, A., & Sifaneck, S. J. (1997). Drug tourists and drug policy on the US-Mexican border: An ethnographic investigation of the acquisition of prescription drugs. *Journal of Drug Issues*, *27*(4), 879–898. doi:10.1177/002204269702700413

Van Der Meer, P. B., Fuentes, J. J., Kaptein, A. A., Schoones, J. W., de Waal, M. M., Goudriaan, A. E., Kramers, K., Schellekens, A., Somers, M., Bossong, M. G., & Batalla, A. (2023). Therapeutic effect of psilocybin in addiction: A systematic review. *Frontiers in Psychiatry*, *14*, 1134454. doi:10.3389/fpsyt.2023.1134454 PMID:36846225

Van Doore, K., Martin, F., & McKeon, A. (2016). International volunteering and child sexual abuse. *Global study on sexual exploitation of children in travel and tourism Expert paper*. New York: Better Care Network. https://bettercarenetwork.org/sites/default/files/Expert%20Paper%20-%20International%20Volunteering%20and%20Child%20Sexual%20Abuse.pdf

Vasilopoulou, C. (2021). The donkeys of Santorini: Workers or slaves? In C. Kline & J. Rickly (Eds.), *Exploring non-human work in tourism: From beasts of burden to animal ambassadors* (pp. 159–180). De Gruyter Oldenbourg. doi:10.1515/9783110664058-010

Vaterlaus, J. M., Patten, E. V., Roche, C., & Young, J. A. (2015). Gettinghealthy: The perceived influence of social media on young adult health behaviors. *Computers in Human Behavior*, *45*, 151–157. doi:10.1016/j.chb.2014.12.013

Verbeke, W. (2005). Agriculture and the food industry in the information age. *European Review of Agriculture Economics*, *32*(3), 347–368. doi:10.1093/eurrag/jbi017

Vidriales, A. L., & Ovies, D. H. (2018). Psychedelic tourism in Mexico, a thriving trend. *Pasos (El Sauzal)*, *16*(4), 1037–1050. doi:10.25145/j.pasos.2018.16.072

Von Essen, E., Lindsjö, J., & Berg, C. (2020). Instagranimal: Animal welfare and animal ethics challenges of animal-based tourism. *Animals (Basel)*, *10*(10), 1830–1847. doi:10.3390/ani10101830 PMID:33050024

Vosko, L. F. (2010). *Managing the margins: Gender, citizenship, and the international regulation of precarious employment*. Oxford University Press.

Voumik, L. C., Nafi, S. M., Majumder, S. C., & Islam, M. A. (2023). The impact of tourism on the women employment in South American and Caribbean countries. *International Journal of Contemporary Hospitality Management*, *35*(9), 3095–3112. doi:10.1108/IJCHM-04-2022-0449

Walton, R. E. (1985). From control to commitment in the workplace. *Harvard Business Review*, (March), 1–9.

Wang, K. C., Hsieh, A. T., & Huan, T. C. (2000). Critical service features in group package tour: An exploratory research. *Tourism Management*, *21*(2), 177–189. doi:10.1016/S0261-5177(99)00047-3

Ward, A. C., & Baker, P. M. (2005). Disabilities and impairments: Strategies for workplace integration. *Behavioral Sciences & the Law*, *23*(1), 143–160. doi:10.1002/bsl.631 PMID:15706600

Warren, C. (2014). Reviewing the skilled migration and 400 series visa programme. *Home Affairs*. https://www.homeaffairs.gov.au/reports-and-pubs/files/carlita-warren.pdf

Wäsche, H., Dickson, G., Woll, A., & Brandes, U. (2017). Social network analysis in sport research: An emerging paradigm. *EJSS European Journal for Sport and Society*, *14*(2), 138–165. doi:10.1080/16138171.2017.1318198

Watts, D. (2003). *Six Degrees: The Science of a Connected Age*. W. W. Norton & Company.

WB-IFC. (2017). *Tourism for development*. WB-IFC. https://documents1.worldbank.org/curated/en/401321508245393514/pdf/120477-WP-PUBLIC-Weds-oct-18-9am-ADD-SERIES-36p-IFCWomenandTourismfinal.pdf

Weber, P., Ludwig, T., Brodesser, S., & Grönewald, L. (2021, May). "It's a kind of art!": Understanding food influencers as influential content creators. In *Proceedings of the 2021 CHI Conference on Human Factors in Computing Systems* (pp. 1-14). 10.1145/3411764.3445607

Weiler, B., & Black, R. (2015). *Tour guiding research: Insights, issues, implications*. Channel View.

Weiler, B., & Ham, S. (2001). Tour guides and interpretation. In D. Weaver (Ed.), *Encyclopedia of ecotourism*. CABI. doi:10.1079/9780851993683.0549

Weiler, B., & Walker, K. (2014). Enhancing the visitor experience: Reconceptualising the tour guide's communicative role. *Journal of Hospitality and Tourism Management*, *21*, 90–99. doi:10.1016/j.jhtm.2014.08.001

Wells, J. C. (2012). Obesity as malnutrition: The role of capitalism in the obesity global epidemic. *American Journal of Human Biology*, *24*(3), 261–276. doi:10.1002/ajhb.22253 PMID:22383142

Werry, M. (2008). Pedagogy of/as/and tourism: Or, shameful lessons. *Review of Education, Pedagogy & Cultural Studies*, *30*(1), 14–42. doi:10.1080/10714410701566207

WFTGA. (n.d.). *What is a professional tourist guide?* WFTGA. https://wftga.org/about/

Wigfield, A. (2001). Post-Fordism, gender and work. Londra: Ashgate Publishing.

Wiggershaus, R. (1994). *The Frankfurt School*. Polity Press.

Wilson, S. L. R. (1993). *Mass media/mass culture: an introduction.* McGraw-Hill, Inc.

Wondirad, A., & Ewnetu, B. (2019). Community participation in tourism development as a tool to foster sustainable land and resource use practices in a national park milieu. *Land Use Policy, 88*, 104155. doi:10.1016/j.landusepol.2019.104155

Wong, C. U. I. (2013). The sanitization of colonial history: Authenticity, heritage interpretation and the case of Macau's tour guides. *Journal of Sustainable Tourism, 21*(6), 915–931. doi:10.1080/09669582.2013.790390

Wright, A. (2004). *Reward Management in Context.* CIPD Publishing.

WSPA. (2021). *The contribution of animal welfare and tourism for sustainable development.* WSPA. https://www.responsibletravel.org/wp-content/uploads/sites/213/2021/03/sustainable-tourism-animal-welfare.pdf

WTO. (2019). *Global report on women in tourism.* World Tourism Organization, UNWTO, Madrid. https://www.e-unwto.org/doi/pdf/10.18111/9789284420384

WTO. (2020). *Regional report on women in tourism in the Middle East.* WTO. https://www.e-unwto.org/doi/epdf/10.18111/9789284422371

WTO. (2022). *Regional report on women in tourism in Asia and the Pacific.* UNWTO, Madrid. https://www.e-unwto.org/doi/epdf/10.18111/9789284423569

WTO-ILO. (2014). Measuring employment in the tourism ındustries – Guide with best practices. World Tourism Organization and International Labour Organization, UNWTO, Madrid.

WTTC. (2019). *Travel & Tourısm: Driving women's success.* WTTC. https://wttc.org/Portals/0/Documents/Reports/2019/Social%20Impact-Driving%20Womens%20Success-Mar%202019.pdf?ver=2021-02-25-182742-097

WTTC. (2022). *Travel & Tourism could grow to $8.6 trillion in 2022, says WTTC.* WTTC. https://wttc.org/news-article/travel-and-tourism-could-grow-to-8-point-6-trillion-usd-in-2022-say-wttc

WTTC. (2022). *Travel & Tourism economic impact 2022, global trends.* WTTC. https://wttc.org/Portals/0/Documents/Reports/2022/EIR2022-Global%20Trends.pdf

Wynn, J. R. (2005). Guiding practices: Storytelling tricks for reproducing the urban landscape. *Qualitative Sociology, 28*(4), 399–417. doi:10.100711133-005-8365-2

Xu (Hazel), Y., & McGehee, N. G. (2017). Tour guides under zero-fare mode: Evidence from China. *Current Issues in Tourism, 20*(10), 1088–1109. doi:10.1080/13683500.2016.1149155

Yahyaoğlu, M. (2022). *Turizmin ve turizm işçilerinin sorunları ve çözüm önerilerimiz.* Devturismis. http://www.devturizmis.org.tr/post/turizmin-ve-turizm-iscilerinin-sorunlari-ve-cozum-onerilerimiz

Yanardağ, M. Ö., & Avcı, M. (2012). Turizm sektöründe istihdam sorunları: Marmaris, Fethiye, Bodrum ilçeleri üzerine ampirik bir inceleme. *Ege Stratejik Araştırmalar Dergisi, 3*(2), 39–62.

Yankholmes, A., & McKercher, B. (2015). Understanding visitors to slavery heritage sites in Ghana. *Tourism Management*, *51*, 22–32. doi:10.1016/j.tourman.2015.04.003

Yarcan, Ş. (2007). Profesyonel turist rehberliğinde mesleki etik üzerine kavramsal bir değerlendirme. *Anatolia: Turizm Araştırmaları Dergisi*, *18*(1), 33–44.

Yayla, Ö., & Ergün, G. S. (2020). Sosyal bilimlerde paradigma ikilemi: Turizm alanında yazılmış lisansüstü tezler üzerine bir inceleme. *Türk Turizm Araştırmaları Dergisi*, *4*(2), 1624–1639. doi:10.26677/TR1010.2020.416

Yetgin Akgün, D., Ay, E., & Çalık, A. Ö. (2022). Customer complaints in package tours and complaint management of tour guides. *Journal of Gastronomy Hospitality and Travel*, *5*(4), 1482–1494. doi:10.33083/joghat.2022.214

Yetgin, D., Yılmaz, A., & Çiftci, G. (2018). Krizlerin turist rehberliği öğrencilerinin kariyer planlamasındaki etkisi. *Journal of Tourism and Gastronomy Studies*, *195*(3), 214. doi:10.21325/jotags.2018.278

Yıldırım, G., & Özbek, Ö. (2019). Profesyonel turist rehberlerine yönelik cinsel taciz ile ilgili bir araştırma. *Turist Rehberliği Dergisi*, *2*(2), 72–87. doi:10.34090/tured.591586

Yıldırım, M. (2021). Tracing the consent, adaptation and resistance practices of an 'unsustainable'workforce: The governmentality of workplaces in tourism industry. *Tourism Management*, *84*, 104260. doi:10.1016/j.tourman.2020.104260

Yin, R. K. (2009). *Case study research: Design and methods* (4th ed.). SAGE.

Yorgun, S. (2013). Türkiye'de turizm sektöründe esnek istihdam uygulamaları ve toplu iş sözleşmelerinde yer alan düzenlemeler. *Turar Turizm & Araştırma Dergisi*, *2*(2), 41–56.

Young, M., & Markham, F. (2020). Tourism, capital, and the commodification of place. *Progress in Human Geography*, *44*(2), 276–296. doi:10.1177/0309132519826679

Yu, X., Weiler, B., & Ham, S. (2001). Intercultural communication and mediation: A framework for analysing the intercultural competence of Chinese tour guides. *Journal of Vacation Marketing*, *8*(1), 75–87. https://journals.sagepub.com/doi/pdf/10.1177/135676670200800108. doi:10.1177/135676670200800108

Zafar, R., Siegel, M., Harding, R., Barba, T., Agnorelli, C., Suseelan, S., Roseman, L., Wall, M., Nutt, D. J., & Erritzoe, D. (2023). Psychedelic therapy in the treatment of addiction: The past, present and future. *Frontiers in Psychiatry*, *14*, 1183740. doi:10.3389/fpsyt.2023.1183740 PMID:37377473

Zamorano, M., & Lizárraga, C. (2022). The Impact of COVID-19 on the Sustainable Development Goals: Achievements and Expectations. *International Journal of Environmental Research and Public Health*, *19*(23), 16266. doi:10.3390/ijerph192316266 PMID:36498340

Zeb, A., Hussain, K., Khan, F., & Safi, A. (2021). The competing value framework model of organizational culture, innovation and performance. *Business Process Management Journal*, *27*(2), 658–683. doi:10.1108/BPMJ-11-2019-0464

Zeifman, R. J., Singhal, N., Breslow, L., & Weissman, C. R. (2021). On the relationship between classic psychedelics and suicidality: A systematic review. *ACS Pharmacology & Translational Science*, *4*(2), 436–451. doi:10.1021/acsptsci.1c00024 PMID:33860173

Zhang, J., & Xu, H. (2023). The power dynamics in local–tourist interactions. *Annals of Tourism Research*. doi:10.1016/j.annals.2022.103509

Zhang, H. Q., & Chow, I. (2004). Application of importance-performance model in tour guides' performance: Evidence from mainland Chinese outbound visitors in Hong Kong. *Tourism Management*, *25*(1), 81–91. doi:10.1016/S0261-5177(03)00064-5

Zhang, J., & Zhang, Y. (2020). Tourism and gender equality: An Asian perspective. *Annals of Tourism Research*, *85*, 103067. doi:10.1016/j.annals.2020.103067

Zhao, S., & Timothy, D. (2017). The dynamics of guiding and interpreting in red tourism. *International Journal of Tourism Cities*, *3*(3), 243–259. doi:10.1108/IJTC-01-2017-0007

Zheng, X., Huang, J., Wu, J., Sun, S., & Wang, S. (2023). Emerging trends in online reviews research in hospitality and tourism: A scientometric update (2000−2020). *Tourism Management Perspectives*, *47*, 101105. doi:10.1016/j.tmp.2023.101105

Zhou, Y. (2010). Festivals, feasts, and gender relations in ancient China and Greece. Cambridge University Press. doi:10.1017/CBO9780511762468

About the Contributors

Barış Çıvak is a faculty member at Anadolu University, Faculty of Tourism, Department of Tourism Management. He completed his Ph.D. in Tourism Management. His academic interests include organization theory, managerial control mechanisms, tourism management, critical issues in tourism, and labor issues.

Engin Bayraktaroğlu received PhD degree in Tourism Management from Anadolu University Graduate School of Social Sciences, Turkey in 2019. He is currently working as Assistant Professor at Anadolu University Faculty of Tourism. His research interest includes destination value, destination development, tourist mobility, philosophy of tourism.

* * *

Bora Alkan is lecturer in Ondokuz Mayıs University.

Hatice Işıl Alkan is Associate Professor in the field of Macroeconomics. She works for Ondokuz Mayıs University.

Feyza Alyu Altinok completed her PhD in Pharmacology at Anadolu University in 2019. During her doctoral studies, she had the opportunity to collaborate with Camerino University, thanks to the support of a scholarship from the European Molecular Biology Organisation (EMBO). Currently, she holds the position of assistant professor in the Department of Pharmacology at the Faculty of Pharmacy, Anadolu University. Her research area focuses on psychopharmacology, a specialized subfield of neuroscience, and she has made significant contributions through various publications related to her research.

Charlotte Bruun is an associate professor at University College of Northern, Denmark, and former associate professor at Aalborg University, Denmark. She holds a PhD in economics and has previously done research on agent-based modelling, monetary theory, stock-flow consistent macroeconomics and welfare theory. Beyond tourism in sub-Saharan Africa her current research interests revolves around business models and sustainability.

Yavuz Demirci is a PhD candidate at Dokuz Eylül University, Institute of Social Sciences. His academic interest includes tourism sociology, culture, tourist behavior, interaction between tourists and residents.

Hamid Derviş is an Associate Professor in the Department of Information Management and Records at Kastamonu University. He holds a Ph.D. in Information Management & Records from Hacettepe University in Ankara, Türkiye. He has a Master's degree in Management Information Systems, a Bachelor's degree, and an Associate's degree in Computer Science from Maryland, United States of America.

S. Emre Dilek is a PhD at the Dept of Gastronomy & Culinary Arts, Tourism and Hotel Management High School, Batman University, Batman, Turkey. He completed his master thesis titled "Green Marketing Applications in Tourism Businesses: A Field Research" in 2012. In this thesis, relationship between tourism and environment was questioned and the concept of green marketing, and green marketing related implementations in tourism were researched. In case of PhD thesis titled "Evaluation of Moral Responsibility of Tourism Regarding Animals in the Context of Meta-Criticism" which is completed in 2016. This study is to make critical analysis by introducing areas that animals are commodified in the tourism industry. He received the title of associate professor in 2021.The topics he is interested are Tourism Sociology, Tourism and Animal Ethics, Philosophy and Green Marketing.

Burak Düz, Ph.D., is a faculty member at Anadolu University, Faculty of Tourism, Department of Tour Guiding. His research interest covers tour guiding within the realm of performance, professional ethics and tour guides' roles.

Anna Hammershøy is an Associate professor at University College of Northern Denmark. She teaches Cultural awareness and customer relations and Destination development at the International Hospitality Management bachelor programme. She holds a Master degree in Culture, Communication and Globalisation from Aalborg University. Her research interests include use of networks in tourism destinations, cooperation and coopetition in tourism businesses and sense of place in tourism destinations.

Emil Nikolov Kotsev is working as an associate professor in Angel Kanchev University of Ruse, Business Development and Innovations Department. His research interests are social management and human resources management.

Nur Kulakoğlu Dilek is a PhD at the Department of Tourism & Hotel Management, Batman University, Batman, Turkey. She completed her master thesis titled "Effectiveness of cultural intelligence through managing cultural differences: A field research on managers of the 'a' class travel agencies" in 2014. PhD thesis titled "An exploratory research on the use and effect of virtual reality technology in tourism sector" which is completed in 2020 at Istanbul University, Department of Tourism Management. In 2014, he started to work as a Research Assistant at Batman University, School of Tourism and Hotel Management, Department of Tourism and Hotel Management. Since 2022, she has been working as an Assistant Professor in the Department of Tourism Management at Batman University, Faculty of Tourism, and her main fields of study are tourism psychology, tourism sociology, virtual reality technology and cultural intelligence.

Pınar Şenel is an assistant professor in the Department of Gastronomy and Culinary Arts at Anadolu University, Turkey. She received her master's degree and Ph.D. in Tourism Management at Anadolu University. She mainly teaches issues related to restaurant design and kitchen planning. Her research focuses on design and atmosphere in gastronomy.

Özgür Devrim Yılmaz is a professor at Dokuz Eylül University, Faculty of Business, Department of Tourism Management, İzmir, Turkey. Her academic interest includes crisis management in tourism, organizational behavior, seasonality in tourism, tourism sociology and smart tourism technologies in tourism enterprises.

Anne Lassen Zakaria is affiliated and has worked with the Department of Business at UCN. Currently she is working at UCN Department of Teacher Education as an assistant professor and coordinator of international relations. Anne's research interests include international education, higher education collaboration between instutions in the global north and the global south including virtual and physical students exchange and the development of communicative and intercultural competences. She has worked with global south issues since 2012 - including tourism, trust, collaboration. In 2016 she defended her PhD on trusting issues and social relations in small business networks dealing with agricultural produce in Ghana.

Index

www.ingramcontent.com/pod-product-compliance
Lightning Source LLC
LaVergne TN
LVHW080644011026
849160LV00014B/1312